MODULATORS, MEDIATORS, AND SPECIFIERS IN BRAIN FUNCTION

Interactions of Neuropeptides, Cyclic
Nucleotides, and Phosphoproteins in
Mechanisms Underlying Neuronal Activity,
Behavior, and Neuropsychiatric Disorders

ADVANCES IN EXPERIMENTAL MEDICINE AND BIOLOGY

Recent Volumes in this Series

MODULATORS, MEDIATORS, AND SPECIFIERS IN BRAIN FUNCTION

Interactions of Neuropeptides, Cyclic
Nucleotides, and Phosphoproteins in
Mechanisms Underlying Neuronal Activity,
Behavior, and Neuropsychiatric Disorders

Edited by

Yigal H. Ehrlich
Jan Volavka
Leonard G. Davis
Eric G. Brunngraber

The Missouri Institute of Psychiatry
University of Missouri - Columbia
School of Medicine
St. Louis, Missouri

PLENUM PRESS • NEW YORK AND LONDON

Library of Congress Cataloging in Publication Data

Main entry under title:

Modulators, mediators, and specifiers in brain function.

(Advances in experimental medicine and biology; v. 116)
"Proceedings of a satellite symposium held in conjunction with the annual
meeting of the Society of Neuroscience, Missouri Institute of Psychiatry, St. Louis,
Missouri, November 4–5, 1978."
Includes index.
1. Brain chemistry – Congresses. 2. Neural receptors – Congresses. 3. Cyclic
nucleotides – Congresses. 4. Neuropsychiatry – Congresses. I. Ehrlich, Yigal H.
II. Society for Neuroscience. III. Series.
QP376.M584   612'.8'042   79-14523
ISBN 978-1-4684-3505-4   ISBN 978-1-4684-3503-0 (eBook)
DOI 10.1007/ 978-1-4684-3503-0

Proceedings of a satellite symposium held in conjunction with the
Annual Meeting of the Society for Neuroscience, held at the Missouri
Institute of Psychiatry, St. Louis, Missouri, November 4–5, 1978.

© 1979 Plenum Press, New York
Softcover reprint of the hardcover 1st edition 1979
A Division of Plenum Publishing Corporation
227 West 17th Street, New York, N.Y. 10011

FOREWORD

While neuroanatomy and neurophysiology were defining
the unique features of the synapse as a site for cell to
cell signaling in the late fifties, neurochemistry was
establishing the identity and studying the biosynthetic
pathways of monoamine neurotransmitters. Meanwhile,
neuropsychiatry was keeping a vigilant eye on the outcome
of this concerted effort with the untold hope that a
genetic defect in neurotransmitter metabolism would account for the pathogenesis of certain psychiatric illnesses. Thus, when neurochemists in the early sixties
began to study the feasibility of measuring the metabolism
of brain neurotransmitters in vivo, clinical biochemists
eagerly adopted these methods to their needs and sought
to verify whether inborn errors of transmitter biogenesis
were a cause for at least certain forms of depression,
mania and schizophrenia.

Undoubtedly, it is still too early to evaluate the
outcome of these studies. However, current opinion holds
that gross inborn errors in transmitter metabolism do
not appear to be operative as a primary cause of psychiatric disorders. Though monoamine metabolism appears to
be defective in certain groups of psychiatric disorders,
the cause of these changes can at best be associated with
changes in patterns of neuronal firing. It is generally
believed that these persistent changes are determined by
a number of unknown factors operative in various psychiatric illnesses. In the attempt to identify the molecular
nature of these unknown factors, the focus of current
research is directed toward transmitter receptors. The
popular questions are whether the number of the receptors
for various transmitters changes in the elderly, in the
manic, and in the depressed patient. There is increasing
evidence that there is a change in the number of certain
receptors or in their affinity for the agonist in two
major neurological diseases: myasthenia gravis and tardive

dyskinesia. In myasthenia gravis, the number of acetylcholine receptors in the neuromuscular junction is
decreased; in tardive dyskinesia the number of dopamine
receptors in the striatum is increased. Such modifications may involve either autoimmune processes, as in
myasthenia, or still poorly understood changes, as in
the processes that regulate dopamine receptor function.
Some of these may involve covalent modification of receptor proteins. Such processes may include phosphorylation,
methylation, and acetylation. The best case for covalent
modification of receptor protein can be made for acetylcholine receptors, and this case is presented in this
volume by A. S. Gordon and I. Diamond in the chapter
entitled "Phosphorylation of the acetylcholine receptor".
Drs. Gordon and Diamond report that some proteins of membrane preparations from the electric organ of T. californica, which contains an extraordinarily high concentration
of acetylcholine receptors, are preferentially phosphorylated. This phosphorylation is stimulated by K+
(0.1 to 0.2 mM) and inhibited by Na+ and 10^{-6} M carbachol.
Perhaps the phosphorylated form of the receptor has
maximal affinity, and the binding of the agonist to the
receptor is reduced when the phosphorylation of the membrane protein is inhibited. Studies utilizing immuno-
chromatography in the presence of phosphorylating conditions have provided evidence that the phosphorylated
component of the membrane binds to specific antibodies to
acetylcholine receptors. In line with this finding are
current investigations in our laboratory (Guidotti, A.,
Toffano, J. and Costa, E. Nature, 275, 553, 1978. Toffano,
J., Guidotti, A. and Costa, E. Proc. Natl. Acad. Sci. US,
75: 4024, 1978) showing that an endogenous protein which
allosterically regulates the binding affinity of GABA
to specific membrane receptors functions as an inhibitor
of protein kinase activity. Though the data on phosphorylation of acetylcholine and GABA receptors are not
yet definitive, the possibility that such a mechanism may
be operative in receptor regulation remains an important
possibility.

Why has the focus of neurochemical research shifted
from transmitter regulation to receptor regulation? As
pointed out by J. L. Barker and T. G. Smith in the chapter
entitled "Three modes of communication in the nervous
system", the present understanding of neuronal communication is no longer restricted to classic synaptic transmission, but also includes neuromodulation and neurohormonal communication. Since receptors are a common

denominator to all the three types of neuronal communication, it is hoped to be able to detect defects in the multiple system for neuronal communication by focusing research efforts on the biochemistry of receptor sites. It is gradually becoming clear that transmitter receptors operate as a supramolecular unit and that other proteins, as well as the binding site for the transmitter, are important. The molecular cooperation that brings about receptor function is facilitated by transient changes that occur in the recognition site as a result of receptor occupancy by the agonist. Thus, a number of specific membrane proteins move in association with other membrane cc ponents within the plane of the fluid mosaic membrane system, when the recognition site of the receptor is unoccupied. When the receptor is occupied, these components may cluster to form a well defined unit. Molecular clustering or collisions are important because they may promote activation of enzymes that modify specific substrates. These substrates may determine the formation of appropriate specifiers of synaptic events. We presently associate two types of response with receptor function: the modification of ion channels and the activation of cyclases. The latter brings about activation of protein kinase and, ultimately, protein phosphorylation. In the chapter entitled "Phosphoproteins as specifiers for mediators and modulators in neuronal function", Y. H. Ehrlich recalls Greengard's suggestion that protein phosphorylation plays an important role as a common pathway used by a diversity of biological regulatory agents. As shown by work discussed in Dr. Ehrlich's chapter, numerous proteins in synaptic plasma membranes can become phosphorylated and, as noted above, recognition sites of receptors may also be phosphorylated. A lack, or an abundance, of phosphorylative activity in disease states may cause an abnormality in metabolic regulation which in turn may become a triggering point for an abnormality in neuronal firing and transmitter metabolism. This is certainly an area of growth for future research. As pointed out by D. H. Boehme, et al., in the chapter entitled "Evidence for the presence of substrates for cGMP dependent protein phosphorylation in human synaptosomal membranes", very few studies exist on alterations in protein phosphorylation in synaptic membranes from human brain. However, before research can boom in this direction, it is necessary to by-pass a few technical obstacles. In fact, as pointed out by D.H. Boehme, et al. in this book, one must first determine whether "rate of protein phosphorylation in membranes depends on ATP/membrane ratios...".

If phosphorylation is an important regulatory mechanism
for synaptic specifiers, then dephosphorylation of
specifiers must occur in situ. As shown by Gordon and
Diamond, the study of the regulation of phosphoprotein
phosphatase activity may be another area of active investigation, particularly if receptor protein phosphorylation is a mechanism to regulate receptor affinity
for transmitters. Since Gordon and Diamond have shown
that photoaffinity label with arylazido-γ-alanyl (γ-32P)-
ATP of the cholinergic receptor shows a specific labelling
of ATP binding sites in three of the eight major polypeptides detected in membrane preparations enriched with
cholinergic receptors, it appears that several proteins
participate in the regulation of receptor phosphorylation.

The chapter by Dr. D. B. Bylund entitled "Regulation
of central adrenergic receptors" addresses two questions
which may have an important role in determining new avenues
for future developments in neurochemical research. One
question concerns the cellular localization of receptors,
the other the participation of the membrane lipids in
receptor regulation. Dr. Bylund discusses data obtained
from Molinoff's laboratory in order to show that β_1-adre-
nergic receptors of rat cerebral cortex are located in
neurons while β_2-adrenergic receptors are located in
glial cells. Since a further classification of receptors
is now extended to other receptors including $GABA_1$ and
$GABA_2$ receptors, it is important to develop precise
methods to determine whether transmitter receptors are
localized on neuronal or glial membranes. This differentiation should not be intended as an attempt to
grade the role of receptors: glial receptors subserving
a function of lesser importance than neuronal receptors.
Actually, we happen to believe that the role of glial
receptors in neuropathology may be as prominent or even
more prominent than that of neuronal receptors. In
our laboratory (Schwartz, J. P., and Costa, E. Neuroscience, 3: 473, 1978 and Naunyn-Schmiedebergs Arch.
Pharmacol., 300: 123, 1978), we stated the belief that
receptors for neurotransmitters which are located in
glial cell membranes may be operative in neuronal-glial
cell communication; in our experiments they appear to
regulate the production and extracellular secretion of
important trophic factors which are taken up by the
axons and are then transported antidromically to the
neuronal soma where they exert their action.

In summary, this book reviews a number of molecular
mechanisms that regulate the function of synaptic receptors. This volume covers various levels of organization and focuses on potential developments of relevance
to neuropathology. Since these mechanisms may become
trends in future basic research related to mental health,
this book can be seen as a compendium of possible ideas,
methods, and research philosophies that may provide
explosive developments in the years to come. It is hoped
that elucidation of receptor function at the molecular
level may establish the area for future discoveries beneficial for mental health. At this moment we have to thank
Drs. Ehrlich, Volavka, Davis, and Brunngraber for having
made available this summary of guiding schemes and having
thereby increased current interest in an area that needs
much attention for future developments.

Erminio Costa, M.D.
Chief, Laboratory of Preclinical
Pharmacology
National Institute of Mental
Health

PREFACE

This volume is an outcome of a satellite symposium
held on November 4th and 5th, 1978, at the Missouri
Institute of Psychiatry, St. Louis, in conjunction with the
eighth annual meeting of the Society for Neuroscience.
The symposium was planned and organized in order to provide information which may help in narrowing the gap
between data obtained by investigators engaged in basic
neuroscience research and the clinical implications of
these findings.

The empirical introduction of psychopharmacological
agents to clinical use in the early 1950's facilitated
investigations aimed at elucidating basic mechanisms
that underlie brain function and behavior. In recent
years, we have witnessed developments in the opposite
direction; discoveries in the laboratories of researchers
in the basic neurosciences have produced new approaches
in treatment of the mentally ill patient. Receptors,
the existence of which had heretofore been largely based
on hypotheses, have been shown as entities which can be
identified, isolated, and studied in the test tube. The
complexity of receptor function has become apparent.
In addition to containing binding sites for a specific
neurotransmitter, the receptor exists as part of a macro-
molecular complex the activity of which may be modulated
by neuroactive peptides and mediated by the generation
of cyclic nucleotides. The cyclic nucleotides appear to
exert their physiological effects by regulating protein
phosphorylation-dephosphorylation cycles. Different
specific phosphoproteins control metabolic events in the
cell interior. In addition, membrane-bound phosphoproteins
may play a role in the regulation of receptor function,
and this process may be regulated by neuropeptides. Thus,
the receptor, when studied as a functional entity, may be
analyzed as a multiplicity of interactions between binding sites, transmitters, modulators, mediators and specifiers in the nervous system.

This book is concerned with recent studies which have sought to define the roles of neuropeptides, cyclic nucleotides, and phosphoproteins in neural function. These subjects are introduced in the first section of the volume, their potential mechanisms of action are discussed in a second section, and a third section is devoted to studies on the relevance of these systems to the etiology and treatment of some neuropsychiatric disorders. The experimental approaches described by the authors include physiological, neurochemical, pharmacological, behavioral, and clinical investigations, and clearly demonstrate the multi-disciplinary nature of current research in neuroscience. We hope that this book will promote interest in studies which focus on the interactions of endorphins, ACTH, cyclic nucleotides, and phosphoproteins in processes that regulate receptor function and may play a role in neuronal adaptation.

We wish to thank the individuals who helped us organize and conduct the symposium; Professor Warren Thompson, Director of the Missouri Institute of Pschiatry, Mr. Weldon Webb of the Division of Continuing Education for the Health Professions-University of Missouri-Columbia, Mr. Bob Wheadon for audiovisual support, our friends at the St. Louis State Hospital-Branch of the Missouri Department of Mental Health, and our co-workers at the Missouri Institute of Psychiatry. Special thanks are due to Ms. Jill Dodds and Mrs. Brenda Fearheiley who helped with the editorial work and typed the manuscripts, and to Mr. Ed Linn who did much of the photographic work for this volume.

April, 1979   Y. H. Ehrlich
                                     J. Volavka
                                     L. G. Davis
                                     E. G. Brunngraber

CONTENTS

SECTION II

Interactions of Neuropeptides, Cyclic
Nucleotides and Phosphoproteins in
Mechanisms Underlying Receptor Function

SECTION III

Clinical Implications

S E C T I O N   I

Mechanisms of Neuronal Communication

INTRODUCTION

This section provides an introduction to the involvement of neuropeptides, cyclic nucleotides, and phosphoproteins in the modulation, mediation, and specification of neuronal function. The first chapter, that by Barker and Smith, serves also as an excellent introduction to this volume. It defines three modes of information transfer in the nervous system: neurotransmission, neuromodulation, and neurohormonal communication, and provides evidence indicating that neuroactive peptides can play a role in each process. Siggins presents criteria which may be used for discrimination between neurotransmitters, neuromodulators, and neuromediators, and focuses on cyclic nucleotides as mediators in the process of synaptic transmission. The role of protein phosphorylation in the events leading from environmental input to persistant alterations in neuronal function is discussed by Ehrlich, who emphasizes the specificity inherent in this system. These three chapters provide an overview of their subject matter, and each is followed by chapters which provide specific examples of studies in this area of research. Thus, Ram discusses the site of action of peptide hormones in the nervous system of invertebrates. Volicer, Mirin, and Meichner present studies concerned with the effects of ethanol on brain cyclic nucleotides. The role of protein phosphorylation in visual function is described by Farber and Lolley. Finally, Boehme, Kosechki, and Marks demonstrate that protein phosphorylative systems can be studied in human (postmortem) brain tissue.

THREE MODES OF COMMUNICATION IN THE NERVOUS SYSTEM

JEFFERY L. BARKER and THOMAS G. SMITH, JR.

Laboratory of Neurophysiology, National Institute
of Neurological and Communicative Disorders and
Stroke, National Institutes of Health
Bethesda, Maryland   20014

INTRODUCTION

How excitable cells communicate with each other has
been the subject of continued study over the past 50 years.
Much of the research done thus far has focused on a form
of communication which occurs at specialized junctions between contiguous elements. The junctions are called "synapses", the communication is known as "synaptic transmission"
and the substance mediating the event is called a "neurotransmitter". The ready availability of preparations appropriate to observe this communication and the development
of techniques to study the details involved have gradually
led to a more complete understanding of synaptic transmission. Since it is the most widely and intensively studied
form of cell-cell signaling in the nervous system, many
investigators have assumed that all intercellular communication is similar and that all substances which can be
extracted from neuronal tissue and which bind to neuronal
membranes or have pharmacological effects are a priori
"neurotransmitters". The logic implicit in this generalization likely reflects both the relatively well-understood
nature of the communication and the natural desire to
simplify neuronal function into an understandable form.
However, our understanding of synaptic transmission is
based almost exclusively on study either of peripheral
synapses between nerve and muscle or nerve and nerve in
vertebrates or of central and peripheral synapses between
nerve and nerve in invertebrates. These preparations
have been chosen owing to their relative simplicity and

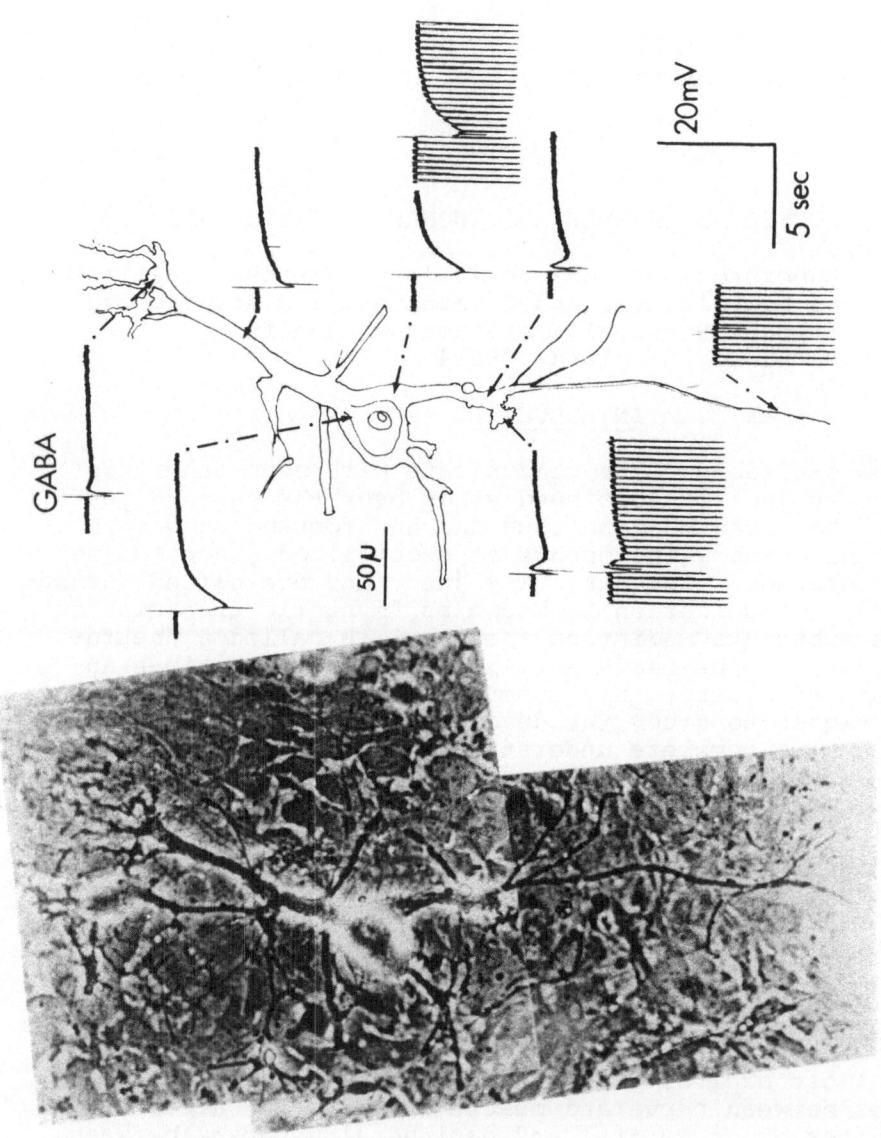

accessibility when compared to the vertebrate central nervous system (CNS). Realistically, these "simple" systems
should be viewed as models of the vertebrate CNS until
synaptic events between vertebrate central neurons can be
studied with the necessary scientific and technical rigor.
There is already sufficient evidence to indicate that synaptic transmission is present in the vertebrate CNS but
this does not necessarily mean that such a method of signaling is the only form of intercellular communication nor
that all endogenous substances function simply as neurotransmitters. Our understanding of synaptic transmission
in model systems may be utilized both as a reference for
examining central synaptic events and for considering other
possible forms of intercellular communication in the nervous system.

In this chapter we will present evidence to suggest
that there are at least three functionally distinct ways
in which neuronal substances may mediate intercellular communication in the nervous system. We will begin by discus-
sing the more familiar synaptic transmission, especially
as it relates to central synaptic events, and then develop
evidence to suggest that there are two other forms of communication distinctly different from conventional synaptic
transmission.

NEUROTRANSMISSION

Synaptic transmission (also called "neurotransmission") has been carefully studied at various neuromuscular
junctions and peripheral vertebrate and central invertebrate ganglia. These preparations have the advantage that
both pre- and post-synaptic elements are freely accessible
to electrophysiological methods, thus allowing study of

Fig. 1. Differential sensitivity of GABA responses on a
    cultured mouse spinal neuron. (Phase photomicrograph
    of living cell on left with outline of cell on right.
    Iontophoresis of identical GABA pulses (24 nA, 100
    msec) onto different regions of spinal neuron shows
    non-uniform topography of sensitivity and response
    complexity. Monophasic, hyperpolarizing responses
    associated with conductance increase are present at
    cell body, while multiphasic responses coupled to
    conductance increase can be seen on processes. Down-
    ward deflections are voltage responses to constant
    current (-1 nA) pulses. Resting membrane potential:
    -55mV. Modified from Barker and Ransom (1978).

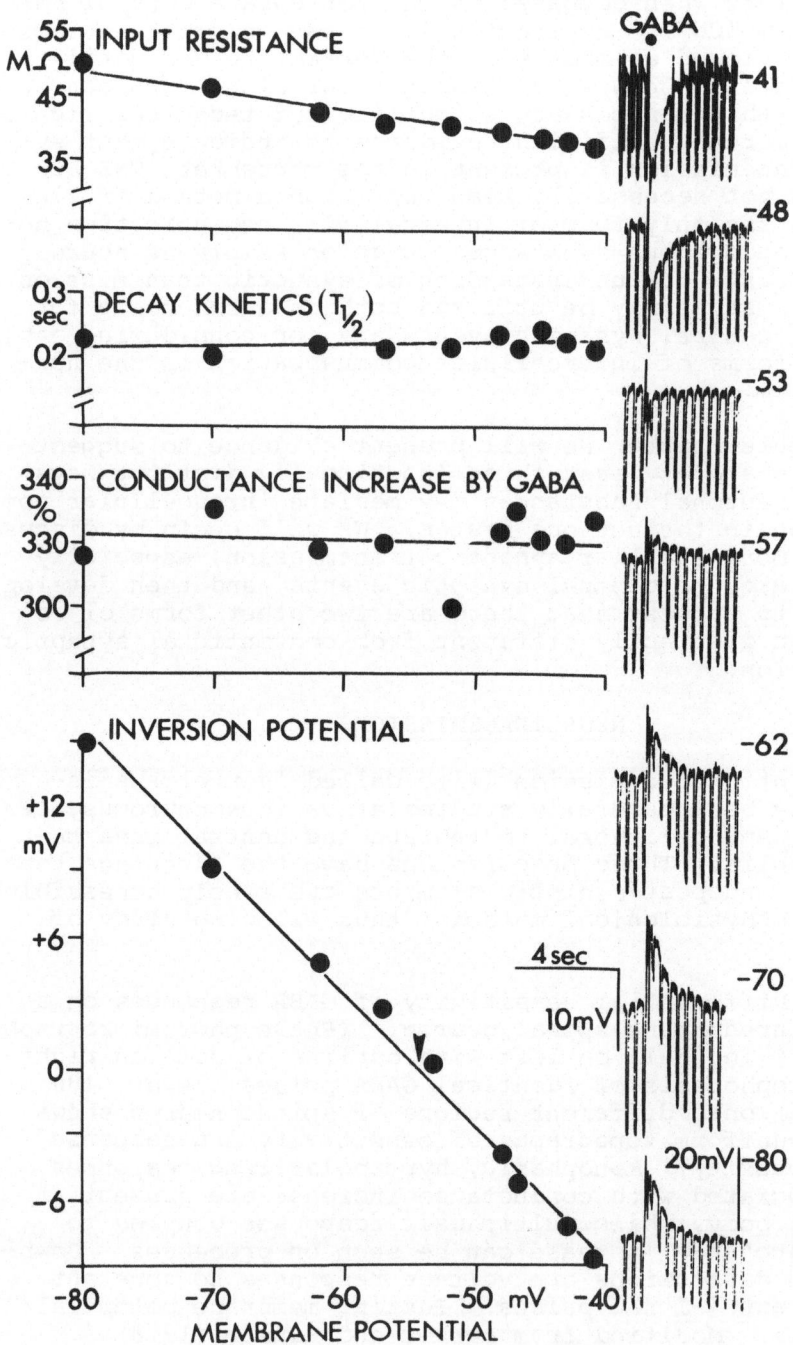

junctional physiology. The observations made with these
preparations have established certain properties of the
signal communicated at synapses. The signal appears to be
derived from the synchronized release of "quanta" of transmitter substance which rapidly diffuse across the synaptic
cleft to engage receptors on the post-synaptic membrane.
Engagement of receptors by transmitter very rapidly leads
to activation or inactivation of post-synaptic membrane
conductance to specific ions. Quanta summate to generate
the post-synaptic potential and current events resulting
from the conductance changes. The rise time of the post-
synaptic response likely reflects the temporal dispersion
of summating quanta. Individual quanta or "miniature" synaptic events are not the most elementary unit involved in
the synaptic events but instead are themselves composed of
many elementary currents, each reflecting activation of a
single receptor-coupled conductance (for review, see NEHER
and STEVENS, 1977). The reaction between neurotransmitter
molecule and membrane receptor occurs relatively independent
of membrane potential. Thus, the elementary conductance
activated during synaptic transmission is constant over the
physiological range of membrane potential. Although the
conductance is constant, the average duration of the elementary events and the decay kinetics of the physiologically
elaborated signal can vary with membrane potential (DIONNE
and STEVENS, 1975; ADAMS, et al., 1976; ASCHER, et al.,
1978). The polarity of the signal is predicated on the
specific ionic conductance(s) activated. Synaptic activation of an ionic conductance whose driving force is depolarized relative to spike threshold will excite the post-
synaptic element, while activation of a conductance with
driving force hyperpolarized relative to spike threshold
will inhibit cell excitability. The brief, but intense
activation of conductance during inhibitory transmission
would also serve to shunt the input resistance of the cell,
thereby attenuating excitatory synaptic events. The

Fig. 2. GABA response properties as a function of mem-
         brane potential on cultured spinal neuron. Membrane
         responses elicited at cell body in response to 40 nA-
         100 msec GABA pulse as membrane potential was varied
         from -80 to -41 mV. The polarity of the response in-
         verts at about -53 mV, while the magnitude of the con-
         ductance increase evoked by GABA and its time course
         of decay appear to change little over the entire 40
         mV range. Resting membrane potential: -57 mV.
         Barker and Ransom, unpublished observations.

synaptic signal is shaped by several factors including membrane potential, diffusion, recent history or receptor activity, cooperativity of both transmitter binding and conductance activation, and temporal and spatial summation.

Some of these aspects of neurotransmitter action can be illustrated by pharmacologic application of exogenous neuronal substances to the surface of central nervous system (CNS) neurons grown in cell culture. Cultured neurons grow as a virtual monolayer and so represent a definite but unknown departure from the normal organization of the nervous system. The cells exhibit many of the physiological properties associated with cells recorded in vivo including exictability, spontaneous synaptic activity and chemosensitivity to a variety of neuronal substances (RANSOM, et al., 1977; BARKER and RANSOM, 1978). Discrete applications of γ-aminobutyric acid (GABA), an amino acid found throughout the nervous system, which is thought to mediate various inhibitory transmissions, revealed both a non-uniform distribution of inhibitory responses as well as the presence of complex excitatory and inhibitory events present at some part of the cell (Fig. 1). The non-uniformity of the responses is likely due to variable densities of GABA receptors on the cell surface which themselves may reflect innervation sites, while the complex nature of some responses may reflect coupling of GABA receptors to different conductances. Similar types of non-uniform and complex response topography have been observed with other neuronal substances, including peptides.

The brief inhibitory response of the cell was associated with an increase in membrane conductance (Fig. 1) which was constant over a wide range of membrane potentials (Fig. 2). A similar constancy of GABA-activated conductance at the single channel level has also been observed (MCBURNEY and BARKER, 1978).

The degree of electrophysiological resolution allowed by the cultured neuronal system has revealed clear differences between the excitatory responses evoked by peptides and acidic amino acids. For example, iontophoresis of the peptide "substance P" leads to a rapid excitatory response whose duration is markedly shorter than that of a response of similar amplitude evoked by the amino acid glutamate (Fig. 3). In addition, the peptide excitation and underlying conductance change desensitize rapidly, preventing summation of responses to closely repeated peptide application (Fig. 3). A short-lived event, apparently identical in properties to that evoked by substance P, has also

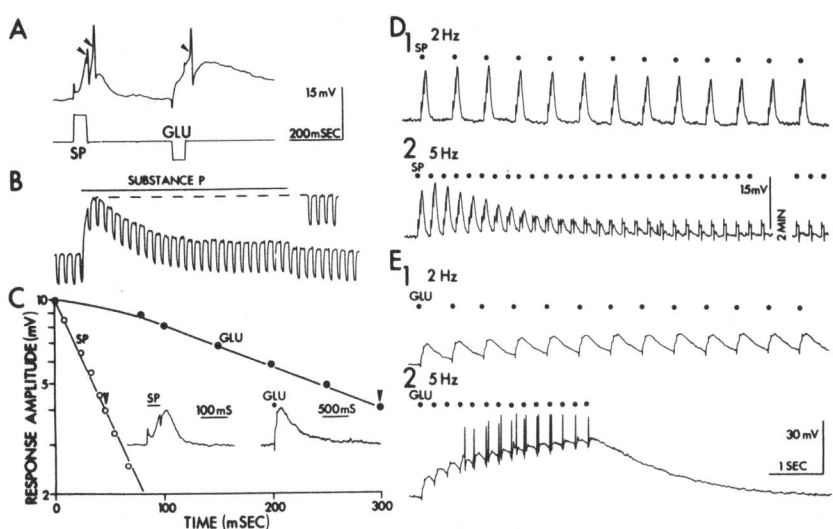

Fig. 3. Excitatory responses to the SP and glutamate compared on the same cultured spinal neuron. A. 50 msec
iontophoretic pulses of SP (25 nA) and glutamate (20
nA) rapidly depolarize the cell above threshold for
generation of spikes (marked by arrowheads), which are
attenuated by the frequency response of the pen-
recorder. Iontophoretic current trace below membrane
potential trace. Resting potential: -52 mV.
B. Sustained application of 40 nA SP (marked by bar
above trace) leads to depolarization and associated
increase in membrane conductance which desensitizes
completely. Downward deflections are voltage responses to -0.2 nA current stimuli. Inset shows voltage responses at same potential as peak of depolarizing response. The abrupt change in potential at end
of the SP iontophoresis is due to the several mVs
coupling artifact between the iontophoretic current
and the voltage recording. Membrane potential: -60
mV. C. Voltage decays of equal-sized SP and glutamate responses are plotted semilogarithmically. Downward arrowheads show time constant of decay which is
about six-fold greater for glutamate response to 25 nA
- 50 msec pulses of SP diminish slightly when applied
at a frequency of 2 Hz. D2. At 5 Hz the responses
desensitize completely and remain desensitized for
2 min. E1. Depolarizing responses to 20 nA - 50
msec pulses of glutamate summate at 2 Hz delivery
rate and produce spikes at 5 Hz (E2). Membrane potential (C-E): -80 mV. (From Vincent and Barker, unpublished observations).

been observed with another peptide, leucine-enkephalin
(GRUOL, et al., 1978; BARKER, et al., 1978c). These pep-
tide-mediated excitations may reflect synaptic transmission of an excitatory signal which is functionally different from glutamate-mediated excitation. Desensitization of
the peptide excitation would effectively limit both the
duration, area and the intensity of excitation yielding a
brief event with a clear refractory period. Peptide-evoked
inhibition of excitation through activation of membrane
conductance appears to differ from amino-acid induced inhibition as well in that the former exhibits little if any,
desensitization (GRUOL, et al., 1978; BARKER, et al.,
1978c; and BARKER and VINCENT, unpublished observations)
while the latter shows desensitization (BARKER and RANSOM,
1978; BARKER and MCBURNEY, unpublished observations).

    These pharmacologically induced membrane responses
may serve as reference for examining physiologically elaborated synaptic events mediated by peptides and other
neuronal substances. In fact, we have begun to apply
fluctuation (or "noise") analysis to the pharmacological
activation of amino acid receptors on cultured spinal neurons. The resolution of the system has allowed us to distinguish between the elementary actions of GABA and glycine
(MCBURNEY and BARKER, 1978; BARKER and MCBURNEY, 1979),
two amino acids which activate a Cl^- conductance on spinal
neurons (BARKER and RANSOM, 1978). We have also been able
to record miniature synaptic currents in cultured spinal
neurons (BARKER and MCBURNEY, unpublished observations)
and to compare the time constant of current decay with the
kinetic properties of the pharmacologically evoked single
channel events. We have found a population of synaptically
evoked currents whose time constant of decay and sensitivity to exogenous drugs correlates well with the behavior
of single channels activated by GABA. This correlational
approach is a useful means of tentatively identifying the
substance(s) mediating specific synaptic events (CRAWFORD
and MCBURNEY, 1976).

    In summary neurotransmission, as defined both in
terms of electrophysiological and anatomical properties,
involves the momentary alteration of membrane conductance
independent of membrane potential through engagement of
receptors at specialized synaptic junctions (see KATZ,
1966 for details). Neurotransmission thus indirectly
changes the excitability of a single excitable element
for a brief period (msec to sec).

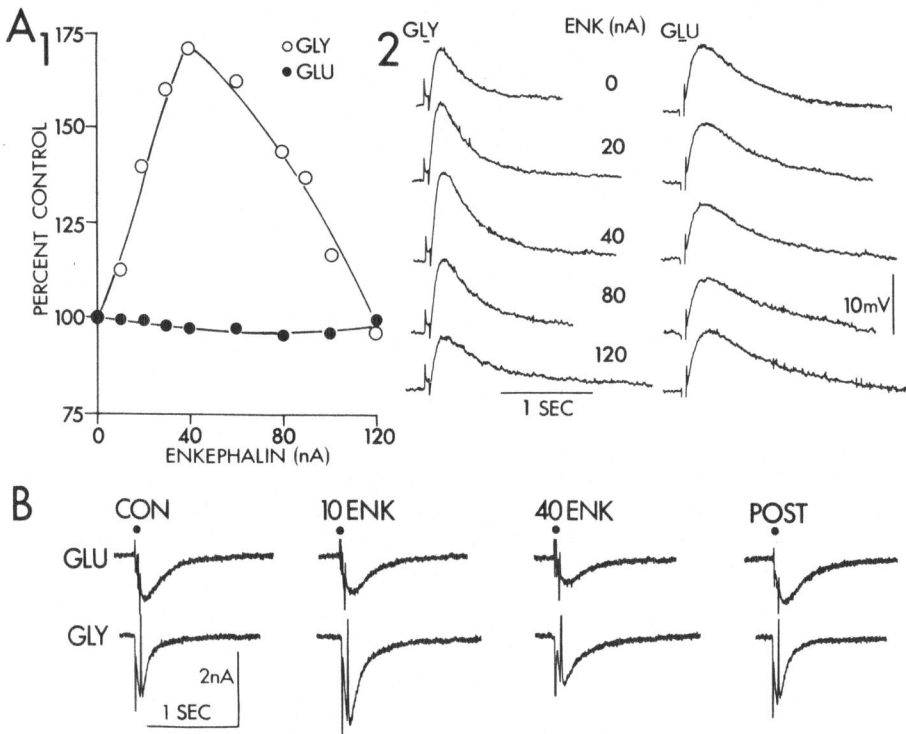

Fig. 4. Enkephalin modulation of amino acid voltage and
current responses on cultured spinal neurons. KCl
recordings from two different cells. A. Modulation
of voltage response to 20 nA - 50 msec glycine pulse
is dependent on ENK iontophoretic current with enhancement at low and moderate ENK current. ENK has
little effect on response to 30 nA - 50 msec glutamate pulse. Normalized data plotted in Al with speci-
men records in A2. Glycine response is depolarizing
when recording with KCl microelectrodes. Membrane
potential: -65 mV. B. Potential clamped to -70 mV
and membrane current responses to 25 nA - 50 msec
glutamate and 40 nA - 50 msec glycine pulses examined
before (CON), during iontophoresis of 10 and 40 nA
ENK and following recovery (POST). The response to
glutamate is depressed in a dose-dependent manner by
ENK, while the response to glycine is enhanced at the
lower, and depressed at the higher ENK current, relative to control (from BARKER, et al., 1978c).

NEUROMODULATION

    We have observed another pharmacological action of
neuronal substances that is distinctly different from
neurotransmission as just defined. This action has been
seen with the peptides substance P and leucine-enkephalin
and involves alteration of the magnitude and kinetics of
the conductance and current responses to putative amino
acid neurotransmitters independent of any other membrane
effects (BARKER, et al., 1978a,b,c,; HUANG, et al., 1978;
VINCENT and BARKER, unpublished observations). We have
tentatively labeled this action "neuromodulation" to
distinguish it from neurotransmission. The peptide alteration in the neurotransmitter-evoked membrane response is
dose-dependent and reversible and independent of membrane
potential. The modulatory effect of the peptides may be
either depression or enhancement. For example, glycine
responses are sometimes increased at low iontophoretic
currents of enkephalin (Fig. 4B) and depressed at higher
enkephalin currents, while glutamate responses are either
uniformly depressed (Fig. 4B) or not affected by enkephalin (Fig. 4A). Enkephalin depression of glutamate
responses evoked on spinal neurons in vivo has been reported in the cat (ZIEGLGANSBERGER, et al., 1976; ZIEGL-
GANSBERGER and FRY, 1976) and rat (SEGAL, 1977). Investigation into the mechanisms of modulation by the peptides
has revealed three probable sites of interaction: 1)
at the receptor level, with the peptide changing the
apparent affinity of the receptor for the neurotransmitter
(HUANG, et al., 1978), 2) at the conductance level, with
peptide affecting the conductance activated by the neurotransmitter but not its binding to its receptor (BARKER,
et al., 1978a) and 3) at the conductance level, with the
peptide altering the apparent ion gradient involved in the
conductance, thereby changing the driving force acting on
the conductance and thus changing the amplitude of the
transmitter-evoked response. An example of this latter
phenomenon is illustrated in Fig. 5. In this example,
enkephalin slightly increases membrane conductance and
potential and depresses the amplitude of the voltage response to a pulse of the inhibitory amino acid β-alanine
(Fig. 5A). The direct effects of the peptide on membrane
properties do not account for all of the depression of the
β-alanine response. Furthermore, the peptide causes changes
in the time course of the depressed response (Fig. 5B).
On closer inspection it is clear that the peptide has also
shifted the inversion potential of the β-alanine response
about 5 mV in the depolarizing direction (Fig. 5C), thus
effectively decreasing the driving force acting on the
neurotransmitter conductance mechanism. A continuous re-

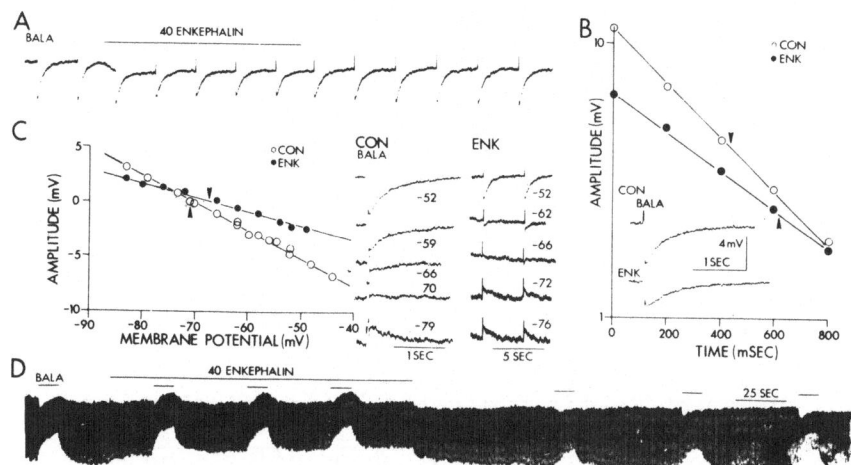

Fig. 5. Enkephalin depresses the membrane response to
β-alanine and alters its inversion potential on a
cultured spinal neuron. A. Continuous pen-recorder
trace of membrane potential showing rapid time course
of depressant effects of 40 nA enkephalin on hyperpolarizing responses to 22 nA - 50 msec pulses of
β-alanine. Enkephalin causes several mVs hyperpolarization. B. Amplitude of β-alanine response
is plotted semi-logarithmically as a function of
time under control conditions ("CON") and in the
presence of 40 nA enkephalin ("ENK"). Arrowheads
indicate time constants of decay. A decrease in decay rate is evident during enkephalin. C. Amplitude of β-alanine responses is graphed as a function
of membrane potential in control ("CON") and during
enkephalin ("ENK"). Arrowheads indicate inversion
potentials of the β-alanine responses. A depolarizing shift is evident during enkephalin. D. Continuous pen-recorder trace showing membrane responses
to applications of 25 nA β-alanine before, during
and after 40 nA enkephalin. A time-dependent loss
of the hyperpolarizing phase of the β-alanine response during enkephalin is evident. Enkephalin
increases membrane conductance by 20% and depresses
the conductance increase evoked by β-alanine.
Membrane potential: -60 mV. Downward deflections
are voltage responses to constant current hyperpolarizing pulses (-0.6 nA, 50 msec).

cord demonstrating the time course of the depolarizing
shift in the voltage response to sustained applications of
β-alanine and the concomitant depression of the conductance
increase to β-alanine during enkephalin is shown in Fig.
5D. β-alanine activates Cl⁻ conductance at the cell body
and the Cl⁻ ion gradient appears to be actively distributed
across spinal neuron membranes since the inversion potential of Cl⁻ - dependent events is typically 10 mV more
hyperpolarized than resting potential (BARKER and RANSOM,
1978). Thus, a 5 mV shift in inversion represents a 50%
change in the driving force relative to resting membrane
potential. Other reports of catecholamines and peptides
affecting the activity of membrane mechanisms involved in
maintaining specific ion gradients have appeared(KOKETSU
and OHTA, 1976; AKASU, et al., 1978; ARECHIGA and ACEVES,
1978).

  Neuromodulation, defined electrophysiologically, involves direct alteration of neurotransmitter-activated
membrane conductance independent of other effects on membrane properties. Neuromodulation, as defined here, has
not yet been observed during the physiological elaboration
of synaptic transmission nor has the anatomical relationship between the neuromodulatory element and its target
cells been described. A wide variety of clinically important drugs can act on the post-synaptic membranes of CNS
neurons to modulate transmitter-activated membrane conductance by altering either receptor affinity for neurotransmitter or the conductance activated by the neurotransmitter
(DOSTROVSKY and POMERANZ, 1973; DUGGAN, et al., 1976;
ZIEGLGANSBERGER and BAYERL, 1976; SEGAL, 1977; CHOI, et al.,
1978; BARKER and RANSOM, 1978b; MACDONALD and BARKER, 1978a;
MACDONALD and BARKER, 1978b). Thus, modulation of transmitter events at post-synaptic sites by endogenous and exogenous substances is likely to be an important aspect of
both the normal physiological function of the CNS and the
cellular mechanism(s) of action of clinically important
CNS-active drugs. Alteration in synaptic transmission by
clinically important drugs acting at pre-synaptic sites in
the CNS has also been reported (e.g., WEAKLY, 1969; MAC-
DONALD and NELSON, 1978), but the membrane mechanisms have
yet to be elucidated.

NEUROHORMONAL COMMUNICATION

  Another type of intercellular communication has recently been described in molluscan nervous systems. The
communication in these invertebrates takes place between
cells not in synaptic contact (COGGESHALL, 1967; FRAZIER,
et al., 1967) and involves long-lasting changes in the

activity of diverse and distant target cells following
stimulation of specific peptidergic neurons called "bag
cells" (BRANTON, et al., 1978; STUART and STRUMWASSER,
1978). The peptide elaborated by the bag cells also induces egg-laying behavior (KUPFERMANN, 1970; ARCH, 1972;
PINSKER and DUDEK, 1977). We have previously examined the
pharmacological effects of peptide extracts of both the
bag cells (BARKER and SMITH, 1977) and other parts of the
molluscan nervous system (IFSHIN, et al., 1975) on the excitability of several of the target neurons in the sea
slug Aplysia californica and the land snail Otala lactea.
In addition, we have applied a variety of peptides derived
from the vertebrate nervous system to the same cells and
found two with actions quite similar to those seen with the
bag cell peptide (BARKER and GAINER, 1974; BARKER, et al.,
1975; BARKER and SMITH, 1976, 1977).

    We have utilized voltage clamp techniques to study the
biophysical mechanisms associated with the peptide actions.
The peptide actions involve regulation of both voltage-
independent and voltage-dependent membrane conductances.
The latter type of conductances are those which underlie
the generation of action potentials and pacemaker potentials. A five-minute application of 10^{-8}M peptide induced
pacemaking which lasted several hours. Whether the prolonged time course reflects prolonged binding of peptide
to receptors or prolonged effect of transiently bound peptide is unknown. The principal actions of the peptide
are 1) depolarization of membrane potential and associated
persistent increase in inward current with little change
in slope conductance at membrane potentials more negative
than the range of the pacemaker potential and 2) induction
or enhancement of the Na^+ and K^+ pacemaker conductances
(Fig. 6), which have been found to underlie bursting pacemaker potential activity (WILSON and WACHTEL, 1974; SMITH,
et al., 1975; BARKER and SMITH, 1978). The Na^+ conductance underlies the negative slope region of the N-shaped
steady state current voltage curve. The peptide also shortens the kinetics of inactivation of the K^+ pacemaker conductance (BARKER and SMITH, 1977). The depolarizing effect
of the peptide (and underlying persistent inward current)
serves to bring membrane potential into the membrane potential region for sequential activation of the two voltage-dependent pacemaker conductances. Threshold concentration for producing these effects is 10^{-9}M. Niether the
"ring" nor the "tail" components of the vasopressin molecule are active and activity is lost upon removal of the
glycineamide terminal. Induction or enhancement of pacemaker activity would serve to re-organize the output of
the cell, producing a facilitated, pulsatile secretion of
product (see BARKER and SMITH, 1977).

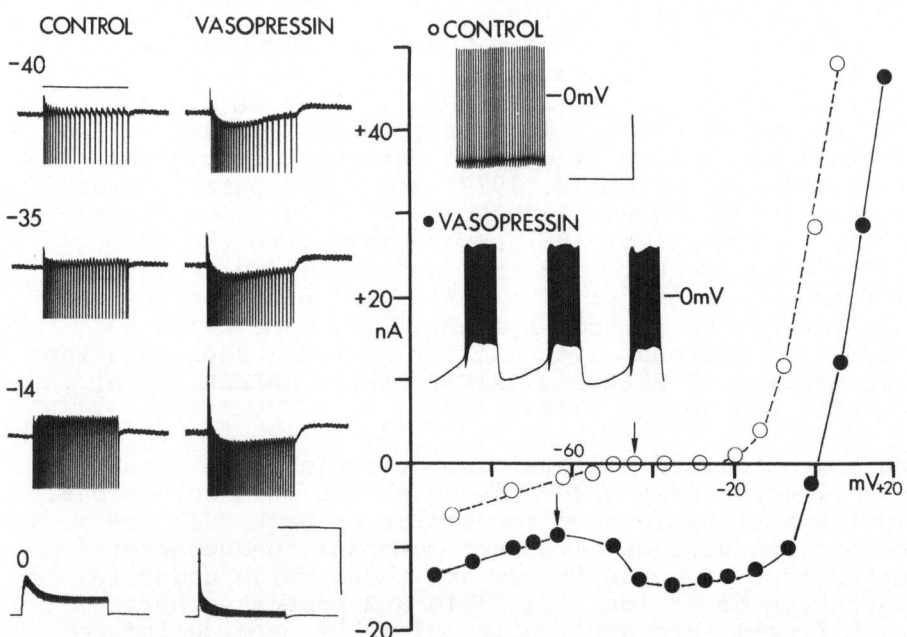

Fig. 6. Vasopressin alters steady-state, voltage-clamp,
     current voltage (I-V) curve. Recordings are from
     peptide-sensitive cell from the snail before (CONTROL)
     and after the bath application of 1 µM vasopressin
     VASOPRESSIN . Membrane potential activity is illu-
     strated in insets of I-V plot on right. Control trace
     shows beating pacemaker activity. Vasopressin in-
     duces bursting pacemaker activity. Zero membrane po-
     tential: "0 mV." Left: Membrane of cell voltage-
     clamped and 5-sec voltage steps imposed (during time
     indicated by bar above current trace marked "-40").
     Currents are shown at different depolarizing voltage
     steps (to membrane potentials indicated by numbers
     above traces) under control conditions and in presence
     of vasopressin. Rapid downward current events re-
     present action potential currents. Presence of slow
     inward current that decreases during the command is
     apparent in the vasopressin-treated membrane. Right:
     I-V curve derived from quasi-steady-state currents
     using most negative or least positive current evoked
     after 1 sec during command. Current axis (nA), vol-
     tage axis (mV). Cell's membrane held at -45 mV in
     control and -50 mV in vasopressin (downward arrows).
     Calibrations: Left, 10 nA (upper three traces) and
     40 nA (lowermost traces), 5 sec; right (inset),
     50 mV, 20s. From BARKER and SMITH (1976).

A similar multiplicity of peptide actions all of which inhibit excitability has been observed on an unidentified cell next to cell 11 in Otala.   The inhibitory actions include 1) hyperpolarization of membrane potential and increase in slope conductance (Fig. 7) and 2)   direct depression of the voltage-dependent inward spike conductance as reflected in a depression of the inward spike currents (BARKER and SMITH, 1977). All of these actions are dependent on the presence of Ca^{++} in the external medium and disappear when Ca^{++} is replaced by an equal concentration of Sr^{++}. Although these actions do not desensitize, they do not long-outlast the period of peptide application, recovery being complete within 5 minutes of washing. The multiple nature of the peptide's effects serve to dampen cell excitability indirectly, by hyperpolarizing membrane potential away from spike threshold and attenuating excitatory synaptic input by shunting membrane resistance, and directly, by depressing the amplitude of action potentials.

The lack of contiguity between the cells involved in this form of communication coupled with the multiple actions of the peptide particularly on voltage-dependent, but also on voltage-independent conductances have prompted us to describe the intercellular signaling as "neurohormonal communication" to distinguish it from neurotransmission and neuromodulation. Based on anatomical and electrophysiological observations, neurohormonal communication may be defined as a regulation of cell excitability through effects on both voltage-dependent and voltage-independent membrane conductances at extra-synaptic sites.

There is a teleological basis for the development and preservation of neurohormonal communication in the nervous system. It represents a class of operations whereby one species of communication molecule initiates and orchestrates the activities of diverse and distant target neurons whose concerted output may lead to an autonomic, endocrine and/or motor behavior of established survival value. In the invertebrate model system the bag cell peptide causes egg-laying behavior, a clearly intricate and crucial pattern necessary for the survival of the species. In the vertebrate, central administration of angiotensin leads to elevation of blood pressure, release of antidiuretic hormone and drinking behavior (see PHILLIPS, 1978 for review). These complementary actions would serve to defend against salt-and-water imbalance. Whether these behaviors reflect a neurohormonal action of the peptide and whether the peptide mediates such events physiologically remain to be studied.

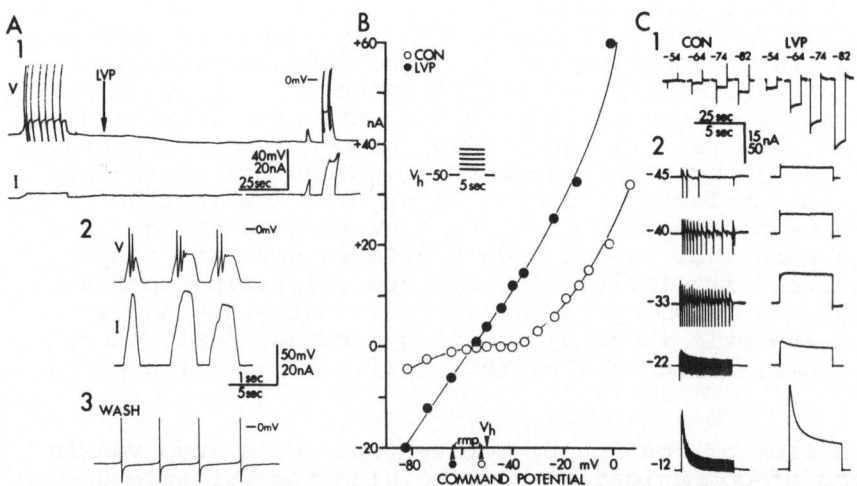

Fig. 7. Inhibitory effects of LVP. Recordings are from
    unidentified cell adjacent to cell (cell 11) excited
    by LVP. (A1) Potential trace (V) above current trace
    (I). Cell is electrically inactive but generates re-
    petitive series of action potentials upon injection
    of depolarizing current, as seen in the early part of
    the trace. Bath application of 0.1 μM LVP (arrow)
    hyperpolarizes membrane and much more current is re-
    quired to generate action potentials: late part of
    trace. Traces were recorded on curvilinear pen re-
    corder. (A2) Relatively fast, large depolarizing
    current pulses evoke small amplitude action potentials
    that rapidly accomodate. Rectilinear traces. (A3)
    Washing in peptide-free solution leads to recovery of
    spikes within minutes. (B). Steady-state I-V curves
    are derived from currents (C) observed at different
    potentials under voltage clamp in control solution
    and in LVP. A holding potential (V_h) of -50 mV and
    5-sec commands were used to generate data. Membrane
    conductance considerably increased and I-V curve
    changed from nonlinear to mainly linear in LVP. Rmp:
    resting membrane potential. (C) Currents evoked dur-
    ing hyperpolarizing (1) and depolarizing commands (2).
    Note the absence of multiple action potential currents
    in LVP. (From BARKER and SMITH, 1977).

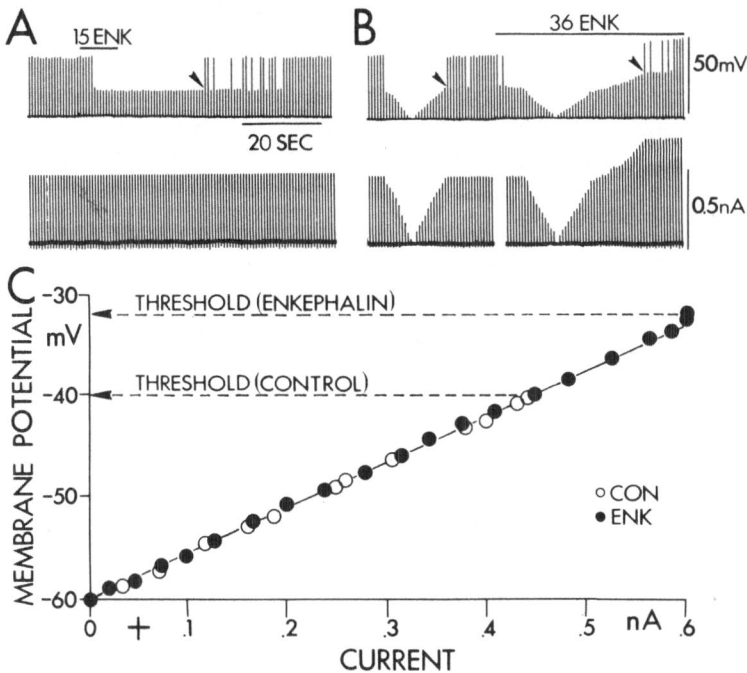

Fig. 8. Enkephalin elevates threshold for action potential generation in a cultured spinal neuron. KAc recording. Cell excitability assessed using suprathreshold 0.4 nA - 50 msec depolarizing current pulses. Voltage traces above current traces. A. Iontophoresis of 15 nA ENK (marked by bar above voltage trace) rapidly blocks action potential generation evoked by current pulses without detectable change in resting membrane potential. The depression clearly outlasts the iontophoretic application. Arrowhead delineates suprathreshold (spike) from subthreshold responses. B. Iontophoresis of 36 nA ENK blocks current-evoked excitation. Increasing amplitude of current pulse reveals elevation in spike threshold without change in the current-voltage relations of the membrane. C. Plot of voltage response to depolarizing currents of increasing amplitude demonstrates similar current-voltage relations and membrane slope conductance in control and during ENK iontophoresis. (From BARKER, et al., 1978c).

TABLE 1. NEURONAL COMMUNICATION

	Neurotransmission	Neuromodulation	Neurohormonal Communication
AVENUE	Between contiguous cells in synaptic contact	? Between contiguous cells	Extra-synaptic
ACTION	Momentary, indirect changes in single cell excitability	Momentary, indirect changes in single cell excitability during neurotransmission	Sustained direct and indirect changes in the excitability of many cells
MECHANISM	Activation, inactivation of membrane conductance independent of voltage	Alteration of synaptically activated voltage-independent conductance	Regulation of voltage-dependent spike or pacemaker conductance
SUBSTANCES	Acetylcholine amino acids, catecholamines, enkephalins, substance P	Enkephalins, substance P	Bag cell peptide acetylcholine, catecholamines, ? enkephalins, ? substance P
ANALOGIES	Telephone conversation	Gain control on telephone conversation	Radio broadcast

We have observed another action of peptides on the
membrane properties of cultured spinal neurons which is
distinct from both conventional neurotransmitter actions
and the aforementioned "neuromodulatory" effects. The
action involves direction depression of excitability
(BARKER, et al., 1978b) and consists of a dose-dependent,
reversible elevation in threshold for spike generation
either without any other detectable changes in membrane
properties or independent of any other membrane effects
(BARKER, et al., 1978c; SMITH, et al., 1978) (Fig. 8).
The depression has been observed with both leucine-enkephalin and substance P, is evident within seconds of apply-
ing either peptide, does not desensitize during prolonged
applications, and wears off within seconds of termination
of the peptide iontophoresis. This action, though subtle,
can effectively attenuate excitability. Since the anatomical relationship between enkephalin- and substance P-
containing nerve terminals and their target cells is not
known, it is unclear whether this pharmacological action
on voltage-dependent spike conductance is mediated by peptides released at synapses, or whether it reflects an
extra-synaptic form of communication between cells.

In fact, regulation of voltage-dependent pacemaker
conductance in invertebrate neurosecretory cells has been
reported to occur following activation of a synaptic pathway and iontophoresis of putative neurotransmitters(WILSON
and WACHTEL, 1978). The distribution of the receptors
activated during the synaptic input is unknown, as are the
morphological details of the "synaptic" innervation in
these examples. Thus, while the observations clearly suggest a form of communication whose functional effect falls
outside the definition of neurotransmission (vide supra),
it is not clear whether this type of signalling should be
considered a form of "neurohormonal communication" between
contiguous elements or whether it represents a new mode of
communication which combines certain aspects of two other
models. Other examples of functional effects of putative
"neurotransmitter" substances on voltage-dependent spike or
pacemaker conductances have been reported (TSIEN, 1974;
BOLTON, 1975; KUBA and KOKETSU, 1975, 1976; GILES and
NOBLE, 1976; DUNLAP and FISCHBACH, 1978; KLEIN and KANDEL,
1978). Some of these actions have been shown to occur on
membranes apparently devoid of synapses (e.g., DUNLAP and
FISCHBACH, 1978) and may be a vertebrate form of "neurohormonal communication". Effects on voltage-dependent conductances which occur following stimulation of a synaptic
pathway and in addition to alteration in voltage-independent conductances (e.g., KUBA and KOKETSU, 1976; WILSON

and WACHTEL, 1978), suggest that multiple forms of communication may exist between contiguous elements in synaptic
contact.

CONCLUSIONS

The recent development and extensive use of central
and peripheral in vitro preparations more complex than the
neuromuscular junction has a new wealth of physiological
and pharmacological observations. Some of these findings
appear to fall within classical definitions of neurotransmitter action, as defined in terms of carefully studied
synaptic events at peripheral or invertebrate central synapses. However, a variety of observations have been reported which clearly do not fall within these definitions,
suggesting that there may be a variety of ways that cells
communicate in the nervous system. Although much of the
evidence to support the notion of multiple forms of intercellular communication is based on data obtained by pharmacological methods, there is reason to consider such data
since some naturally occurring synaptic events can be
closely mimicked by discrete pharmacological applications
(e.g., KUFFLER and YOSHIKAMI, 1975). Such mimickry is
possible in the case of synaptic potentials because they
are brief and clearly detectable signals, properties which
have led to the development of experimental techniques
designed to mimic the natural membrane events. Since
the physiological events occurring during "neurohormonal
communication" can be also mimicked by pharmacological
application of bag cell peptide, other membrane actions
induced by pharmacological application of neuronal substances (e.g., neuromodulation) might well reflect naturally occurring events.

From the foregoing it is reasonable to suggest that
intercellular communication in the nervous system may not
be restricted to "neurotransmission" occurring at specialized synaptic junctions between contiguous elements, but
rather that intercellular communication in the nervous
system may be more complex, utilizing a variety of signals
and messengers to generate different patterns of excitability. The three types of intercellular signalling
discussed above may be likened to several forms of modern
communication. For example, synaptic transmission is
similar to a telephone conversation between two people in
that the conversation is private and requires hard-wiring,
just as neurotransmission is a form of single cell-to-
single cell communication which requires contiguous neurons
in contact at specialized synaptic junctions. The signal

generated during neurotransmission is brief and shaped by
temporal, spatial, cooperative, desensitizing, diffusion
and metabolic factors. Ostensibly the signal functions to
momentarily alter cell excitability, both by bringing membrane potential closer to, or further away from spike threshold, and by increasing or decreasing membrane resistance,
thereby enhancing or attenuating the efficacy of other
synaptically evoked events. Neuromodulation, as defined
above, would be analogous to gain or volume control over a
telephone conversation. Alteration in the time course of
the synaptic conductance by peptides might be likened to
a "tuning" of the transmitted signal. Whether neuromodulation requires immediate proximity of the controlling
element with subsynaptic receptors needs to be studied.
Neurohormonal communication would be analogous to a radio
broadcast which involves public transmission and private
reception between remote elements. During this type of
communication only those receivers properly tuned to the
appropriate frequency will sense the signal. Thus, only
those target cells with the proper receptors (possibly
designed to engage different parts of the humorally con- ·
veyed molecule) will be engaged. The neurohormone-recep-
tor interaction, like the coupling of neurotransmitter
receptors to functionally different conductances, may
cause different effects in different target cells, leading
to excitation in some and inhibition in others. Thus, the
same signal may have a different meaning or consequence to
different receivers. The neurohormonal mode of intercellular communication would allow one cell type to regulate the
excitability of diverse and distant target neurons so as to
produce a concerted pattern of autonomic, endocrine and
motor behavior. This form of communication has not yet
been demonstrated in the vertebrate CNS. The three forms
of communication are outlined in Table 1.

REFERENCES

ADAMS, P. R., GAGE, P. W. and HAMMILL, D. P. (1976) Brain
    Res. 115, 506-511.
AKASU, T., OHTA, Y. and KOKETSU, K. (1978) Experientia,
    34, 488-490.
ARCH, S. (1972) J. Gen. Physiol. 60, 102-119.
ARECHIGA, H. and ACEVES, J. (1978) Soc. Neurosci. Abst.
    4, 505.
ASCHER, P., MARTY, A. and NEILD, T.O. (1978) J. Physiol.
    278, 177-206.
BARKER, J. L. and GAINER, H. (1974) Science, 184, 1371-
    1373.
BARKER, J. L. and MCBURNEY, R. N. Nature (in press).

BARKER, J. L. and RANSOM, B. R.   (1978a)   J. Physiol.
    280, 331-354.
BARKER, J. L. and RANSOM, B. R.   (1978b)   J. Physiol.
    280, 355-372.
BARKER, J. L. and SMITH, T. G. (1976)   Brain Res.   103,
    167-170.
BARKER, J. L. and SMITH, T. G. (1977)   in Society for
    Neuroscience Symposia, Vol. II: Approaches to the
    Cell Biology of Neurons (Cowan, W.M. and Ferrendelli,
    J.A., eds.) p. 340-373. Soc. Neurosci., Bethesda.
BARKER, J. L. and SMITH, T. G. (1978)   in Abnormal Neu-
    ronal Discharges, (Chalazonitis, N. and Boisson, M.,
    eds.)   Raven, New York. pp. 359-388.
BARKER, J. L. IFSHIN, M. and GAINER, H.   (1975)   Brain
    Res.   84, 501-513.
BARKER, J. L. NEALE, J. H., SMITH, T. G. and MACDONALD,
    R. L. (1978a)   Science 199, 1451-1453.
BARKER, J. L., NEALE, J. H. and SMITH, T. G.   (1978b),
    Brain Res. 154, 153-158.
BARKER, J. L., GRUOL, D. L., HUANG, L. M., NEALE, J. H.
    and SMITH, T. G.   (1978) in Characteristics and Func-
    tion of Opioids (Van Ree, J. and Terenius, L.)pp.
    87-98. Elsevier, North-Holland.
BOLTON, T. B. (1975)   J. Physiol.   250, 175-202.
BRANTON, W. D., MAYERI, E., BROWNELL, P. and SIMON, S. B.
    (1978)   Nature. 274, 70-72.
CHOI, D. W., FARB, D. H. and FISCHBACH, G. D. (1978)
    Nature, 269, 342-343.
COGGESHALL, R. E. (1967)   J. Neurophysiol.   30, 1263-1287.
CRAWFORD, A. C. and MCBURNEY, R. N. (1976)   J. Physiol.
    250, 205-226.
DIONNE, V. and STEVENS, C. F. (1975).   J. Physiol.   251,
    245-270.
DOSTROVSKY, J. and POMERANZ, B. (1973)   Nature New Biol.
    246, 222-224.
DUGGAN, A. W., HALL, T. G. and HEADLEY, P.M., (1976),
    Nature New Biol., 264, 456-458.
DUNLAP, K. and FISCHBACH, G. D. Nature   (in press)
FRAZIER, N. T., KANDEL, E. R., KUPFERMANN, I., WAZIRI, R.
    and COGGESHALL, R. E. (1967)   J. Neurophysiol.   30,
    1288-1351.
GILES, W. and NOBLE, S. J. (1976)   J. Physiol. 261, 103-
    123.
GRUOL, D. L., HUANG, L. M., BARKER, J. L. and SMITH, T. G.
    (1978)   Neurosci. Abst.   4, 408.
HUANG, L. M., GRUOL, D. L., BARKER, J. L. and SMITH, T. G.
    (1978)   Neurosci. Abst.   4, 410.
IFSHIN, M., GAINER, H. and BARKER, J. L. (1975)   Nature
    254, 72-74.

KATZ, B. (1966) Nerve Muscle and Synapse (McGraw-Hill, New York).

KLEIN, M. and KANDEL, E. R. (1978) Proc. Natl. Acad. Sci. 75, 3512-3516.

KOKETSU, K. and OHTA, Y. (1976) Life Sci. 19, 1009-1014.

KUBA, K. and KOKETSU, K. (1975) Brain Res. 89, 166-169.

KUBA, K. and KOKETSU, K. (1976) Jap. J. Physiol. 26, 703-716.

KUFFLER, S. W. and YOSHIKAMI, D. (1975) J. Physiol. 244, 703-730.

KUPFERMANN, I. (1970) J. Neurophysiol. 28, 865-876.

MACDONALD, R. L. and BARKER, J. L. (1978a) Nature, 271, 563-564.

MACDONALD, R. L. and BARKER, J. L. (1978b) Neurology, 28, 325-333.

MACDONALD, R. L. and NELSON, P. G. (1978) Science, 199, 1449-1451.

MCBURNEY, R. N. and BARKER, J. L. (1978) Nature, 274, 596-597.

NEHER, E. and STEVENS, C. F. (1977) Ann. Rev. Biophys. Bioeng. 6, 345-401.

PHILLIPS, M. I. (1978) Neuroendocrinol. 25, 354-377.

PINSKER, H. and DUDEK, F. E. (1977) Science 197, 490-493.

RANSOM, B. R., et al. (1977) J. Neurophysiol. 40, 1132-1150.

SEGAL, M. (1977) Neuropharmacology, 16, 587-592.

SMITH, T. G., BARKER, J. L., GRUOL, D. L. and HUANG, L. M. (1978) Neurosci. Abst. 4, 415.

SMITH, T. G., BARKER, J. L. and GAINER, H. (1975) Nature, 253, 450-452.

STUART, D. G. and STRUMWASSER, F. (1978) Neurosci. Abst. 4, 207.

TSIEN, R. (1974) J. Gen. Physiol. 64, 293-305.

WEAKY, J. N. (1969) J. Physiol. 204, 63-77.

WILSON, W. A. and WACHTEL, H. (1974) Science 186, 932-934.

WILSON, W. A. and WACHTEL, H. (1978) Science 202, 772-775.

ZIEGLGANSBERGER, W. and BAYERL, H. (1976) Brain Res. 115, 111-128.

ZIEGLGANSBERGER, W., FRY, J. P., HERZ, A., MORODER, L. and WURSCH, E. (1976) Brain Res. 115, 160-164.

ZIEGLGANSBERGER, W. and FRY, J. P. (1976) in Opiates and Endogenous Opioid Peptides (Kosterlitz, H. W., ed.) pp. 231-238.

DO BEHAVIORALLY ACTIVE POLYPEPTIDE HORMONES ACT AT
CRUCIAL "COMMAND" SITES OR AT MANY SITES, FROM "COMMAND"
DOWN TO "FINAL COMMON PATHS"?

Jeffrey L. Ram

Department of Physiology
Wayne State University, School of Medicine
Detroit, Michigan 48201

Mechanisms by which polypeptides interact with neurons in higher animals may have their resolution suggested
by analogy with results in simpler systems, such as in
cell culture systems (BARKER, et al., 1978), in isolated
parts of the nervous system, such as in brain slices
(BLOOM, et al., 1978), or in nearly intact, but simpler,
nervous systems of invertebrates (see below). For questions concerning organizational aspects of the nervous
system, cell culture and brain slice preparations lack
the requisite organization to provide an appropriate
analogy, but invertebrate systems offer technically accessible but nevertheless sufficiently structured preparations for studying these problems. For example, one question of interest to us is whether polypeptide hormones
which affect behavior exert their effects by acting at a
few crucial "command" sites in the nervous system or whether they interact at many levels, from "command" down to
"final common paths" residing in motor neurons and their
target muscles.

The concept of "command" cells in the nervous system
was first suggested by experiments in the crayfish (e.g.
WIERSMA, 1952) which showed that stimulation of one or a
few specific interneurons--the "command" neurons--could
bring into play a complex and coordinated sequence of activation of motor neurons leading to a recognizable motor
behavior (ATWOOD & WIERSMA, 1967). Analogous command functions have been described in other invertebrates (e.g.
in Pleurobranchaea, DAVIS, et al., 1974a; GILLETTE, et al.,
1978), and similar concepts have been invoked in explana-

tions of vertebrate behavior, particularly with regard to
locomotion (GRILLNER, 1975).

In order to analyze where in the "command" structure
of the nervous system that a hormone may be exerting its
effects, at least three things must be studied:
    (1)   behavioral effects of the hormone,
    (2)   neuronal circuits underlying the affected be-
          haviors, and
    (3)   effects of the hormone on identified neurons in
          the circuits at all levels of the "command"
          structure.
In no case has there been a complete analysis of the effects of a hormone with respect to the items listed above.
In the following discussion, progress towards achieving
these goals in several systems is reviewed.

Polypeptide hormones having dramatic and specific
effects on behavior have been studied in both vertebrates
and invertebrates. Probably, the best established and
most selective behavioral effect of a polypeptide hormone
in a mammal is the effect of angiotensin II (A II), which
drives animals to drink (SIMPSON & ROUTTENBERG, 1973).
The site of action of plasma A II for this effect is
thought to be the subfornical organ (SFO; SIMPSON, et al,
1975; SIMPSON & ROUTTENBERG, 1973). Unfortunately, little
detailed information is available concerning the role of
the SFO in the neuronal circuits controlling drinking behavior, and the exclusive role of the SFO as a locus of
A II effects on drinking behavior has also been challenged
(HOFFMANN & PHILLIPS, 1976). Moreover, the neuronal circuitry of the SFO is not well understood: With iontophoretic application of A II, saralasin (an A II antagonist),
and acetylcholine to SFO neurons (PHILLIPS & FELIX, 1976)
at least four pharmacologically distinguishable classes
of neurons in the SFO can be identified. Morphologically,
three areas of the SFO can be distinguished (PHILLIPS, et
al., 1974), and with the electron microscope, four classes
of neurons have been described (DELLMANN and SIMPSON,
1975). The relationships of these various types of neurons and the serotonergic (LICHTENSTEIGER, 1967) and cholinergic (LEWIS & SHUTE, 1967) innervation of the SFO to
drinking behavior is unknown. While there is therefore
little chance, at the present time, to determine on a
cellular basis the organizational levels at which A II
has this effect on behavior, continuing research in this
area will certainly be of great interest.

In invertebrates, Truman and co-workers have shown
in a silk-moth that a hormone which is most likely a

polypeptide causes the complex series of behaviors involved in eclosion from the pupal case (TRUMAN & SOKOLOVE,
1972; TRUMAN, 1978).   In the gastropod mollusc Pleurobranchaea, a polypeptide hormone (*) causes a stereotyped
egg-laying behavior (RAM, et al., 1977) which is accompanied by a suppression of feeding behavior (DAVIS, et
al., 1974b).   Similarly, in Aplysia, another gastropod,
a polypeptide egg-laying hormone is also present (KUPFER-
MANN, 1967; TOEVS & BRACKENBURY, 1969; STRUMWASSER, et
al., 1969; ARCH, 1976) which causes, as in Pleurobranchaea,
an accompanying suppression of feeding behavior (STUART
& STRUMWASSER, 1978).

    All of the above invertebrate systems are electro-
physiologically accessbile in that they have relatively
large neurons (some are up to 1 mm in diameter in Pleurobranchaea) which are reidentifiable from preparation to
preparation.   In Pleurobranchaea and Aplysia, neuronal
circuit analysis have identifed many of the neurons involved in feeding behavior, including command neurons,
coordinating neurons, motor neurons, and sensory neurons
(DAVIS, et al., 1973; SIEGLER, et al., 1974; LEE & LIE-
GEOIS, 1974; GILLETTE & DAVIS, 1977; GILLETTE, et al.,
1978; WEISS, et al., 1975; KANDEL, 1976).   Since feeding
behavior is one of the behaviors affected during egg-
laying episodes, circuit analysis of feeding behavior
provides a necessary prerequisite for determining whether
egg-laying hormone acts directly on the feeding circuits,
and if so, at what sites.

    The isolated nervous system of Pleurobranchaea is
illustrated in Figure 1.   The nervous system consists of
several discrete ganglia which can be easily dissected
out of this marine slug and can be maintained in an electrophysiologically viable condition for more than a day
in a simple medium consisting only of sea water.   Bioassay experiments on homogenates of various ganglia from
the central nervous system show that the egg-laying hormone is found in two symmetrically located ganglia, the
pedal ganglia (Table I).

    *Throughout this paper, egg-laying hormone will be
    referred to for convenience in the singular, al-
    though present evidence does not exclude the possi-
    bility that several molecular species, working
    either alone or synergistically could be responsible
    of egg-laying and accompanying behaviors (RAM, et
    al., 1977).

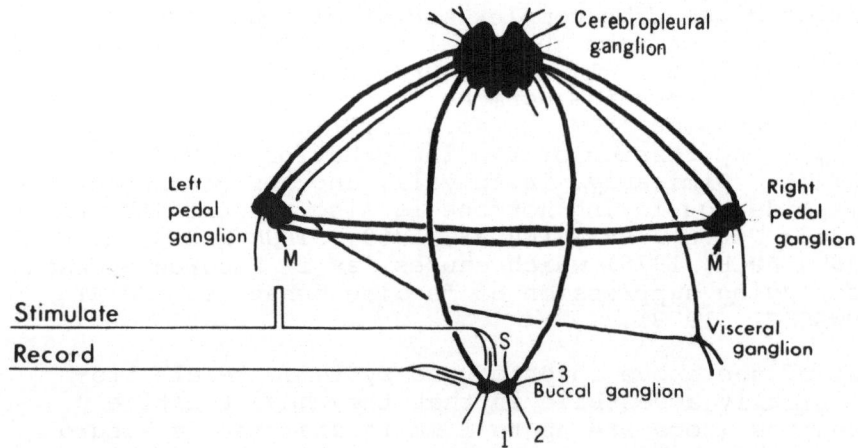

Fig. 1.   Schematic diagram of the isolated nervous system
          of Pleurobranchaea. Stimulation of one or both
          stomatogastric nerves causes coordinated output of
          action potentials from nerve roots of buccal gan-
          glia, as is illustrated in Figure 3. Buccal gan-
          glion nerve roots: S, Stomatogastric nerve; 1,
          first root; 2, second root; 3, third root. M,
          medial lobe.

     The egg-laying hormone has been partially purified
from pedal ganglia, and its molecular weight estimated
by means of gel filtration on Sephadex G-50. Egg-laying
behavior was caused by column eluates having an average
molecular weight of about 6000 daltons (RAM, et al.,
1977). In subsequent experiments, pedal ganglia were
homogenized in water, centrifuged at 8000 x g for 25 min.,
and fractionated on Sephadex G-50 in 0.2 M ammonium acetate (pH 7.0). Fractions were freeze-dried and run on
SDS polyacrylamide gels. Fractions eluting where egg-
laying hormone would have eluted showed a major band at
the low molecular weight end of the gel, where a 6000
dalton polypeptide should run (Figure 2). Heterogeneity
of the material run on the gel is indicated by the small
shoulder at a higher molecular weight, although the pattern for this fraction is comparatively simple compared to
that for higher molecular weight eluates (Figure 2).

Additional analytical techniques for assessing heterogeneity of hormone containing fractions and for purifying the hormone further are presently under investigation.

TABLE I.   BIOASSAY RESULTS (LARGE NON-LAYING DONORS)

Test sample (homogenate in 1.0 ml sea water)	Animals laying	Samples tested
LEFT PEDAL GANGLION	9	11
RIGHT PEDAL GANGLION	9	9
CEREBROPLEURAL GANGLION	0	6*
BUCCAL AND VISCERAL GANGLIA COMBINED	0	5*

*Subsequent injection of the recipients of these
samples with pedal ganglion homogenates gave egg-
laying in 5 of 6 cerebropleural recipients and 4 of
5 of the buccal/visceral ganglia recipients.

To localize the egg-laying hormone within the pedal
ganglia, protein patterns of various parts of the pedal
ganglia were analyzed on SDS polyacrylamide gels. Only
the medial lobes (Figure 1) of these ganglia gave a staining pattern with a band at the low molecular weight end,
corresponding to the band in the 6.0 fraction illustrated
in Figure 2 (RAM, et al., 1977). This result, along with
bioassay experiments, staining characteristics of medial
lobes in sections, and the incorporation of ^{3}H-leucine
into low molecular protein in the medial lobe suggest
that the medial lobes are the main locus for synthesis
and storage of egg-laying hormone in the pedal ganglia
(RAM, et al., 1977).

Electrophysiological investigations on the feeding
motor system in the Pleurobranchaea nervous system have
been done by Davis, Mpitsos, Siegler, Gillette, Lee, and
others, including myself (Figure 3; DAVIS, et al., 1973;
SIEGLER, et al., 1974; LEE & LIEGEOIS, 1974; GILLETTE &

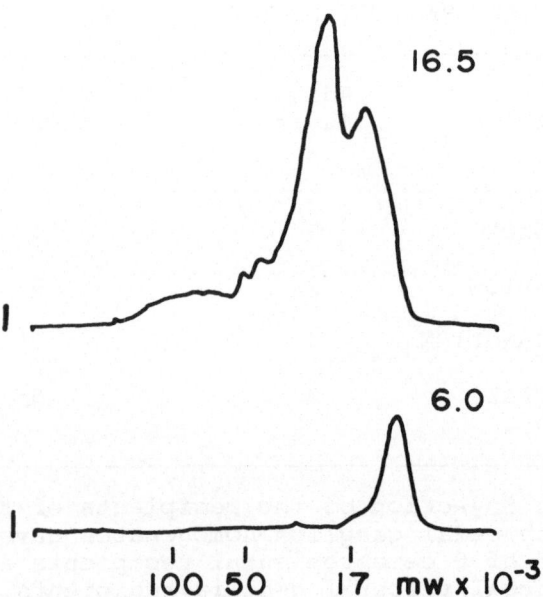

Fig. 2. SDS gel electrophoresis patterns of pedal ganglia
   fractions separated on Sephadex G-50. Thirteen pedal
   ganglia were homogenized in 2 ml double distilled
   water. Bioassay of .15 ml of the homogenate caused
   egg-laying. Remainder was frozen, and upon thawing
   was centrifuged at 8000 x g for 25 min. Sucrose
   (3 mg) and blue dextran (3 mg) were added, and the
   sample was put on a 1.5 x 60 cm Sephadex G-50 Fine
   column (equilibrated with 0.2 M ammonium acetate,
   pH 7.0) from which it was eluted with the column
   buffer. Fractions were freeze-dried and then run on
   SDS polyacrylamide gels (RAM, et al., 1977). Gels
   were stained with Coomassie Blue and densitometer
   tracings made. The patterns in the figure are for
   fractions eluting where proteins of average molecular
   weight of 16,500 daltons (16.5) and 6,000 daltons
   (6.0) would have eluted according to calibration
   with a set of molecular weight standard proteins
   (as in RAM, et al., 1977).

DAVIS, 1977; GILLETTE, et al., 1978). These investigations have demonstrated that in the isolated nervous
system, stimulation of the stomatogastric nerves of the
buccal ganglia produces a coordinated output from buccal
ganglia nerve roots (Figure 3). These nerves innervate
the musculature of the buccal mass, and their coordinated
output can produce the oscillatory movements of the buccal
mass which occur during feeding (DAVIS, et al., 1973).
Recently, Davis and Villet (submitted) have observed that
this output in the isolated nervous system is subject to
modulation by pedal extracts containing the egg-laying
hormone.

     In Davis & Villet's experiments, isolated nervous
system preparations consisted of the cerebropleural and
buccal ganglia connected by the cerebrobuccal connective
nerves. Bursting activity associated with the feeding
motor program in the small oral veil nerves and buccal
third root nerves in response to stomatogastric nerve
stimulation (5 - 10 V, 1 Hz, for 2 min.) was examined.
Addition of pedal ganglia extracts to the medium caused
a suppression of feeding bursts in 12 of 16 preparations.
The four preparations in which no change was observed had
been dissected from animals with immature gonads, thereby
suggesting that the effect seen was related to the sexual
maturity of the animal. Most of the animals which responded had mature gonads (greater than 0.8 - 1.0 g, RAM,
et al., 1977 and analysis by DAVIS & VILLET), and the
average suppression seen in these animals was to less than
30% of the pre-extract number of bursts in response to
the stomatogastric nerve stimulus. Upon washing out the
extract, feeding output returned to control levels for
10 of the 12 suppressed preparations, but usually declined
again to a low level within two hours after the washout.
The effect appears to be specific to pedal ganglia extracts, as is egg-laying hormone, since extracts of other
ganglia did not cause a similar suppression (N = 3).
Experiments with purified egg-laying hormone are obviously
necessary to determine whether this effect can be attributed to the egg-laying hormone; however, the experiments
certainly provide an experimentally produced suppression
of a feeding motor program for which analysis of the
organizational level or levels at which the effect is
produced may be investigated.

     Experiments relevant to the question of how egg-
laying hormones act on molluscan feeding is also available
from studies on Aplysia. Stuart & Strumwasser (1978)
report that Aplysia egg-laying hormone (which is not

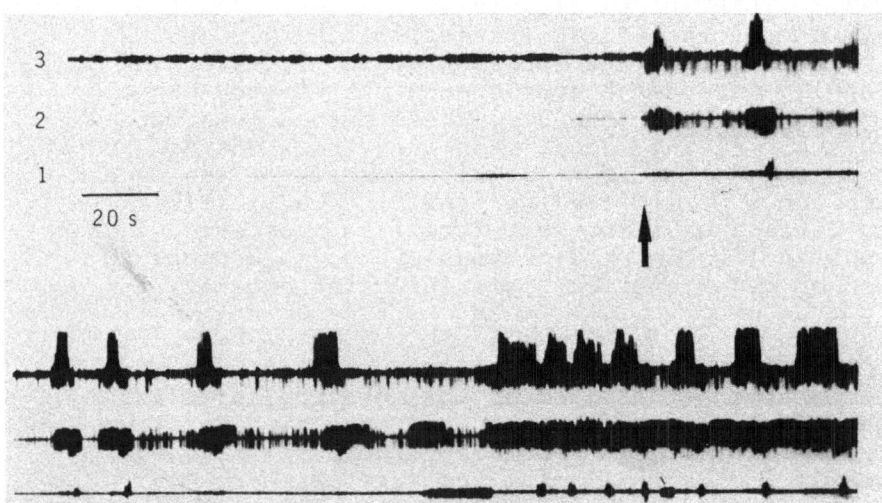

Fig. 3. Electrophysiological activity recorded in an
         isolated buccal ganglion preparation. Activity is
         recorded from buccal ganglia nerve roots: third
         root, 3; second root, 2; and first root, 1. Stimu-
         lation of stomatogastric nerve (5 Hz, 5 V) was begun
         at the arrow. A break of approximately 10 sec occurs
         between the top three lines in the figure and the
         bottom three lines.

identical to Pleurobranchaea egg-laying hormone (RAM, et
al., 1977)), partially purified by ammonium sulfate pre-
cipitation, ion exchange chromatography, and gel filtration, caused an increase in the tonic firing rate of at
least two pairs of neurons in an isolated buccal ganglion
preparation. The onset of this effect resembles in time
course the suppression of feeding behavior in intact animals after injection of the egg-laying hormone. Since
one of these pairs of neurons has axons that go into the
buccal musculature, Stuart & Strumwasser (1978) suggest
that this pair of tonically firing neurons may innervate
muscles in a way that antagonizes feeding behavior. The
other pair of buccal neurons excited by the egg-laying
hormone has axons in the nerves connecting the buccal
ganglia to the cerebral ganglion and could conceivably

play a coordinating role, modulating the activity of feeding arousal cells in the cerebral ganglion (e.g. the meta-
cerebral cells, WEISS, et al., 1975). With regard to
sites at which the egg-laying hormone may be producing
these effects, several alternatives are possible: The
hormone may have a direct and independent effect on each
pair of neurons; the hormone may excite one pair which
then excites the other; or both pairs may be excited by a
common presynaptic neuron on which the hormone actually
acts. Each possibility would, of course, have different
implications for the question of the hierarchical level(s)
at which polypeptide hormones act.

It should also be noted that the phenomena reported
as possible effects of egg-laying hormones in Aplysia and
Pleurobranchaea on feeding circuitry are qualitatively
different. It remains to be seen whether both types of
effects are found in both animals (and not seen because
different aspects of buccal ganglia output were chosen
for study), whether the responses to egg-laying hormones
are indeed different, or whether the differences are due
to the drastically different degree of hormone purity
and concentration between the two studies.

Effects of the Aplysia egg-laying hormone in the
abdominal ganglion and elsewhere in the animal are also
under investigation. The egg-laying hormone of Aplysia
is synthesized and stored in two clusters of neurons in
the abdominal ganglion. These neurons send their processes, among other places, into the connective tissue
sheath of the ganglion, from which egg-laying hormone
presumably can be released into the interstitial spaces
of the ganglion (COGGESHALL, 1967; FRAZIER, et al., 1967).
Mayeri, et al., (1978a) have shown that activation of the
egg-laying hormone secreting cells is frequently followed
by excitation of cell L10, an interneuron which acts as a
command neuron for heart acceleration (Figure 4). Although it is not known whether this effect is due to a
direct effect of the egg-laying hormone on L10, this result takes on added significance in light of the recent
observation of Smock, et al., (1978) that the heart of
Aplysia can be excited directly by a polypeptide, which
by all criteria thus far tested, including protease sensitivity, anatomical localization, and gel filtration, is
identical to egg-laying hormone. If it can be shown that
the effects seen by Mayeri, et al., (1978a) on L10 are due
to the polypeptide studied by Smock, et al., (1978) acting
directly on L10, then this will demonstrate that a hormone
can exert direct control on a motor system both at the
command neuron level and also at the target of that com-

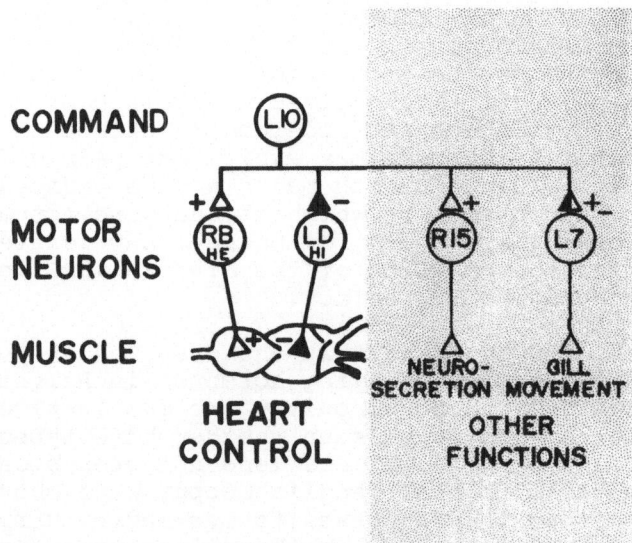

Fig. 4. L10, an identified cell in the abdominal ganglion
        of Aplysia, has synaptic connections which enable it
        to act as a command cell for heart rate and to co-
        ordinate gill movements and neurosecretion with heart
        rate changes. The figure shows some of the many con-
        nections which L10 has in the abdominal ganglion
        (adapted from KANDEL, 1976). Possible sites of hor-
        mone action are at many levels in this command hier-
        archy, including L10, R15, and the heart (see text).

mand, i.e. the heart. This story lacks other elements,
among them the behavioral confirmation that the heart is
in fact accelerated during egg-laying and also an assess-
ment of the relative importance of the central nervous
system effect and the direct cardioaccelerating effect of
the hormone, if in fact it normally occurs.

     Mayeri, et al., (1978b) have also shown that activa-
tion of egg-laying hormone secreting cells is accompanied
by excitation of cell R15, a neurosecretory cell believed
to be involved in water balance in Aplysia (KUPFERMANN &
WEISS, 1976; STINNAKRE & TAUC, 1969; studies in a homo-
logous cell, GAINER, 1972). In collaboration with Arch
(BRANTON, et al., 1978), Mayeri's laboratory has now shown
that application of partially purified (gel filtration,
isoelectric focusing, and dialysis) egg-laying hormone
directly to R15 causes similar excitatory effects. Though

it remains possible that excitation of R15 could nevertheless have been due to an indirect effect through diffusion of the hormone to another target site in the ganglion, the most parsimonious interpretation would be that
egg-laying hormone does indeed directly excite R15. Since
L10 is also presynaptic to Rl5 (Figure 4), it appears that
this may be an example of a hormone acting directly on two
neurons, one of which is presynaptic to the other. In
this case, the "final common path" is not a motor neuron
or a muscle but a neurosecretory cell. As with the work
on the heart and the heart circuits, this story also lacks
the "behavioral" confirmation, that the effect of activating R15, presumably water uptake, accompanies normal egg-
laying behavior.

In summary, the following observations may be made:
(1) Pleurobranchaea and Aplysia have polypeptide hormones
which have behavioral effects; (2) the sites of action of
these hormones can be studied at a cellular level, and the
organizational levels at which they act can be determined
in these systems; (3) the weight of evidence so far argues for multiple sites of action and for action at several
levels in the hierarchy of command; however, (4) much
work remains to be done to adequately analyze this question.

ACKNOWLEDGEMENTS

Research on Pleurobranchaea by the author described
herein was supported by NIH Grant Number 05384-17 and
NIH post-doctoral fellowship #1 F32 NS05238-01, and was
carried out in part in collaboration with W. J. Davis
and S. R. Salpeter and with the support of NIH Research
Grants to Davis (NS09050 and MH23254). The author grate-
fully acknowledges receiving several unpublished manuscripts from W. J. Davis and E. Mayeri, to which references are made in the text.

REFERENCES

ARCH, S. (1976) Amer. Zool. 16, 167-175.

ATWOOD, H. L. & WIERSMA, C. A. G. (1967) J. Exp. Biol.,
    46, 249-261.

BARKER, J. L., NEALE, J. H., SMITH, T. G., JR., & MAC-
    DONALD, R. L. (1978) Science, 199, 1451-1453.

BRANTON, W. D., ARCH, S., SMOCK, T., & MAYERI, E. (1978)
    Proc. Nat. Acad. Sci., 75, (in press).

COGGESHALL, R. E. (1967) J. Neurophysiol., 30, 1263-1287.

DAVIS, W. J., MPITSOS, G. J., & PINNEO, J. M. (1974b)
    J. Comp. Physiol., 90, 225-243.

DAVIS, W. J., MPITSOS, G. J., SIEGLER, M. V. S., PINNEO,
    J. M., & DAVIS, K. B. (1974a) Amer. Zool., 14,
    1037-1050.

DAVIS, W. J., SIEGLER, M. V. S., & MPITSOS, G. J. (1973)
    J. Neurophysiol., 36, 258-274.

DELLMANN, H. D. & SIMPSON, J. B. (1975) Brain-Endocrine
    Interaction II. The Ventricular System (Karger,
    Basel), pp. 166-189.

FRAZIER, W. T., KANDEL, E. R., KUPFERMANN, I., WAZIRI, R.,
    & COGGESHALL (1967) J. Neurophysiol., 30, 1288-1351.

GAINER, H. (1972) Brain Res., 39, 387-402.

GILLETTE, R. & DAVIS, W. J., (1977) J. Comp. Physiol.,
    116, 129-159.

GILLETTE, R., KOVAC, M. P., & DAVIS, W. J. (1978) Science,
    199, 798-801.

GRILLNER, S. (1975) Physiol. Rev., 55, 247-304.

HOFFMANN, W. E. & PHILLIPS, M. I., (1976) Brain Res.,
    108, 59-73.

KANDEL, E. R. (1976) Cellular Basis of Behavior (San
    Francisco: W. H. Freeman).

KUPFERMANN, I. (1967) Nature, 216, 814-815.

KUPFERMANN, I. & WEISS, K. R. (1976) J. Gen. Physiol.,
    67, 113-123.

LEE, R. M. & LIEGEOIS, R. J. (1974) J. Neurobiol., 5,
    157-164.

LEWIS, P. R., & SHUTE, C. C. D. (1967) Brain, 90, 521-
    540.

LICHTENSTEIGER, W. (1967) Brain Res., 4, 52-59.

MAYERI, E., BROWNELL, P. & BRANTON, W. D. (1978a) J.
    Neurophysiol. (in press).

MAYERI, E., BROWNELL, W. D., BRANTON, W. D., & SIMON, S. B.
    (1978b) J. Neurophysiol. (in press).

PHILLIPS, M. I. & FELIX, D. (1976) Brain Res., 109, 531-
    540.

PHILLIPS, M. I., BALHORN, L., LEAVITT, M. & HOFFMANN, W.
    (1974) Brain Res., 80, 95-110.

RAM, J. L., SALPETER, S. R., & DAVIS, W. J. (1977) J. Comp. Physiol., 119, 171-194.

SIEGLER, M. V. S., MPITSOS, G. J., & DAVIS, W. J. (1974) J. Neurophysiol., 37, 1173-1196.

SIMPSON, J. B. & ROUTTENBERG, A. (1973) Science, 181, 1172-1175.

SIMPSON, J. B., SAAD, W. A., & EPSTEIN, A. N. (1975) in Onesti, G., Fernandes, M. & Kim, K. E. (ed.) Regulation of Blood Pressure by the Central Nervous System (New York: Grune & Stratton) pp. 191-202.

SMOCK, T., ARCH, S., & LLOYD, P. (1978) Soc. Neurosci. Abst., 4, 206.

STINNAKRE, J. & TAUC, L. (1969) J. Exp. Biol., 51, 347-361.

STRUMWASSER, F., JACKLET, J. W., & ALVAREZ, R. F. (1969) Comp. Bioch. Physiol., 29, 197-206.

STUART, D. K. & STRUMWASSER, F. (1978) Soc. Neurosci. Abst., 4, 207.

TOEVS, L. A. & BRACKENBURY, R. W. (1969) Comp. Bioch. Physiol., 29, 207-216.

TRUMAN, J. W. (1978) J. Exp. Biol. 74, 151-173.

TRUMAN, J. W. & SOKOLOVE, P. G. (1972), Science, 175, 1491-1493.

WEISS, K. R., COHEN, J. & KUPFERMANN, I. (1975) Brain Res., 99, 381-386.

WIERSMA, C. A. G. (1952) Cold Spring Harbor Symp. Quant. Biol., 17, 155-163.

NEUROTRANSMITTERS AND NEUROMODULATORS AND THEIR MEDIATION

BY CYCLIC NUCLEOTIDES

G.R. SIGGINS

The Salk Institute,
La Jolla, California

INTRODUCTION

In a symposium devoted to the discussion of modulators, mediators and specifiers in brain, it seems propi-
tious first to address the issue of what these agents are
and what criteria may be developed to define them. This
task seems particularly timely now because of the bur-
geoning number of substances found in brain which seem
not to strictly fit our preconceived notions of a neurotransmitter. Therefore, after developing criteria for
neurotransmitters, mediators and modulators, I will provide examples of substances that could satisfy the criteria for these classifications, with special emphasis on
norepinephrine (NE), cyclic AMP and the opioid peptides.

The hypothesis of chemical neurotransmission was
first developed for the peripheral nervous system, which
was thought to include nerve terminals that communicated
to muscle tissue via release of acetylcholine or catecholamines (ELLIOT, 1904: DALE, 1906; DALE, 1914; LOEWI, 1921).
This chemical doctrine was later generalized to the central nervous system, although not without vigorous discussion along the way (cf. DALE, 1935; BROOKS & ECCLES,
1947). Arising out of the controversy and reseach in
this area was the principle that different chemicals (and
different neurons) transmit different messages (excitation or inhibition). Most of the resistance to the doctrine of central chemical transmission seems to have arisen
out of consideration for the speed by which spinal cord

synapses transmitted their messages. Sir John Eccles,
then a critic of the doctrine, seemed to feel that central messages demanded rapid electrical transmission, although later studies have shown that the original peripheral models for the chemical doctrine (the autonomic
nervous system) showed both rapid and slow forms of chemical communication (BURNSTOCK & HOLMAN, 1963; SPEDEN, 1964).
Although the ECCLES group performed the experiments that
eventually proved the chemical doctrine for the spinal
cord (COOMBS et al., 1955; ECCLES et al., 1954), the
narrow view that persisted until recently was that neurotransmission operated in only two modes, fast excitation
via excitatory postsynaptic potentials (EPSPs) and fast
inhibition via inhibitory postsynaptic potentials (ISPSs),
both caused by the opening of conductance channels in the
post-synaptic membrane (see ECCLES, 1964; KATZ, 1966)
brought about by the release of fast-acting transmitters
from large, rapidly conducting axons.

However, results from more recent electrophysiological and ultrastructural studies on autonomically innervated smooth muscle systems point up the fact that neurotransmission need not always involve fast signals transferred across narrow, specialized synaptic clefts. Features common to many neuro-effector smooth muscle systems
(see BURNSTOCK & HOLMAN, 1963; CAESAR et al., 1957;
RICHARDSON, 1958; LEVER et al., 1967) are: 1) long junctional delays; 2) long time courses of post-junctional
potential changes: 3) the requirement for repetitive
nerve activation to produce detectable, summated responses;
4) wide junctional gaps of up to thousands of angstroms,
without synaptic specializations: 5) transmitter release from boutons or varicosities "en passage".

It is now becoming apparent that nerve cells also
communicate with one another by means of a much more complex vocabulary than a simple rapid "yes" or "no". Ink-
lings of the idea that neuronal transmission might not
always require the rapid opening of conductance channels
was seen in the non-synaptic models of the photoreceptor
(BAYLOR & FUORTES, 1970) and the stretch receptor (NAKA-
JIMA & TAKAHASHI, 1966), where "leaky" ion channels seemed
to be closed or "inactivated" to produce the physiological
response. The possibility that such a novel form of
response might apply also to vertebrate central neurotransmission gained credence with results of studies on
mammalian and amphibian sympathetic ganglia. Here, a new
form of slow transmission, the slow IPSP (sIPSP) and the
slow EPSP (sEPSP), occurred without an increase in membrane
conductance, and co-existed with the more usual form of

rapid nicotinic depolarization (EPSP) accompanied by a
conductance increase (ECCLES & LIBET, 1961; NISHI & KOKET-
SU, 1968). Although a controversy still centers on whether
the sIPSP is generated by a decrease in ionic conductance
(WEIGHT & PADJEN, 1973a) or by activation of an electrogenic pump (see NISHI & KOKETSU, 1968), the important
point is that investigators began to ask if such a new
form of transmission might not also apply to the central
nervous system.

Shortly after the discovery of the sIPSP and the
sEPSP, it became apparent that similar slow mechanisms
might apply also to certain forms of synaptic transmission
in brain. Thus, exogenous norepinephrine (NE) and activation of NE-containing fibers was found to evoke slow
hyperpolarizations in cerebellar Purkinje cells associated
with either a decrease or no change in ionic conductance
(HOFFER et al., 1973; see below), while muscarinic, cholinergic depolarizing responses associated with a decreased
conductance were reported in a cortical area thought to
possess cholinergic fibers (KRNJEVIC et al., 1971). These
responses in brain neurons thus appear to parallel those
in sympathetic ganglia, where the sEPSP is thought to
arise from activation of muscarinic receptors (ECCLES &
LIBIT, 1961; KOKETSU, 1969), while the sIPSP may be generated by release of a catecholamine (ECCLES & LIBET, 1961,
but see WEIGHT & PADJEN, 1973b).

However, at about this time researchers began to anticipate the presence in brain of even more complicated
forms of neuronal communication. For example, hypotheses
were advanced that the post-synaptic action of certain
neurotransmitters (NE, ACh) might be mediated by the generation of cyclic nucleotides (cyclic AMP and cyclic GMP;
see below). In addition, with the discovery of a wide
array of neuroactive peptides in brain came the likelihood
that several of these might alter the responses of neurons
to other neurotransmitters without having a direct action
of their own. Examples of this putative form of neuronal
communication were seen in iontophoresis experiments
showing that TRH could alter the excitatory action of ACh
(YARBROUGH, 1976), and that opioid peptides could reduce
synaptic efficacy and glutamate and ACh responses (ZIEGL-
GANSBERGER & FRY, 1976; ZIEGLGANSBERGER & BAYERL, 1976;
BARKER et al., 1978); in both cases interactions occurred
without an apparent direct peptide effect on neuronal excitability.

It is against this brief historical backdrop that an

attempt will be made to develop criteria to be used for
identifying neurotransmitters, neuromodulators and mediators in nervous tissue.   Where possible I will apply broad
criteria with sufficient latitude for inclusion of possible undiscovered new forms of neuronal communication,
rather than later be forced to generate new terms for a
process that might still be best described as, for example,
neurotransmission.   In this effort, the features of peripheral neurotransmission will be kept in mind as continuing
models for central communication.

Neurotransmitter Criteria

Analysis of synaptic function in the nervous system
labors against a frustratingly complex system of hetero-
genous interconnected neurons and associated neuroglia.
Therefore, the analysis of how modulators and mediators

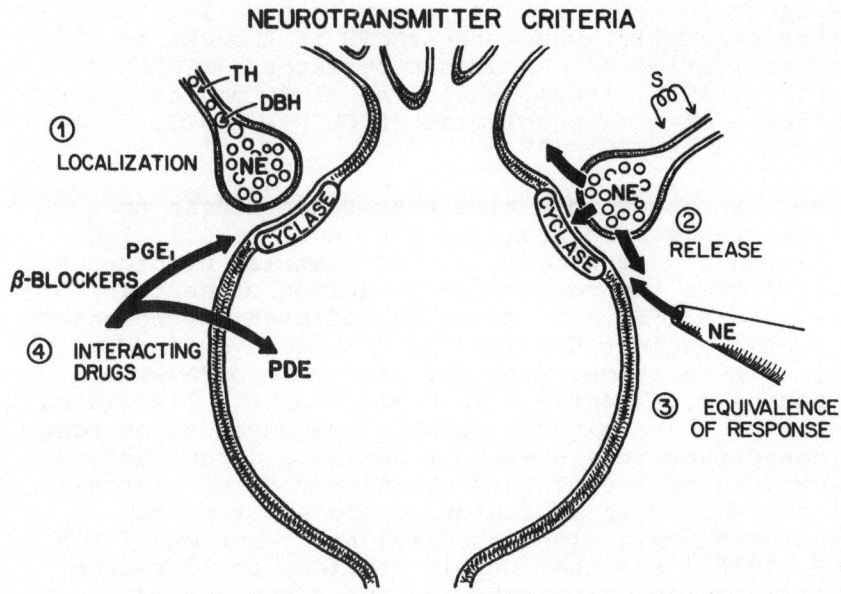

Fig. 1: Schematization of the major criteria for establishing a substance as a neurotransmitter, using norepinephrine (NE) as an example of a central inhibitory transmitter operating through β-adrenergic receptors and activation of adenylate cyclase.

might participate in neuronal communication might be best
undertaken after discussion of the criteria for the identification of a synaptic transmitter, which might be considered a first order messenger.

My personalized criteria (Fig. 1) for identification
of a neurotransmitter may be paraphrased and condensed
from the several criteria previously suggested (cf. WERMAN,
1966):

1. Neuronal localization of the substance and its
enzymes of synthesis and degradation.
2. Release of the substance upon selective activation of a specific neuronal pathway.
3. Identical physiological response to exogenously
applied transmitter and to activation of the pathway.
4. Identical action of pharmacological agents
(antagonists, etc.).

To some, these criteria may show a lack of concern
for the ultrastructural bases of neurotransmission (e.g.,
synaptic "specializations"). However, such considerations
were omitted because they are still very controversial,
because they have been discussed in detail elsewhere
(DISMUKES, 1977; BLOOM, 1979a, b), and because of continued difficulties in documenting correlations between
structure and physiology. Moreover, the possibilities
for a broad array of morphological forms of neurotransmission in CNS may be indicated in the peripheral nervous
system: some sympathetic boutons transmit direct messages
without synaptic specializations to their smooth muscle
contacts in an "en passage" fashion, but do show them at
intraganglionic nerve-nerve contacts (HOKFELT, 1967).

It will be noted that there is no mention in the
stated criteria of the speed of transmission. Thus, as
in peripheral systems, a slowly acting substance is just
as much a neurotransmitter as a fast acting one. Moreover, a substance is not excluded as a neurotransmitter
because it alters (or "modulates") the response of another
neurotransmitter. If this were the case, most currently-
conceived neurotransmitters would be excluded because they
alter membrane conductance (or resistance), and the result
of this would likely be altered currents generated by
other synapses or transmitters. For example, GABA and
classical IPSPs inhibit cells by increasing conductance
to Cl^- ions: such an increase in conductance would cause
a decrease in the membrane resistance such that a given
synaptic current generated by another input to the cell

would generate a smaller potentiate shift across the
reduced resistance. Thus, GABA might be expected to
inhibit firing by two mechanisms, hyperpolarization and
reduction of other synaptic potentials. By the same
reasoning, application of glutamate or activation of a
fast EPSP (increasing conductance to Na^+ and reducing
membrane resistance) might be expected to excite cells
by depolarization, but reduction of other synaptic potentials should also be considered.

The word "response" in criterion number 3 above was
purposely chosen as a broad term, in order to cover a wide
spectrum of possible direct mechanisms of transmitter
action on cell excitability. Thus, substances which
hyperpolarize or depolarize by increased or decreased conductance, or by activating or inactivating electrogenic
pumps, or which alter spike thresholds (with or without
potential changes), will all be considered neurotransmitters by these criteria because they directly affect
neuronal excitability.

Mediator Criteria

A mediator of neuronal communication might be best
exemplified by the role of cyclic nucleotides as "second
messengers". The second messenger concept as currently
applied to brain has evolved from the mediator role of
cyclic AMP in peripheral hormonal responses, as first
suggested by SUTHERLAND et al. (1965). Modified for
neuronal transmission or local modulation (see below),
this concept may be summarized (Fig. 2) as follows: a
synaptically released neurotransmitter or locally released
neuroactive agent could act at certain pre- or postsynaptic
receptors to activate adenylate cyclase and the synthesis
of cyclic AMP. Intracellular cAMP would then initiate
subsequent enzymatic or molecular events, which, among
other actions (e.g., long-term trophic effects) could
result in changes in membrane potential and cell discharge
rate. Four major criteria may be adapted from the criteria
for hormones, to establish that the action of a transmitter is mediated by a cyclic nucleotide (BLOOM, 1975,
1976: SIGGINS, 1978).

1. Exogenous neurotransmitter substance and activation
of the synaptic pathway both regulate intracellular levels
of cyclic nucleotide in the postsynaptic cell.
2. The change in intracellular cyclic nucleotide
content should precede "the biological event" triggered
by the transmitter or nerve pathway.

3.   Responses to the transmitter or nerve pathway
should be logically altered by drugs that specifically
interact with the nucleotide cyclase or that inhibit the
appropriate phosphodiesterase.
4.   Exogenous cyclic nucleotides (and analogues which
activate protein kinase) should elicit the biological
event caused by the transmitter or nerve pathway.

These criteria are schematized in Fig. 2.

Unfortunately, attempts to satisfy the second messenger criteria for central neurons meets with considerable
technical obstacles, such as the indirect actions of
systemic drugs, blood-brain barriers to systemic agents,
slow nucleotide sampling and measurement times compared
to fast synaptic events, and relative impermeability of
cyclic nucleotides into target cells. Several of these
obstacles can be partially overcome in the central nervous
system by the techniques of microiontophoresis and

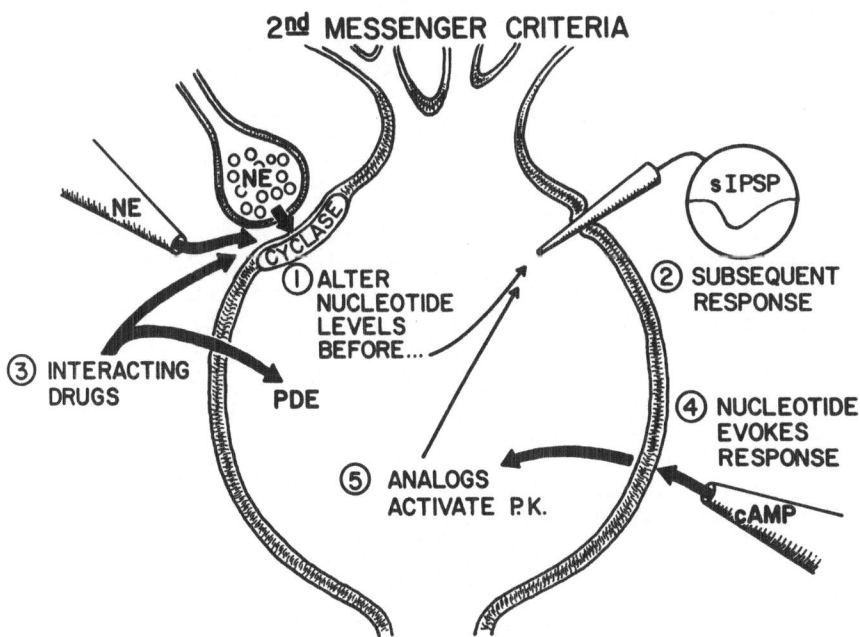

Fig. 2: Schematization of criteria for second messenger
    mediation, using cyclic AMP as an example of a
    second messenger mediating norepinephrine first
    messages postsynaptically. Reproduced from SIGGINS,
    (1978).

electrophysiology, as has been applied to several brain
areas (see below).

In spite of these difficulties, an important new
consideration with respect to the likelihood that cyclic
nucleotides mediate the function of neurotransmitters is
the idea that a neurotransmitter can generate potential
and conductance changes not by altering passive membrane
properties but by altering neuronal metabolism (or energy
function) which in turn alters transmembrane properties.
This constitutes the "energy" dimension of neurotransmission as seen by BLOOM (1979a).

Modulator Criteria

The term modulator has received increasing attention
recently as a catch-all category capable of including the
action on all substances whose actions differ from those
of the classically conceived spinal cord transmitters.
Indeed, references to "modulators" in the literature seem
to be increasing at a frequency in direct correlation to
the discovery of new neuroactive substances; the variety
of definitions of the term "modulator" seems almost as
numerous as the number of new brain substances. These
definitions range from an emphasis on a long time-course
of action (FREDERICKSON, 1977), to any substance (e.g.,
CO_2 or NH_3) released from non-synaptic sources (FLOREY,
1967). However, a criterion common to most definitions
is the notion that a modulator should have no direct effect
of its own, but that it can alter the excitability of the
post-synaptic neuron in response to other synaptic afferents (BARKER & SMITH, 1977; BARKER et al., 1978).
Here again, I will apply a broad definition of modulator,
using primarily the latter criteria (schematized in Fig. 3).
In my view, a modulator should:

    1. Alter responses of other neurotransmitters
    2. Have no direct effect on spiking, membrane poten-
tial or conductance.
    (3. Be localized in the vicinity.)

No distinction is made here between those modulators
released from neurons and those not, since such agents
may be sub-classified as "neuromodulators" and "local
modulators", respectively, without loss of clarification.
Again, no consideration is given to the distance over
which the substance must diffuse (i.e., the synaptic
"gap") or whether synaptic specializations are present

(see BLOOM, 1979b); such considerations would have assigned
catecholamines and acetylcholine of the autonomic system
as "modulators" rather than transmitters. Nor is strict
adherence paid to the criteria of BARKER & SMITH (1977),
and BARKER et al. (1978), who define a neuromodulator as
a substance that "alters synaptic receptor-coupled conductances without direct activation of such conductances".
The two criteria above are drawn in a less restrictive
manner, so as to exclude neurotransmitters which directly
alter cell excitability by means other than strict activation of conductances (such as by alteration of electrogenic pumps, spike thresholds and inactivation of conductances; see "Transmitter Criteria" above), and so as not
to exclude alterations of other synaptic receptors that
could operate by these other means.

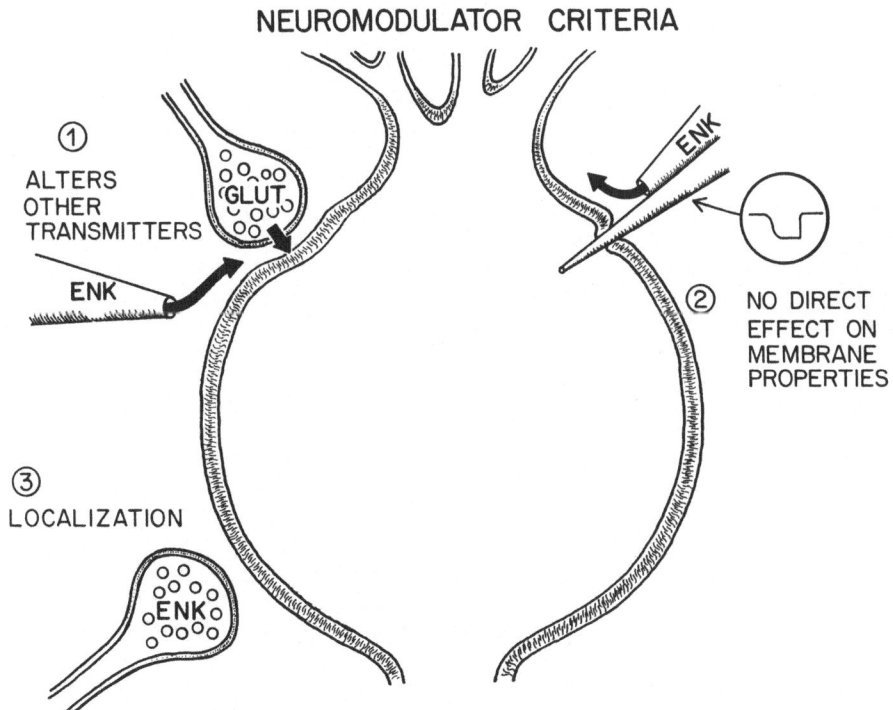

Fig. 3:   Schematization of the two major criteria for a
          neuromodulator, using enkephalin (ENK) as an example
          of a substance that reduces excitatory synaptic trans-
          mission (via glutamate; GLUT) yet has no direct action
          of its own on membrane excitability.

However, even with such a broadly-drawn definition
of "modulator", formidable technical difficulties are
encountered in proving their existence. Proof of modulation of other synaptic afferents or receptors is difficult
enough, and usually relies upon sophisticated electrophysiological and iontophoretic techniques. Much more
difficult, however, is strict documentation of a lack
of direct effect of a substance. This usually requires
intracellular recording. Moreover, negative results in
iontophoretic and electrophysiological techniques are
always suspect especially when little is known of the
spatial geometry on a neuron of the receptors for a
putative "modulator". Thus, if the receptors are located
on dendrites some distance from the soma, where the intracellular electrode is usually inserted, potential changes
produced by diffusion of the substance to these remote
sites may be decremented to undetectable levels before they
are conducted to the soma. Likewise, receptor-induced
changes in ionic conductance on distant dendrites are
likely not to be detected by the usual intracellular
(intrasomatic) stimulation method. Therefore, strict
proof of neuromodulation requires methodologies for tracking
receptor sensitivity along the entire surface of the neuron.

Having now applied broad criteria for classification
of transmitters, modulators and mediators, and presented
some caveats for experimental efforts to document the
existence of these messengers in brain, we can now turn
to specific examples of chemicals that appear to fulfill
at least a majority of these criteria.

Central Norepinephrine and its Mediation by Cyclic AMP

With the possible exception of acetylcholine, criteria
for a central transmitter seem most completely satisfied
by norepinephrine. This state of affairs arises in part
because histochemical methods for localizing NE (cf. MOORE
& BLOOM, 1978) are less ambiguous than for other putative
central transmitters, thus more easily satisfying criterion
number 1 (see above). Fulfillment of criterion number 2,
release of activation, is less directly fulfilled although
it is possible to detect metabolites of NE and diminished
endogenous NE levels in cortex after activation of the
Locus Coeruleus (LC), the source of NE-containing fibers
to the cortex (KORF et al., 1973). Satisfaction of the
last two criteria have been the subject of intensive investigations, carried out on the cerebellar Purkinje cell
and the hippocampal Pyramidal cell, both targets of the
LC NE-containing pathway.

Coeruleo-Cerebellar Pathway. In the cerebellar
Purkinje cell, extracellular iontophoretic application
of NE or stimulation of the LC was found to produce equivalent effects on the Purkinje cell, namely, inhibition
of spontaneous discharge with extracellular recording
(SIGGINS et al., 1969; HOFFER et al., 1973), and a unique
hyperpolarization associated with either no change or a
decrease in membrane conductance with intracellular recording (SIGGINS et al., 1971 b, c; HOFFER et al., 1973).
This latter effect was in direct contrast to changes seen
with classical IPSP's or with iontophoresis of a putative
inhibitory transmitter like GABA (SIGGINS et al., 1971a),
in which the hyperpolarization is associated with increased membrane conductance.

In contrast again to classical spinal cord IPSP's,
the inhibitory response to single LC stimuli showed long
latencies (including conduction time and synaptic delay)
averaging about 125 msec and responses lasting 0.5 sec
or more. Trains of pulses to LC at 10/sec, which is
nearly the optimal frequency for both the LC pathway and
for peripheral sympathetic nerves (see FOLKOW, 1952;
CARPENTER & TANKERSLEY, 1959), could evoke inhibitions
outlasting the stimulus for up to a minute or more (HOFFER
et al., 1973). Similar responses with long latencies and
durations of action had been previously reported for both
sympathetic target tissues (e.g., smooth muscle; see
"Introduction") and for the sIPSP of sympathetic ganglia
(LIBET, 1967). Since the response to exogenous NE has
similar long latencies and durations, and is accompanied
by an unusual conductance decrease, it would appear that
transmitter criterion number 3 is satisfied.

Unfortunately, the existence of long latencies in
response to LC stimulation confounds attempts to prove that
the pathway responsible for inhibition is monosynaptic.
However, results of iontophoretic experiments with local
application of NE antagonists provide indirect support for
a monosynaptic inhibitory projection from LC to the Purkinje cell, and in addition, help to satisfy transmitter
criterion number 4. Thus, both effects of NE iontophoresis
and LC stimulations are comparably antagonized by phenothiazines (FREEDMAN & HOFFER, 1975; SIGGINS et al., 1976a),
by cobalt or lead (FREEDMAN et al., 1975), by lithium
(SIGGINS & HENRIKSEN, in preparation) as well as by
prostaglandins and beta-adrenergic antagonists (HOFFER
et al., 1973). Furthermore, the effects of LC stimulation
are also abolished when the synthesis and storage of NE
are blocked pharmacologically and when the pathway is

destroyed with 6-OHDA (HOFFER et al., 1973). These pharmacological manipulations were selective in that other
pathways or neurotransmitters were not affected.

Coeruleo-Hippocampal Projection. More recently, the
properties of the LC-cerebellar circuit have been generalized to other LC target areas. Stimulation of LC in the
awake rat produces a two-fold rise in hippocampal cyclic
AMP with no change in cyclic GMP (SEGAL & GUIDOTTI, in
preparation). Physiological studies indicate that the
hippocampal NE projection from LC produces cellular effects
virtually identical to those of the locus coeruleus on
cerebellar Purkinje cells: in hippocampus, both LC and
NE slow pyramidal cell discharge with long latencies and
long durations of actions. Both responses are also blocked
with beta-blockers, by prostaglandins of the E series,
by lithium and by phenothiazines (SEGAL & BLOOM, 1974;
SEGAL, 1974; BLOOM, 1975). Moreover, the action of the
LC pathway is blocked by chronic pretreatment with 6-OHDA
or acute pretreatment with reserpine and alpha methyl
tyrosine or with inhibitors of dopamine beta hydroxylase
(SEGAL & BLOOM, 1974).

Septum. SEGAL (1976) has recently reported studies
on neurons in the medial septal nuclei and in the diagonal
band of Broca, which also receive a projection from the
locus coeruleus. LC stimulation was found to produce, as
in the other areas already described, an inhibition of
long latency (30-100 msec) and long duration (100-300
msec). These inhibitory effects were blocked by depletion
of NE stores or by pretreatment with 6-OHDA.

Cyclic AMP as a Mediator of Central NE Effects.
With respect to second messenger mediation by cyclic
nucleotides, the central catecholamine-containing pathways merit investigation because they satisfy three practical considerations: 1) catecholamines meet most or all
of the criteria above for a synaptic transmitter; 2) biochemically, catecholamines are known to activate adenylate
cyclase or elevate cyclic AMP levels in many peripheral
systems and in various discrete regions of the central
nervous system by definable receptors; and 3) as shown
above, the source neurons and target neurons of the central
catecholamine pathways have been sufficiently characterized
so that their effects can be determined and related to the
effects of cyclic nucleotides and related substances.

From the electrophysiologist's point of view, the
cerebellar Purkinje cell (P-cell) is still the best candidate for a target neuron that receives a noradrenergic

input capable of generating cyclic AMP postsynaptically.
The data reinforcing this notion have been reviewed in
detail elsewhere (BLOOM, 1975; SIGGINS, 1977, 1978) but
may be summarized (in order of mediator criteria) as
follows: 1) catecholamines elevate cyclic AMP levels
and increase adenylate cyclase activity in cerebellum
in vitro, and exogenous NE and stimulation of the LC increase cyclic AMP histochemical immuno-reactivity in
P-cells in vivo; 2) the increase in cyclic AMP immunoreactivity is detectable at least by the time the electrophysiological effects of LC stimulation are apparent;
3) the inhibitory effects on P-cells of LC stimulation
or of NE iontophoresis are potentiated by several phosphodiesterase (PDE) inhibitors and antagonized by agents
(e.g., PGE_1 and E_2, MJ-1999, fluphenazine, Lathanum) known
to block NE-elevated cyclic AMP levels in vitro; 4) with
extra- and intracellular recording, responses to iontophoresis of cyclic AMP and several more potent synthetic
analogues generally mimic the inhibitory hyperpolarizing
action of iontophoretic NE and LC stimulation, as well
as displaying the novel increases in membrane resistance
usually seen with such noradrenergic stimuli.

Thus, the criteria for cyclic AMP mediation appear
to be largely satisfied for the inhibitory NE input to
Purkinje cells, except for the technical inability to
detect postsynaptic increases in cyclic AMP at a time prior
to the "biological event" triggered by NE. Although dis-
agreement exists as to the exact percentage of P-cells
inhibited by iontophoretic cyclic AMP (GODFRAIND & PUMAIN,
1971; LAKE & JORDAN, 1974), this discrepancy can be explained by poor cell penetrability and other technical
considerations (BLOOM et al., 1974; SHOEMAKER et al., 1975).
Moreover, iontophoresis of derivatives of cyclic AMP
(e.g., 8-p-Chlorophenyl-cyclic AMP) known to have a greater
action on the protein kinase enzyme (the intracellular
"receptor" for cyclic AMP) than the parent compound, can
depress the activity of up to 90% of P-cells (SIGGINS &
HENRIKSEN, 1975). In addition, the strong correlation
between percentage of P-cells depressed and the potency
of several derivatives in activating protein kinase argues
for an involvement of cyclic AMP-dependent protein kinase
in the depressant responses.

In studies of cultured Purkinje neurons, GAHWILER(1976)
has reported potentiation of NE and cyclic AMP-induced
depressions with phosphodiesterase inhibitors. He also
observed that the thresholds for inhibitory responses to
NE were 100-1000 times lower than for cyclic AMP applied

by superfusion, in keeping with predictions from several
studies on peripheral tissues where second messenger mediation has been proven (see RALL, 1972; BLOOM, 1975).

The suggestion that actions of extracellularly-
applied cyclic AMP are mediated by conversion to adenosine
and activation of an adenosine receptor is disproven by
the observations in cerebellum and the cerebral cortex
that methylxanthine-type phosphodiesterase inhibitors
potentiate NE and cyclic AMP, yet block the effects of
adenosine or 5'AMP (SIGGINS et al., 1971a; HOFFER et al.,
1973; STONE & TAYLOR, 1977). At any rate, since the
effects of adenosine are also inhibitory and proposed by
the PHILLIS group to be mediated by cyclic AMP (PHILLIS
et al., 1974· KOSTOPOULOS et al., 1975), the implication
is that the physiological action of cyclic AMP is also
inhibitory, thus confirming the original findings with
iontophoresis of cyclic AMP (SIGGINS et al., 1969; SIGGINS
et al., 1971 a, b).

Evidence similar to that found in the cerebellum
exists for cyclic AMP mediation of the inhibitory norepinephrine input to hippocampal and cerebral cortical
pyramidal cells (SEGAL & BLOOM, 1974; STONE & TAYLOR,
1977). In brief, exogenous catecholamine elevates cyclic
AMP in vitro, the inhibitory effects of LC activation or
NE iontophoresis are affected in a predictable way by drugs
which interact with the cyclic AMP system, and iontophoresis
of cyclic AMP generally mimics the inhibitory action of
LC stimulation and iontophoretic NE. However, immunohistochemical methods for in vivo localization of cyclic
AMP to the pyramidal cells have not yet been applied to
the hippocampus or cortex, nor has it been possible to
detect cyclic AMP in these structures prior to the NE-
induced biological event. Research suggesting that the
inhibitory responses of caudate neurons to dopamine are
also mediated by cyclic AMP is reviewed elsewhere (SIGGINS,
1977, 1978).

Other more indirect data also support the hypothesis
of mediation of NE-induced inhibition by cyclic AMP. Thus,
lithium and lanthanum, each of which have been shown to
block NE-induced elevations of cyclic AMP in vitro (NATHAN-
SON et al., 1976; BLOOM, 1975; NATHANSON, 1977), also both
antagonize NE-evoked inhibitions of spontaneous activity
in cerebellar Purkinje cells and hippocampal pyramidal
cells (NATHANSON et al., 1976; SEGAL, 1974). Lithium also
antagonizes LC-induced depression in hippocampus (SEGAL,
1974) and cerebellum (SIGGINS & HENRIKSEN, in preparation).

Modulation of Adenylate Cyclase by Altering Catecholamines

Biochemical studies in the last decade have projected the interesting concept that the catecholamine/cyclic AMP system is not a static entity, but can be modulated reciprocally by changes in the functionality of the endogenous catecholamine systems. In general, chronic reductions in catecholamine innervation or release (e.g., by 6-hydroxydopamine or reserpine) bring about an increase in responsivity of brain adenylate cyclase to β-adrenergic or dopaminergic stimulation, while pharmacological manipulations designed to elevate synaptic catecholamines (e.g., by re-uptake blockers such as desipramine; DMI) reduce sensitivity to these agonists (HUANG et al., 1973; VETULANI & SULSER, 1975; SCHULTZ, 1976).

Electrophysiological methods have also been brought to bear on this "homeostatic" phenomenon. Previous studies showed that disruption of the dopamine input to the caudate nucleus by 6-OHDA results in a supersensitivity of caudate neurons to iontophoretically applied dopamine and apomorphine (SIGGINS et al., 1976b), thus presenting a cellular correlation of the behavioral (rotatory) supersensitivity to systemically injected apomorphine (UNGERSTEDT, 1971). Recent studies in our laboratory suggest that the converse electrophysiological experiments also support the principle of cyclic AMP homeostatis. Acute application of DMI by iontophoresis potentiates the effects of NE on P-cells (HOFFER et al., 1971) presumably by blocking NE uptake into noradrenergic terminals. In contrast, chronic treatment of rats with DMI (30-50 mg/kg/day for 8-10 days) significantly reduces the responsivity of cerebellar P-cells to iontophoretically applied NE (Table 1) as determined by the elevation of the average ejection current required to evoke threshold depressions of spontaneous activity (SIGGINS & SCHULTZ, 1978). These biochemical and iontophoretic findings thus represent another confirmation of the mediator hypothesis for cyclic AMP in NE responses.

Norepinephrine as a Neuromodulator

In spite of the strong evidence favoring a transmitter role for NE, there is data suggesting that NE could also be a modulator. Indeed, those predisposed to define modulators on the basis of speed or duration of action, on the basis of lack of induced increase in ionic conductance, or on the basis of mediation through an energetic event, have already applied the modulator label to NE (see review

TABLE I

DMI on Purkinje Cell Thresholds
(nA Current#) to Norepinephrine

	NE Thresholds
Controls (26)	24 $\pm$ 3.7 nA
DMI Acute (18)	16 $\pm$ 2.9
DMI Chronic (28)	43 $\pm$ 3.6**
DMI 5-D Withdrawal (22)	31 $\pm$ 5.9*

#Mean iontophoretic current values $\pm$ S.E.M.
One-Way Analysis of Variance; Newman-Keuls Analysis
(N) = Number of Purkinje Cells studied
*p = .05;   **p < .01

by DISMUKES, 1977).   However, to my mind there is only one
compelling piece of evidence to suggest that NE has the
major quality of a modulator: it can clearly influence
the responsiveness of a given target cell to its other
afferents.   Studies in the awake squirrel monkey auditory
cortex (FOOTE et al., 1975), in the cerebellum of the
anesthetized rat (FREEDMAN et al., 1976, 1977), and in the
hippocampus of the awake rat (SEGAL & BLOOM, 1976) suggest
a heterosynaptic influence of the LC which should be considered additional to the classical mode of synaptic opera-
tion termed inhibition.   For example, in the cerebellar
Purkinje cell (FREEDMAN et al., 1976, 1977) and the hippocampal pyramidal cell (SEGAL & BLOOM, 1976), conditioning
stimuli in the LC or iontophoretic application of NE will
potentiate the effects of nonadrenergic inhibitory inputs
for considerable periods. Moreover, in the cerebellum,
the excitatory afferents of parallel fibers and climbing
fibers are also potentiated by NE and LC activation.   Such
potentiated responses may be similar to the enhanced evoked
responses seen following LC conditioning stimuli in lateral
geniculate (NAKAI & TAKAORI, 1974) and on the increased
ratio of evoked responses to spontaneous activity seen
during iontophoresis of NE to acoustically reactive units
of the squirrel monkey auditory cortex (FOOTE et al., 1975).

As stated above ("Neurotransmitter Criteria") one
might expect that a substance which decreases conductance

(or increases membrane resistance) might potentiate other
inputs, by virtue of its ability to enlarge a potential
produced by a given synaptic current across an increased
resistance. In verification of this, climbing fibers
EPSP's are increased in size by NE and LC activation
(SIGGINS et al., 1971b, c). Although it is suggested
that the potentiating effects of NE outlast the direct
inhibitory action on extracellularly recorded spontaneous
activity (FREEDMAN et al., 1976, 1977), intracellular
recording is needed to verify that the membrane potential
and conductance is still not altered when the spike frequency (seen extracellularly) has returned to normal.

Opioid Peptides: Neurotransmitters or Neuromodulators?

The discovery of endorphins, the endogenous peptide
ligands for the stereospecific, naloxone sensitive opiate
receptor (see TERENIUS & WAHLSTROM, this volume), has
created a flurry of activity attempting to delineate the
physiological action and possible function of these peptides. As with the catecholamines, research to date has
suggested a role for these peptides more complex than
can be conceptualized either as strictly a neurotransmitter
or a neuromodulator. Electrophysiologic research on
opiates and opioid peptides has been primarily through
extracellular single unit recordings directed at CNS areas
with high density opiate receptors or involved with noci-
ception.

In extracellular recording, most of the stereospecific,
naloxone-antagonizable actions of opioid peptides are inhibitions of single unit discharge (spontaneous, glutamate
or ACh-evoked activity) which are qualitatively similar
throughout the mammalian central and peripheral nervous
system (see review by ZIEGLGANSBERGER & FRY, 1978).
However, some major exceptions exist: naloxone-reversible
excitatory responses are seen in pyramidal cells in the
hippocampus (NICOLL et al., 1977; HILL et al., 1977), the
Renshaw cell in the spinal cord (see DAVIES & DRAY, 1978;
DUGGAN et al., 1976) and some less well-identified cells
in various parts of the CNS (see NICOLL et al., 1977).
However, recent studies using GABA blockade by bicuculline
and blockade of transmitter release by Mg ions indicate
that the excitatory responses of hippocampal pyramidal
neurons may now be viewed as a primary inhibitory effect
on neighboring inhibitory interneurons, resulting in excitation by disinhibition (SIGGINS et al., 1978).

However, intracellular recordings are required to
measure synaptically- and chemically-induced changes in
membrane potential and membrane conductance for elucidation of mechanisms of opiate action, and recently a few
such studies have been completed.

Early studies in the spinal cord showed that intravenous administration of morphine agonists depressed
polysynaptic EPSPs, an effect antagonized by opiate antagonists (JURNA et al., 1973). Recent studies employing
intracellular recording of spinal cord neurons with simul-
taneous extracellular microiontophoretic application revealed that morphine and opioid peptides do not appear to
change membrane potential or resting membrane resistance,
but decrease the rate of rise of the EPSPs (ZIEGLGANS-
BERGER & FRY, 1976: ZIEGLGANSBERGER & BAYERL, 1976).
Opiates and opioid peptides also depress the glutamate-
induced depolarization in addition to synaptic activation.
With respect to cat spinal neurones, microiontophoretically
applied glutamate is considered to increase the postsynaptic membrane conductance to sodium ions (ZIEGLGANSBERGER &
PUIL, 1972). ZIEGLGANSBERGER and co-workers (1976) therefore postulate that the opiates interfere with the chemically excitable sodium channel comparable to those also
operated by synaptically released excitatory transmitters.
The fact that the depolarizing responses to glutamate are
post-synaptic and antagonized by opiate agonists suggests
that the opiate receptors involved in this effect are also
located on the post-synaptic membrane.

The anti-glutamate action of the opioid peptides in
vivo have recently been confirmed with spinal neurons grown
in culture (BARKER et al., 1978). Kinetic analysis indicates that the inhibitory action of the opiate peptides
are brought about by a non-competitive mechanism on the
postsynaptically located sodium-ionophore. However, these
postsynaptic modulator-like actions of the opiates and
opioid peptides are just a few of the many types described.
For example, several reports of pre-synaptic influence
have appeared, in which it would appear that the opiates
depress transmitter release (CAVILLO et al., 1974; MAC-
DONALD & NELSON, 1978; HILLER et al., 1978), perhaps by
involving mechanisms similar to the modulation of post-
synaptic ionic processes. Furthermore, the study of
BARKER et al. (1978) shows a multitude of presumably post-
synaptic actions of enkephalin on the cultured mouse
spinal-neuron preparation, including direct positive and
negative effects on membrane excitability associated with
potential and/or conductance changes, abrupt naloxone-

resistant depolarizations, enhancement of amino acid-
induced actions without change in membrane properties, as
well as the depression of glutamate actions described
above. However, the possibility should be considered that
some of these diverse opiate actions may arise as a pharmacological "curiosity" in cells that do not ordinarily
ever receive endorphinergic influence in vivo.

     With regard to multiple actions of opioid peptides,
the case of myenteric plexus neurons is an interesting
one. Here, the analgesic potency of opiates correlates
with their depressant effect upon the electrically induced
twitch of the guinea pig ileum and is accurately reflected
in single unit studies of myenteric plexus neurons (NORTH
& HENDERSON, 1975; NORTH, 1976). The stereospecific
depressant effect of the opiates here can be seen in Ca^{++}-
free/high Mg^{++} solution, indicating a postsynaptic effect
(NORTH & WILLIAMS, 1976, 1977). Interestingly, with intracellular recording, many myenteric neurons show no change
in membrane potential or conductance in response to superfusion or iontophoresis of the peptides. However, other
cells are seen which show hyperpolarizing actions with
increased membrane conductance (NORTH et al., 1979). Although this effect seems at variance to the data obtained
in central neurons cited above, a different ionic mechanism
in these neurons cannot be excluded. However, a common
mechanism may still be operative since the hyperpolarizing
response to enkephalin becomes smaller as the basal membrane potential of myenteric plexus cells becomes more
negative (less injured?) and since non-linearity in the
current/voltage characteristic (anomalous rectification)
could then account for the apparent increase in membrane
conductance. However, a different technical problem
applies to the studies on spinal neurons in vivo or in
vitro, namely the possibility of direct but remote actions
of the peptides on membrane properties which are not conducted to or picked up by a somatically placed intracellular
electrode (see discussion under "Neurotransmitter Criteria").

                            SUMMARY

     An effort has been made here to devise criteria
allowing discrimination between neurotransmitters, modulators and mediators. However, after consideration of several
technical pitfalls in studies of these criteria, and examination of the properties of two examples of neuroactive
agents (norepinephrine and endorphins) often referred to

as "modulators", it is still difficult to classify these
agents in all cases. Thus, in most central targets where
NE-fibers are known to terminate, the synaptic actions of
NE appear to have properties of both a neuromodulator and
a neurotransmitter. Although much more research needs to
be pursued, the opioid peptides may be neuromodulators
for some neurons (spinal cord neurons) and neurotransmitters for others (myenteric plexus and spinal cord
neurons). It may be that classification of such peptide
agonists will need to be done on a cell-by-cell basis,
with the endogenous peptides subserving a multi-faceted
role in central and peripheral neuronal communication.

As more and more endogenous ligands and transmitter-
like substances are extracted from brain, it begins to
appear that the language of neuronal communication is much
richer than originally imagined from responses of spinal
neurons to the fast-acting classical neurotransmitters.
Indeed, it may evolve that the "deviant" forms of com-.
munication or transmission are more the rule than the
exception. In the final analysis, each neurotransmitter
may possess its own "fingerprint" of holistic actions
attesting to the unique individuality of neuron types and
their neurotransmitters. Such individualities might be
expected to acoomplish more sophisticated integrative
operations, and hence behaviors, than could simple rapid
"yes" or "no" messages.

ACKNOWLEDGEMENTS

I am indebted to Drs. Floyd Bloom, Roy Wise and
Leonard Koda for their critical evaluation of ideas in
this discussion, and to Ms. Nancy Callahan for her secre-
tarial skills.

REFERENCES

BARKER, J.L., NEALE, J.H., SMITH, T.G. JR. & MACDONALD,
     R.L. (1978) Science 199, 1451-1453.
BARKER, J.L. & SMITH, T.G. JR. (1977) Neurosci. Symp. 2,
     340-373.
BAYLOR, D.A. & FUORTES, M.G.F. (1970) J. Physiol. (Lond.)
     207, 77-92.
BLOOM, F.E. (1975) Rev. Physiol. Biochem. Pharmacol. 74,
     1-103.
BLOOM, F.E. (1976) in Advances in Biochemical Pharma-
     cology, Vol. 15, pp. 273-282, Raven Press, New York.

BLOOM, F.E. (1979a) in The Peptidergic Neuron (Fuxe, K.,
    Hokfelt, T. & Luft, R., eds), Pergamon Press, New
    York (in press).
BLOOM, F.E. (1979b) in Catecholamines: Basic and Clinical
    Frontiers (Usdin, E., ed.) (in press).
BLOOM, F.E., SIGGINS, G.R. & HOFFER, B.J. (1974) Science
    185, 627-629.
BROOKS, C. & ECCLES, J.C. (1947) Nature 159, 760-764.
BURNSTOCK, G. & HOLMAN, M.E. (1963) Ann. Rev. Physiol.
    25, 61-90.
CAESAR, R., EDWARDS, G. & RUSKA, H. (1957) J. Biophys.
    Biochem. Cytol. 3, 867-877.
CALVILLO, O., HENRY, J.L. & NEUMAN, R.S. (1974) Canad.
    J. Physiol. Pharmacol. 52, 1207-1211.
CARPENTER, F.G. & TANKERSLEY, J.C. (1959) Am. J. Physiol.
    196, 1185-1188.
COOMBS, J.S., ECCLES, J.C. & FATT, P. (1955) J. Physiol.
    130, 326-373.
DALE, H.H. (1906) J. Physiol. 34, 163-206.
DALE, H.H. (1914) J. Physiol. 48, 3iii-3iv.
DALE, H.H. (1935) Proc. R. Soc. Med. 28, 319-332.
DAVIES, J. & DRAY, A. (1978) Brit. J. Pharmacol. 63,
    87-96.
DISMUKES, K. (1977) Nature 269, 557-558.
DUGGAN, A.W., DAVIES, J. & HALL, J.G. (1976) J. Pharmacol.
    Exp. Ther. 196, 107-120.
ECCLES, J.C. (1964) The Physiology of Synapses, Springer-
    Verlag, Berlin.
ECCLES, J.C., FATT, P. & KOKETSU, K. (1954) J. Physiol.
    126, 524-562.
ECCLES, R.M. & LIBET, B. (1961) J. Physiol. (Lond.) 157,
    484-503.
ELLIOTT, T.R. (1904) J. Physiol. 31, 20.
FLOREY, E. (1967) Fed. Proc. 26, 1164-1178.
FOLKOW, B. (1952) Acta Physiol. Scand. 25, 49-76.
FOOTE, S.L., FREEDMAN, R. & OLIVER, A.P. (1975) Brain
    Res. 86, 229-242.
FREDERICKSON, R.C.A. (1977) Life Sci. 21, 23-40.
FREEDMAN, R. & HOFFER, B.J. (1975) J. Neurobiol. 6, 277-
    288.
FREEDMAN, R., HOFFER, B.J. & WOODWARD, D.J. (1975) Br.
    J. Pharmacol. 54, 529-539.
FREEDMAN, R., HOFFER, B.J., PURO, D. & WOODWARD, D.J.
    (1976) Brit. J. Pharmac. 57, 603-605.
FREEDMAN, R., HOFFER, B.J., WOODWARD, D.J. & PURO, D.
    (1977) Exp. Neurol. 55, 269-288.
GAHWILER, B.H. (1976) Nature 259, 483-484.
GODFRAIND, J.M. & PUMAIN, R. (1971) Science 174, 1257.
HILL, R.G., MITCHELL, J.F. & PEPPER, C.M. (1977) J.

Physiol. 272, 50-51.
HILLER, J.M., SIMON, E.J., CRAIN, S.H., PETERSON, E.R.
(1978) Brain Res. 145, 396-400.
HOFFER, B.J., SIGGINS, G.R. & BLOOM, F.E. (1971) Brain
Res. 25, 523-534.
HOFFER, B.J., SIGGINS, G.R., OLIVER, A.P. & BLOOM, F.E.
(1973) J. Pharmacol. Exp. Ther. 184, 553-569.
HOKFELT, T. (1968) Zeit. f. Zellforsch. 91, 1-74.
HUANG, M., HO, A.K.S. & DALY, J.W. (1973) Mol. Pharmacol.
9, 711-717.
JURNA, I., GROSSMANN, W. & THERES, C. (1973) Neuropharma-
cology 12, 983-993.
KATZ, B. (1966) Nerve Muscle and Synapse, p. 193, McGraw-
Hill, Inc., New York.
KOBAYASHI, H. & LIBET, B. (1968) Proc. Natl. Acad. Sci.
60, 1304-1311.
KOKETSU, K. (1969) Fed. Proc. 28, 101-112.
KORF, J., ROTH, R.H. & AGHAJANIAN, G.K. (1973) Eur. J.
Pharmacol. 23, 276-282.
KOSTOPOULOS, G.K., LIMACHER, J.J. & PHILLIS, J.W. (1975)
Brain Res. 88, 162-165.
KRNJEVIC, K., PUMAIN, R., & RENAUD, L. (1971) J. Physiol.
215, 247-268.
LAKE, N. & JORDAN, L.M. (1974) Science 183, 663-664.
LEVER, J.D., GRAHAM, J.D.P. & SPRIGGS, T.L.B. (1967)
Bibl. Anat. 8, 51-55.
LIBET, B. (1967) J. Neurophysiol. 30, 494-514.
LOEWI, O. (1921) Pflugers Arch. 189, 238-242.
MACDONALD, R.L. & NELSON, P.G. (1978) Science 199, 1449-
1451.
MOORE, R.Y. & BLOOM, F.E. (1978) Ann. Rev. Neurosci. 2,
in press.
NAKAI, Y. & TAKAORI, S. (1974) Brain Res. 71, 47-60.
NAKAJIMA, S. & TAKAHASHI, K. (1966) J. Physiol. 187, 105.
NATHANSON, J. (1977) Physiol. Rev. 57, 158-256.
NATHANSON, J., FREEDMAN, R. & HOFFER, B.J. (1976) Nature
261, 330-331.
NICOLL, R.A., SIGGINS, G.R., LING, N., BLOOM, F.E. &
GUILLEMIN, R. (1977) Proc. Natl. Acad. Sci. USA 74,
2584-2588.
NISHI, S. & KOKETSU, K. (1968) J. Neurophysiol. 31, 717-
728.
NORTH, R.A. (1976) Neuropharmacology 15, 1-9.
NORTH, R.A. & HENDERSON, G. (1975) Life Sci. 17, 63-66.
NORTH, R.A., KATAYAMA, Y. & WILLIAMS, J.T. (1979) Brain
Res., in press.
NORTH, R.A. & WILLIAMS, J.T. (1976) Nature (Lond.) 264,
460-461.
NORTH, R.A. & WILLIAMS, J.T. (1977) Fed. Proc. 36, 3667.

PHILLIS, J.W., KOSTOPOULOS, G.K. & LIMACHER, J.J. (1974)
    Can. J. Physiol. Pharmacol. 52, 1227-1229.
RALL, T.W. (1972) Pharmacol. Rev. 24, 399-409.
RICHARDSON, K.C. (1958) Am. J. Anat. 103, 99-135.
SCHULTZ, J. (1976) Nature 261, 417-418.
SEGAL, M. (1974) Nature 250, 71-73.
SEGAL, M. (1976) J. Physiol. 261, 617-631.
SEGAL, M. & BLOOM, F.E. (1974) Brain Res. 72, 99-114.
SEGAL, M. & BLOOM, F.E. (1976) Brain Res. 107, 513-525.
SHOEMAKER, W.J., BALENTINE, L.T., SIGGINS, G.R., HOFFER,
    B.J., HENRIKSEN, S.J. & BLOOM, F.E. (1975) J. Cyclic
    Nucleotide Res. 1, 97-106.
SIGGINS, G.R. (1977) in Cyclic Nucleotides: Mechanisms
    of Action (Cramer, H. & Schultz, J., eds.), John
    Wiley and Sons, Ltd., London-New York.
SIGGINS, G.R. (1978) in Neuronal Information Transfer
    (Karlin, A., Tennyson, V.M. and Vogel, H.J., eds.),
    p. 339, Academic Press, New York.
SIGGINS, G.R. & HENRIKSEN, S.J. (1975) Science 189,
    559-561.
SIGGINS, G.R., HOFFER, B.J. & BLOOM, F.E. (1969) Science
    165, 1018-1020.
SIGGINS, G.R., HOFFER, B.J. & BLOOM, F.E. (1971a) Brain
    Res. 25, 535-553.
SIGGINS, G.R., OLIVER, A.P., HOFFER, B.J. & BLOOM, F.E.
    (1971b) Science 171, 192.
SIGGINS, G.R., HOFFER, B.J., OLIVER, A.P. & BLOOM, F.E.
    (1971c) Nature 233, 481-483.
SIGGINS, G.R., HENRIKSEN, S.J. & LANDIS, S.C. (1976a)
    Brain Res. 114, 53-65.
SIGGINS, G.R., HOFFER, B.J., BLOOM, F.E. & UNGERSTEDT, U.
    (1976b) in The Basal Ganglia (Yahr, M.D., ed.), pp.
    227-248, Raven Press, New York.
SIGGINS, G. & SCHULTZ, J. (1978) Abstracts of 7th Inter.
    Cong. of Pharmacol. (IUPHAR), 346.
SIGGINS, G.R., ZIEGLGANSBERGER, W., FRENCH, E., LING, N.
    & BLOOM, F. (1978) Neurosci. Abstracts 4, 414.
SPEDEN, R. (1964) Nature 202, 193-194.
STONE, T.W. & TAYLOR, D.A. (1977) J. Physiol. 266, 523-
    543.
SUTHERLAND, E.W., OYE, I. & BUTCHER, R.W. (1965) Rec.
    Progr. Hormone Res. 21, 623-642.
UNGERSTEDT, U. (1971) Acta Physiol. Scand. Suppl. 367, 69.
VETULANI, J. & SULSER, F. (1975) Nature 257, 495.
WEIGHT, F.F. & PADJEN, A. (1973a) Brain Res. 55, 219-224.
WEIGHT, F.F. & PADJEN, A. (1973b) Brain Res. 55, 225-228.
WERMAN, R. (1966) Comp. Biochem. Physiol. 18, 745-766.
YARBROUGH, G.G. (1976) Nature 263, 523-524.
ZIEGLGANSBERGER, W. & BAYERL, J. (1976) Brain Res. 115,
    111-128.

ZIEGLGANSBERGER, W. & FRY, J.P. (1976) in Opiates and En-
    dogenous Opioid Peptides (Kosterlitz, H.W., ed.),
    pp. 213-238, Elsevier/North-Holland, Biomedical Press,
    Amsterdam.
ZIEGLGANSBERGER, W. & FRY, J.P. (1978) in Development in
    Opiate Research (Herz, A., ed.), Marcel Dekker,
    New York, New York.
ZIEGLGANSBERGER, W. & PUIL, E.A. (1972) Exp. Brain Res.
    17, 35-49.

METABOLIC FACTORS AFFECTING BRAIN CYCLIC NUCLEOTIDES

L. VOLICER, R. MIRIN and R. MEICHNER

Boston University School of Medicine
Department of Pharmacology
Boston, Mass. 02118

Initial investigations of cyclic nucleotides indicated that they act as second messengers mediating hormonal stimulation (ROBISON et al., 1971). More recently it was recognized that in the central nervous system cyclic nucleotides might play a more general role as modulators of nerve transmission (IVERSEN et al., 1975; MCILWAIN, 1976). Thus changes of cyclic nucleotide levels might significantly affect the function of specific brain structures. In this paper we would like to demonstrate that some drugs affect cyclic nucleotide levels not by changes of neurotransmitter release but by their effects on brain metabolism.

Several investigators demonstrated that brain levels of cyclic AMP are increased by asphyxia (SWANSON, 1969; SATTIN, 1971). This increase is probably mediated by formation and release of adenosine because it is inhibited by methylxanthines (SATTIN, 1971) which block stimulation of adenylate cyclase by adenosine (SATTIN & RALL, 1970). Brain cyclic AMP levels are also increased after decapitation of experimental animals (KAKIUCHI & RALL, 1968; DITZION et al., 1970; UZUNOV & WEISS, 1971). This increase is due to both anoxia and depolarization of cell membranes. Cell membrane depolarization leads to release of adenosine (SHIMIZU et al., 1970) and therefore the effect of both anoxia and depolarization is probably mediated by adenosine. In agreement with that, the post-decapitation rise of cyclic AMP is inhibited by pretreatment with theophylline (PAUL et al., 1970).

We have observed that the postdecapitation rise of
cyclic AMP levels in the cerebellum is inhibited by pretreatment with ethanol (VOLICER & GOLD, 1973) and have
studied the mechanism of this ethanol effect.   Brain enzymes were inactivated by microwave irradiation at various times after decapitation in Sprague-Dawley rats and
levels of ATP, cyclic AMP and adenosine in the hind brain
were measured (VOLICER et al., 1977).

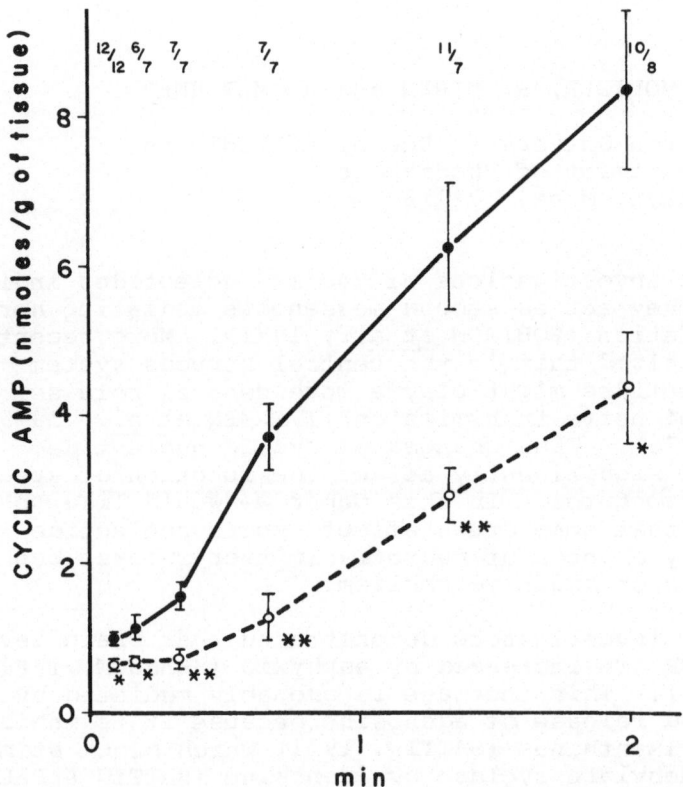

Fig. 1: Changes of cyclic AMP levels induced by decapi-
        tation in the hind brain of control (o——o) and etha-
        nol-treated (o---o) rats.   The bars indicate S.E.M.
        and the fractions indicate number of determinations in
        control animals over the number of determinations in
        ethanol-treated rats.   Significance; *P< 0.025, **P <
        0.005.   Reprinted from Volicer et al., 1977.

Cyclic AMP levels in control animals increased more than 8-fold during two minutes after decapitation (Fig. 1). In contrast to that, in animals pretreated with 6 g/kg of ethanol 1 hr. before sacrifice the increase was smaller than in controls.

Adenosine levels increased after decapitation in control animals but did not change significantly in ethanol-treated animals (Fig. 2). This indicates that ethanol prevents the increase of adenosine induced by anoxia and depolarization and this effect might be responsible for the inhibition of the post-decapitation rise of cyclic AMP observed in ethanol-treated rats.

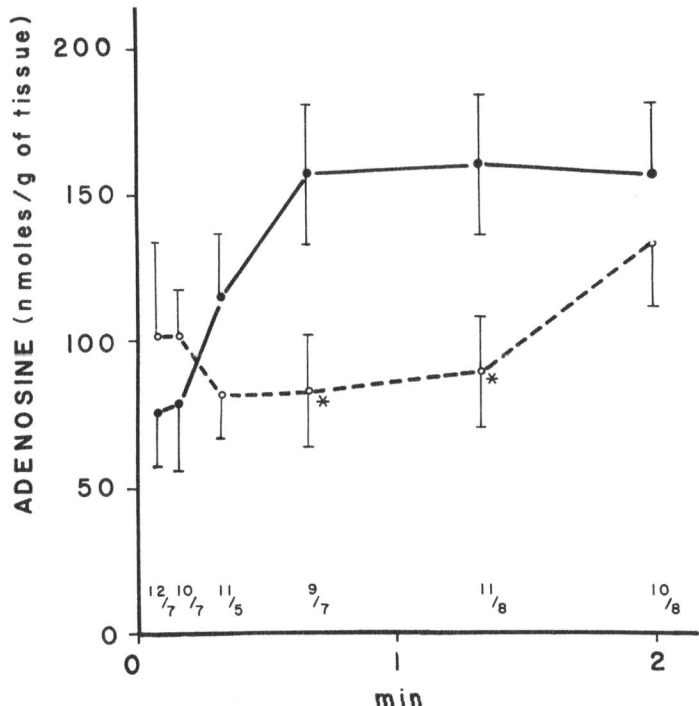

Fig. 2: Changes of adenosine levels induced by decapitation in the hind brain of control (o—o) and ethanol-treated (o---o) rats. The bars indicate S.E.M. and the fractions indicate number of determinations in control animals over number of determinations in ethanol-treated rats. Significance; *P < 0.05. Reprinted from Volicer, et al., 1977.

Since it has been reported that ethanol inhibits Na,
K-activated ATPase (ISRAEL & SALAZAR, 1967; SUN &
SAMORAJSKI, 1972; GRISHAM & BARNETT, 1973) it was possible that the inhibition of adenosine formation was due
to decreased ATP metabolism. We therefore measured ATP
levels in the hind brain at various times after decapitation and found that the decrease of ATP levels was
much slower in ethanol-treated animals than in controls
(Fig. 3). These results therefore indicate that ethanol
affects formation of cyclic AMP by inhibition of ATP
degradation and not through a change of neurotransmitter
release.

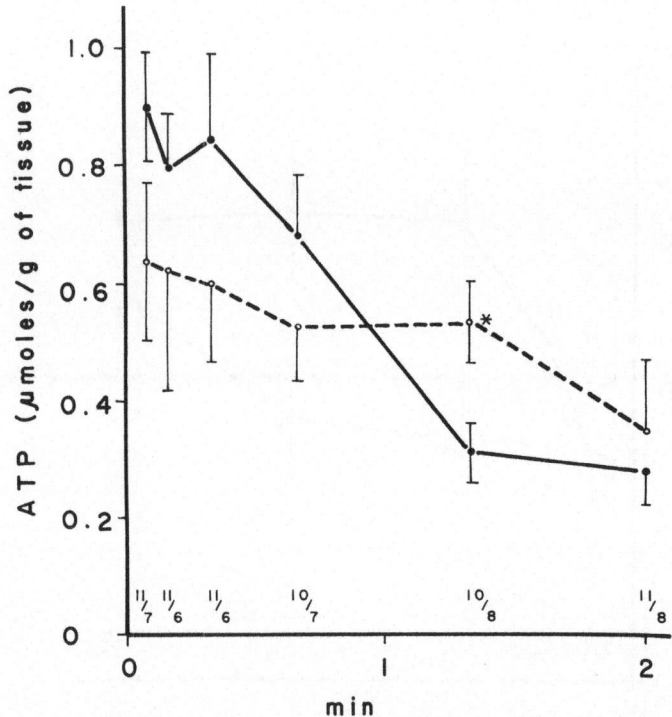

Fig. 3: Changes of ATP levels induced by decapitation
in the hind brain of control (o—o) and ethanol-treated (o---o) rats. The bars indicate S.E.M. and the
fractions indicate number of determinations in control
animals over number of determinations in ethanol-treated rats. Significance; *P< 0.05. Reprinted from Volicer, et al., 1977.

Administration of ethanol also affects the second
cyclic nucleotide, cyclic GMP. As shown in Fig. 4,
brain cyclic GMP levels are decreased in a dose-dependent
manner one hour after ethanol administration (VOLICER &
HURTER, 1977). There is some evidence that cyclic GMP
might act as a second messenger mediating the effect of
cholinergic stimulation or responding to increased intracellular calcium levels (GOLDBERG & HADDOX, 1977). However, the effect of ethanol on cyclic GMP is unaffected
by cholinergic agonists and antagonists (VOLICER &
KLOSOWICZ, 1978) and by intracerebral calcium fusion
(SCHMIDT & VOLICER, in preparation).

Guanylate cyclase is also activated by oxidation
(MITTAL & MURAD, 1977) and it is possible that its activity in vivo depends on a redox potential of the tissue.
It was reported (RAWAT & KURIYAMA, 1972; RAWAT et al.,
1973) that ethanol administration affects the redox

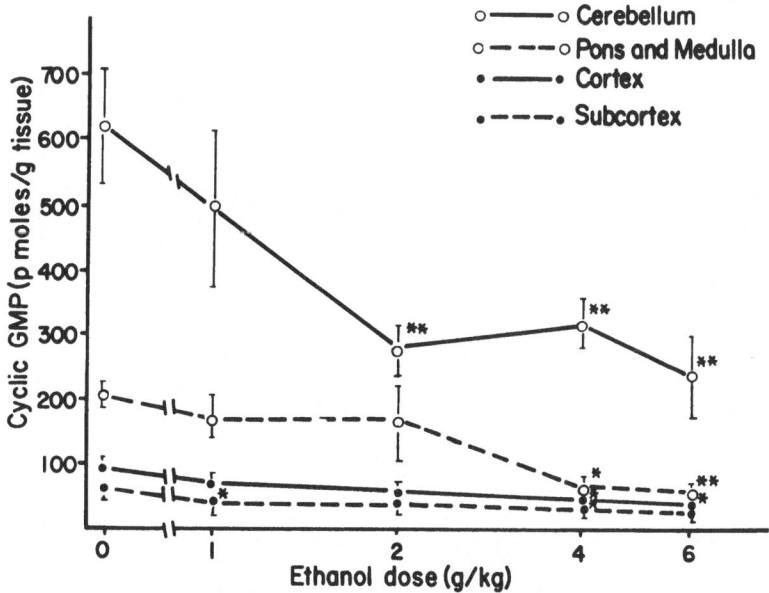

Fig. 4: Cyclic GMP levels in brains of control rats and
         rats treated with various doses of ethanol. The
         points are means of values obtained in six animals and
         the bars indicate values which are significantly dif-
         ferent from water-treated controls. *p< 0.05; **P<
         0.01. Reprinted from Volicer and Hurter, 1977.

potential of brain tissue resulting in an increased ratio
of reduced to oxidated metabolites. Although this finding was not supported by the results of VELOSO et al.,
(1972), it is possible that a shift of redox potential
toward reduction might lead to a decreased guanylate
cyclase activity and decreased cyclic GMP levels.

In order to test this hypothesis we inhibited ethanol metabolism by pretreatment with an alcohol dehydro-
genase inhibitor, pyrazole, which was shown to prevent
changes of brain redox potential after ethanol administration (RAWAT & KURIYAMA, 1972). Pyrazole itself in
a dose of 225 mg/kg decreased cyclic GMP levels in the
cerebellum but it prevented the ethanol-induced decrease
(Fig. 5). Virtually identical changes of cyclic GMP
levels after ethanol and pyrazole administration were
observed in the cerebral cortex. (Fig. 5).

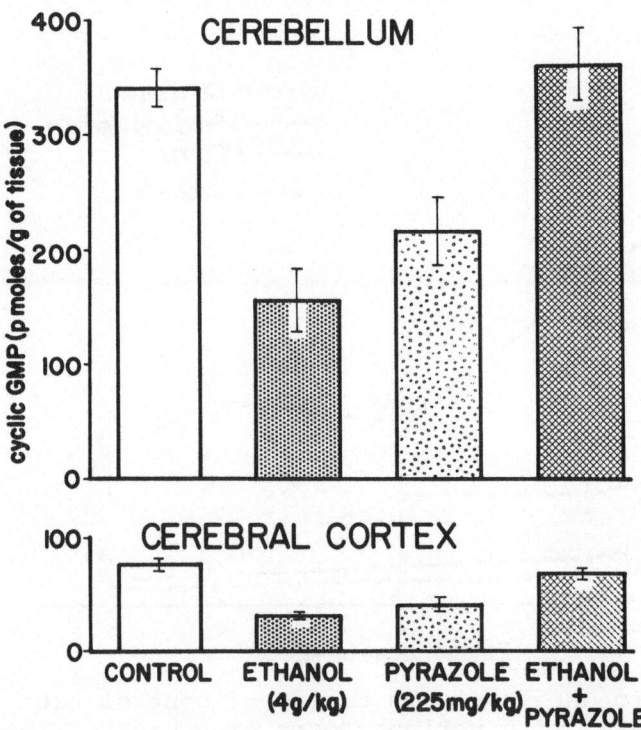

Fig. 5:   Effect of pyrazole pretreatment on the cyclic
          GMP-lowering effect of ethanol in rat brain. Columns
          represent means(6-7 determinations)and the bars re-
          present S.E.M.

In contrast to the effects of pyrazole and ethanol on cyclic GMP levels cyclic AMP levels were unchanged after their administration (Table 1). The lack of effect of ethanol on cyclic AMP levels in this experiment might be due to different origins of these Sprague-Dawley rats because genetic differences in the effect of ethanol on cyclic AMP levels were recently described (CHURCH & FELLER, 1978).

TABLE 1

Effect of pyrazole and ethanol on cyclic AMP levels in rat brain.

Treatment	Cerebellum	Cerebral Cortex
Control	684 + 86	623 + 46
Ethanol (4 g/kg)	596 + 84	744 + 91
Pyrazole (225 mg/kg)	564 + 73	700 + 72
Ethanol & Pyrazole	500 + 69	674 + 51

Pyrazole was administered i.p. one hr. before administration of ethanol by intubation and the rats were sacrificed one hr. later. Cyclic AMP levels are expressed in p moles/g of tissue + S.E.M.

TABLE 2

Effect of pyrazole pretreatment on the cyclic GMP-lowering action of ethanol in the rat cerebellum.

	Control	Ethanol 4 g/kg
Control	699 + 125	173 + 40**
Pyrazole (450 mg/kg)	273 + 80*	1615 + 190**

Pyrazole was administered i.p. one hr. before administration of ethanol by intubation and the rats were sacrificed one hr. later. Cyclic GMP levels are expressed in p moles/g of tissue + S.E.M. *$P < 0.05$ and **$P < 0.01$ compared to controls.

A higher dose of pyrazole (450 mg/kg) not only prevented the ethanol-induced decrease of cyclic GMP levels
but it actually reversed its effect. Administration of
ethanol one hr. after pyrazole pretreatment led to cyclic
GMP levels significantly higher than control levels in
both cerebellum (Table 2) and cerebral cortex.

At the same time this dose of pyrazole produced
some degree of sedation and potentiated the ethanol-
induced sedation (Table 3). Thus, although the biochemical effect of ethanol was prevented or even reversed,
the behavioral effect of ethanol was not diminished.
These results indicate that the decrease of cyclic GMP
levels after ethanol administration might be secondary
to ethanol metabolism and probably does not mediate the
behavioral depression. In addition to that, these results provided evidence that changes of redox potential
are affecting the cyclic GMP levels in vivo.

Ethanol is a unique drug because it acts as a metabolic substrate and therefore leads to significant shifts
of intermediary metabolism. However, many other drugs
are also metabolized or may lead to secondary changes
of intermediary metabolism. Therefore metabolic changes
have to be considered when a possible role of cyclic
nucleotides in pharmacological effects of centrally
acting drugs is being investigated.

TABLE 3

Effect of pyrazole pretreatment on the ethanol-induced
depression of motor behavior.

	Control	Pyrazole (450 mg/kg)
Control	0/6	2/6
Ethanol (4 g/kg)	2/6	6/6

Pyrazole was administered i.p. an hr. before administration of ethanol by intubation. Rats were tested an hr.
later on an inclined plane (VOLICER & GOLD, 1973). The
rats were placed head down and observed for 1 minute.
Numbers in the table are rats sliding off the plane/
total number of rats tested.

Acknowledgement

This work was supported by USPHS grant AA-00189. The expert assistance of Mrs. Jenifer Woodworth in preparation of the manuscript is greatly appreciated.

REFERENCES

CHURCH, A.C. & FELLER, D. (1978) The influence of genetics on ethanol-induced changes in brain cyclic nucleotide levels. Alcoholism 2, 189.

DIZION, B.R., PAUL, M.I. & PAUK, G.L. (1970) Measurement of adenosine 3',5'-monophosphate in brain. Pharmacology 3, 25-31.

GOLDBERG, N.D. & HADDOX, M.K. (1977) Cyclic GMP metabolism and involvement in biological regulation. Ann. Rev. Biochem. 46, 823-896.

GRISHAM, C.M. & BARNETT, R.E. (1973) The effects of long-chain alcohols on membrane lipids and the (Na^+-K^+)-ATPase. Biochim. Biophys. Acta. 311, 417-422.

ISRAEL, Y. & SALAZAR, I. (1967) Inhibition of brain microsomal adenosine triphosphatases by general depressants. Arch. Biochem. & Biophys.122, 310-317.

IVERSEN, L.L., IVERSEN, S.D. & SNYDER, S.H. (1975 Synaptic Modulators in Handbook of Psychopharmacology, Vol. 5, Plenum Press, New York.

KAKIUCHI, S. & RALL, T.W. (1968) The influence of chemical agents on the accumulation of adenosine 3',5'-phosphate in slices of rabbit cerebellum. Mol. Pharmacol. 4, 367-378.

MCILWAIN, H. (1976) Translocation of neural modulators - 2nd category of nerve signal. Neurochem. J. 1, 351-368.

MITTAL, C.K. & MURAD, F. (1977) Properties and oxidative regulation of guanylate cyclase - mini-review. J. Cyc. Nucl. Res. 3, 381-391.

PAUL, M.I., PAUK, G.L. & DITZION, B.R. (1970) The effect of centrally acting drugs on the concentration of brain adenosine 3',5'-monophosphate. Pharmacology, 3, 148-154

RAWAT, A.K. & KURIYAMA, K. (1972) Ethanol oxidation: effect on the redox state of brain in mouse. Science, 176, 1133-1135.

RAWAT, A.K., KURIYAMA, K. & MOSE, J. (1973) Metabolic consequences of ethanol oxidation in brains from mice chronically fed ethanol. J. Neurochem. 20, 23-33.

ROBISON, G.A., BUTCHER, R.W. & SUTHERLAND, E.W. (1971)
    Cyclic AMP, Academic Press, New York and London.
SATTIN, A. (1971)   Increase in the content of adenosine
    3',5'-monophosphate in mouse forebrain during
    seizures and prevention of the increase by methyl-
    xanthines. J. Neurochem. 18, 1087-1096.
SATTIN, A. & RALL, T.W. (1970)   The effect of adenosine
    and adenine nucleotides on the cyclic adenosine 3',
    5'-phosphate content of guinea pig cerebral cortex
    slices. Molec. Pharmacol. 6, 13-23.
SHIMIZU, H., CREVELING, C.R. & DALY, J. (1970)   Stimu-
    lated formation of adenosine 3',5'-cyclic phosphate
    in cerebral cortex: synergism between electrical
    activity and biogenic amines. Proc. Nat. Acad.
    Sci. 65, 1033-1040.
SUN, A.Y. & SAMORAJSKI, T. (1972)   Effects of ethanol on
    the activity of adenosine triphosphatase and acetyl-
    cholinesterase in synaptosomes isolated from guinea-
    pig brain. J. Neurochem. 17, 1365-1372.
SWANSON, P.D. (1969)   The effects of oxygen deprivation
    on electrically stimulated cerebral cortex slices.
    J. Neurochem. 16, 35-45.
UZUNOV, P. & WEISS, B. (1971) Effects of phenothiazine
    tranquilizers on the cyclic 3',5'-adenosine mono-
    phosphate system of rat brain. Neuropharmacology,
    10, 697-708.
VELOSO, D., PASSONNEAU, J.V. & VEECH, R.L. (1972)   The
    effect of intoxicating doses of ethanol upon inter-
    mediary metabolism in rat brain. J. Neurochem.
    19, 2679-2686.
VOLICER, L. & GOLD, B.I. (1973)   Effect of ethanol on
    cyclic AMP levels in rat brain. Life Sci. 13,
    269-280.
VOLICER, L., MIRIN, R. & GOLD, B.I. (1977)   Effect of
    acute ethanol administration on the cyclic AMP
    system in rat brain. J. Stud. Alcohol 38, 11-24.
VOLICER, L. & HURTER, B.P. (1977)   Effects of acute and
    chronic ethanol administration and withdrawal on
    adenosine 3',5'-monophosphate and guanosine 3',5'-
    monophosphate levels in the rat brain. J. Pharmacol.
    Exp. Ther. 200, 298-305.
VOLICER, L. & KLOSOWICZ, B.A. (1978)   Effect of drugs
    which increase cyclic GMP levels on ethanol-induced
    depression of acoustic startle response. Alcoholism
    2, 189.

PHOSPHOPROTEINS AS SPECIFIERS FOR MEDIATORS AND MODULATORS

IN NEURONAL FUNCTION

Yigal H. Ehrlich

The Missouri Institute of Psychiatry
University of Missouri-Columbia, School of
Medicine, St. Louis, Missouri 63139

INTRODUCTION

A large body of evidence has accumulated indicating
that protein phosphorylation plays an important role in the
regulation of metabolic processes in eukaryotic cells.
Thus, glycogenolysis, glycogen synthesis and lipolysis are
controlled, respectively, by the phosphorylation of glycogen phosphorylase-kinase, glycogen-synthetase and trigly-
ceride-lipase (for review see RUBIN and ROSEN, 1975).
Subsequent to these initial discoveries, it was found that
the phosphorylation of proteins is involved in the regulation of a diversity of physiological and metabolic processes occuring in many different organs. In addition, it
has been demonstrated that in each of these organs the process of protein phosphorylation is influenced by physiological affectors that regulate cellular functions in the
respective target tissue (e.g. polypeptide hormones, steroid
hormones, cyclic nucleotides, calcium ions, etc.). The
integration of all these findings has led to the suggestion
(GREENGARD, 1978a) that the phosphorylation of proteins may
be a final common pathway for a diversity of biological
regulatory agents.

The process of protein phosphorylation consists of
the transfer of the gamma phosphate from adenosine tri-
phosphate (ATP) to serine and/or threonine hydroxyl groups
in protein substrates. This transfer is catalysed by
the enzyme protein kinase (Figure 1). The enzyme phosphoprotein phosphatase catalyses the hydrolysis of the

phosphoester bond and the protein then resumes its original
structure.   The phosphorylation and dephosphorylation of
proteins thus provide the means to carry out reversible
structural alterations that may mediate functional changes.
This system does not require de-novo synthesis nor degradation of pre-existing molecules.   It represents, therefore,
an "ideal" target for regulatory agents that produce rapid
and transient changes in cellular activity.   On the other
hand, phosphorylative activity may serve also as a site of
long-term molecular adaptation, since a modification induced in the process of protein phosphorylation could result in long-lasting alterations in cellular function.

    This chapter will briefly review studies that have
implicated protein phosphorylation in neuronal function,
will focus on the element of specificity provided by this
system, and summarize studies which have indicated the
potential role of certain specific phosphoproteins in
processes whereby environmental and pharmacological inputs
induce long-lasting alterations in brain function.

Protein phosphorylation in neuronal function.

    The first study to implicate protein phosphorylation
in neural function was reported by Heald (1975).   He demonstrated that a brief depolarization of respiring slices of
cerebral cortex by electrical pulses caused a significant

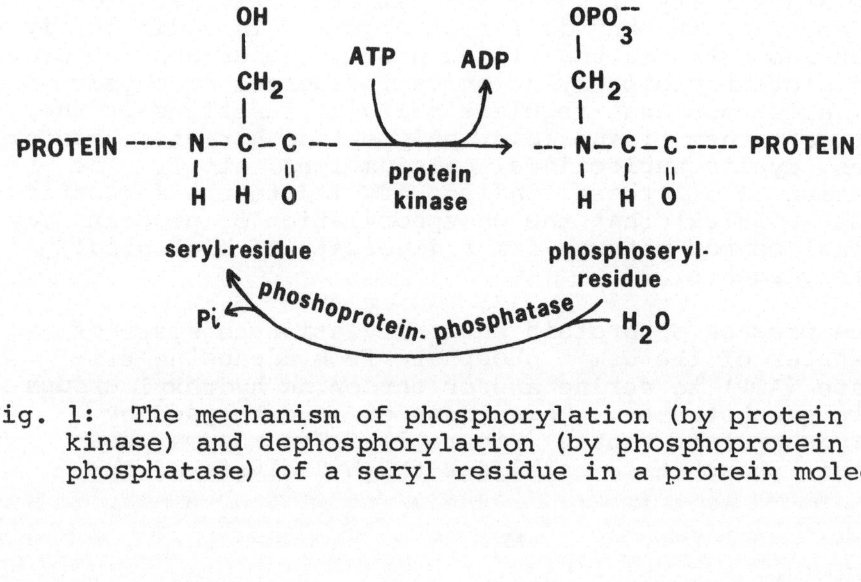

Fig. 1:   The mechanism of phosphorylation (by protein
    kinase) and dephosphorylation (by phosphoprotein
    phosphatase) of a seryl residue in a protein molecule.

increase in the amount of protein-bound phosphate in the
tissue. Subsequent studies, carried out in the laboratory
of R. Rodnight in London, showed that this increase occurs
primarily in membrane-bound proteins, is mediated by the
action of neurotransmitters, and may involve activation of
protein kinases by cyclic AMP (reviewed by RODNIGHT, 1975;
WILLIAMS and RODNIGHT, 1977). Attempts to determine the
location of the affected membranes (WILLIAMS, et al, 1974)
showed that noradrenaline induced a significant increase
in protein phosphorylation in neuronal fractions but not
in crude glial preparations, while an increase in protein
phosphorylation induced by histamine and 5-hydroxytrytamine
was significant in the glial but not in the neuronal cell
body fractions. A recent study, using hippocampal slices,
demonstrated that repetitive stimulation produces changes
in the phosphorylation of a specific protein in synaptic
membranes (BROWNING, et al, 1979). Both this latter study
and the earlier reports by Rodnight's group, have suggested
that the observed changes in phosphorylation are related
to stimulus-dependent potentiation of synaptic transmission
such as demonstrated by Libet and Tosaka (1970).

     A different approach used in studying the role of pro-
tein phosphorylation in neuronal function consists of the
isolation of subcellular fractions from brain, and characterization of the protein kinases, phosphoprotein phosphatases and phosphoprotein substrates present in each
fraction. Early work in this area (RODNIGHT and LAVIN,
1966) showed that brain membranes are capable of carrying
out an enzymatic transfer of phosphate from ATP to protein
substrates which constitute part of the structure of these
membranes. This activity will be referred to hereinafter
as endogenous phosphorylation. Later studies from the
same laboratory demonstrated that this endogenous phosphorylation activity is stimulated by cyclic AMP (WELLER
and RODNIGHT, 1970, 1973). The cyclic AMP-stimulated endogenous phosphorylative activity was found to be enriched
in preparations of synaptic membranes (RODNIGHT, 1975) and
within them, in subfractions containing synaptic junctions
(WELLER and MORGAN, 1976).

     A major contribution to our present knowledge of the
role of protein phosphorylation in neuronal function has
been provided by studies from the laboratory of P. Greengard. Following the discovery of cyclic AMP and the for-
mulation of the "second messenger concept", it was suggested that cyclic AMP acts as a second messenger in the
post-synaptic action of neurotransmitters (see also
SIGGINS, this volume). Greengard's group discovered

cyclic AMP-dependent protein kinase activity in brain and
characterized it using exogenous substrates, such as histones (MIYAMOTO, et al, 1969). They have found that this
enzyme, endogenous substrates for its activity and phosphoprotein phosphatase are present in high levels in preparations of synaptic membranes (JOHNSON, et al, 1971).
Studying the mechanisms of synaptic transmission in the
mamalian superior cervical ganglion, Greengard and Kebabian
(1974) obtained experimental data suggesting that cyclic
AMP mediates dopaminergic transmission, leading to the
production of slow inhibitory post-synaptic potentials,
while cyclic GMP mediates muscarinic cholinergic transmission leading to the production of slow excitatory post-
synaptic potentials. Based on these findings, P. Greengard
(1976) has suggested that rapid changes in the phosphorylation state of synaptic proteins, brought about by cyclic
nucleotides, produce altered membrane permeability and
consequent changes of membrane polarization. Subsequent
studies in Greengard's laboratory, including the identification, characterization and isolation of a specific substrate for cyclic AMP-dependent protein kinase in synaptic
membranes, provided further support for this suggestion.
A monograph presenting a detailed and comprehensive review
of the studies by Greengard and his colleagues has been
published recently (GREENGARD, 1978b).

The study of protein phosphorylation in neural membranes, using either tissue slices or isolated synaptic
membranes, has provided evidence that cyclic AMP-dependent
protein kinase is involved in the post-synaptic action of
neurotransmitters, and has suggested that this activity may
play a role in the slow electrogenic response of neurons
to neurohumoral stimulation. Another line of investigation
has implicated protein phosphorylation in additional aspects
of neuronal function. In this investigative approach, the
ability of various components in the neuron to serve as
phosphate acceptors for protein kinase is tested, and the
effects of phosphorylating conditions on their activity is
examined. Such studies have implicated protein phosphorylation in the regulation of tyrosine hydroxylase activity.
This enzyme, which catalyzes the rate limiting step in the
biosynthesis of catecholamines, requires for its activity
a reduced pterine cofactor which serves as an electron
donor. Lovenberg, et al (1975) have demonstrated that subjecting a crude preparation of tyrosine hydroxylase to
phosphorylating conditions decreases the enzyme's Km for
the pterine cofactor and increases its Ki for dopamine.
Similar findings were reported by Goldstein, et al (1976).
Another enzyme whose intimate involvement in neuronal

function is well established is dopamine-sensitive adenylate cyclase. The activity of this enzyme is also
affected by protein phosphorylation. Chueng, et al (1975)
have shown that the calcium dependent regulator (CDR) of
phosphodiesterase activity also regulates membrane-bound
adenylate cyclase activity. Gnegy, et al (1976a) reported
that CDR is present in synaptic membranes from brain, and
have shown that its release from the membrane increases
when membrane-bound proteins in the preparation are phosphorylated by an exogenous, cyclic AMP-dependent protein
kinase. A further study (GNEGY, et al, 1976b) demonstrated
that phosphorylation-dependent interaction of CDR with
neostriatal membranes controls the sensitivity of adenylate
cyclase to stimulation by dopamine. Using immunoprecipi-
tation methods, this study has shown that adenylate cyclase
itself is not the phosphorylated substrate in this process.
Nonetheless, since the phosphorylation and dephosphorylation of proteins has been shown to be a mechanism involved
in the structural re-arrangement of membrane components
(GAZITT, et al, 1976), it is possible that the alterations
in membrane-bound adenylate cyclase are reflective of
general changes in membrane fluidity, effected by a phosphorylation process. Similar mechanisms may be operating
in the process whereby protein phosphorylation alters the
movement of calcium ions in neuronal systems (WELLER and
MORGAN, 1977). The latter study has demonstrated that incubation of synaptosomes under phosphorylating conditions
lowers the rate of calcium efflux and influx, without
affecting the equilibrium binding of calcium to the membranes.

     Using a similar approach, the phosphorylation of
proteins has been implicated in the control of calcium-
dependent neurotransmitter release, microtubular function,
RNA and protein synthesis and the regulation of various
soluble and membrane-bound enzymes in nervous tissue (for
recent reviews see RUBIN and ROSEN, 1975; GREENGARD, 1976;
WILLIAMS and RODNIGHT, 1977; GREENGARD, 1978). In these
processes, phosphorylation systems may act as intermediaries
for the action of various neurotransmitters, peptide and
steroid hormones, cyclic nucleotides and calcium ions.
Thus, protein phosphorylation may play a crucial role in
the series of events that occur in processes whereby cells
in the nervous system respond to changes in the internal
and external environment. The presumed relationships
between environmental input, protein phosphorylation and
neuronal activity are illustrated in Figure 2. As pointed
out by Williams and Rodnight (1977) and by Greengard (1978a),

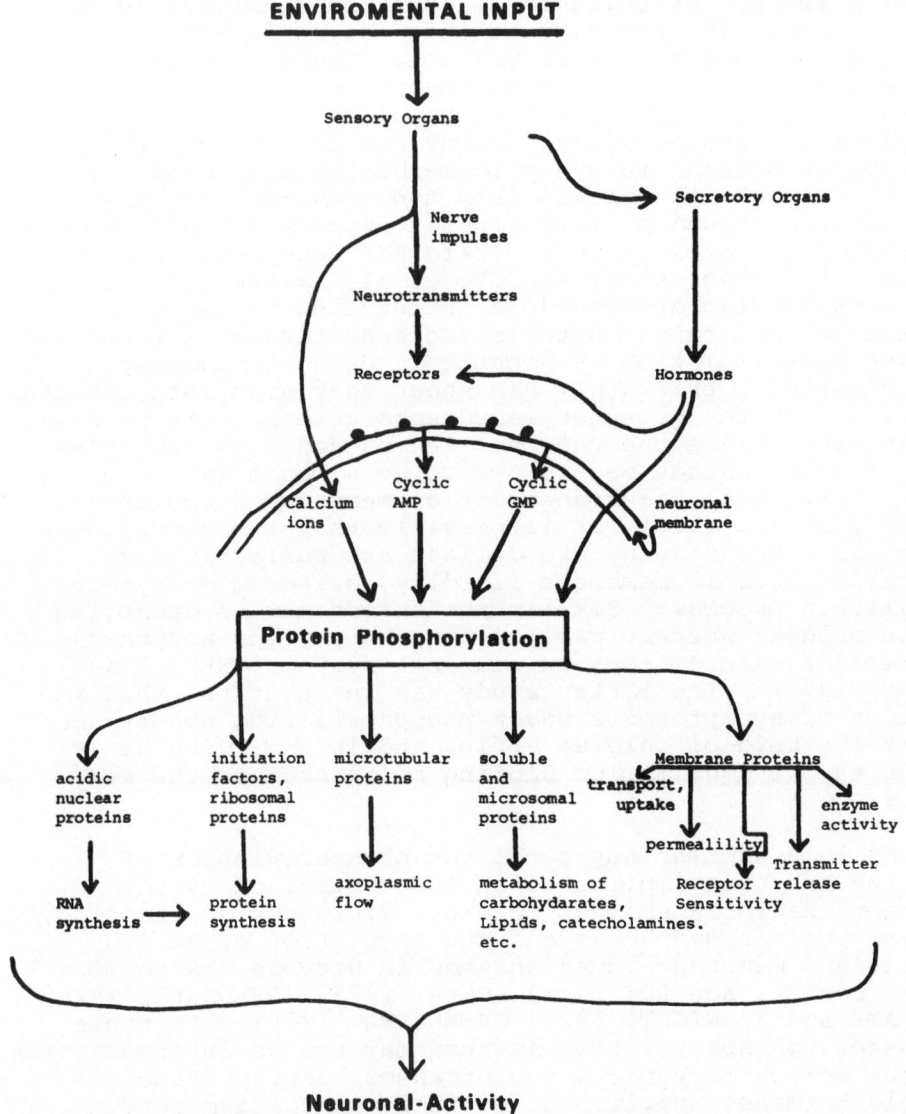

Fig. 2: The series of events that lead from environmental
input to altered neuronal activity, in which protein
phosphorylation systems act as mediators and specifiers for the influences of neurotransmitters, hormones,
cyclic nucleotides and calcium ions. Based on a survey of the literature cited in the text.

there is good evidence that the phosphorylation of proteins is indeed regulated, either directly or indirectly,
by the various physiological affectors indicated in Figure
2. However, the evidence for involvement of phosphorylated
proteins in many aspects of neuronal function is still
largely circumstantial. In a review of the early literature in this field, Rubin and Rosen (1975) indicated that
the establishment of direct causal relationships between
protein phosphorylation and defined cellular functions must
await the identification, isolation and purification of
proteins that serve as specific substrates for protein
kinases. The studies described below were carried out with
this purpose in mind.

Phosphoproteins as specifiers for neurotransmitters and neuromodulators.

As early as 1969, Kuo and Greengard (1969) advanced
the hypothesis that all the biological effects of cyclic
AMP are mediated and specified by protein kinases. In
view of the great diversity of physiological and metabolic
events regulated by cyclic AMP in the nervous system, and
the demonstration that only two types of protein kinase
can be isolated from membranes or cytosol of nervous tissue
(UNO, et al, 1976), the question of the element which provides the specificity in this mode of regulation remained
open. Moreover, recent findings ascribe multiple functions
also for cyclic nucleotide-independent protein kinase, and
in particular to calcium-sensitive protein kinase. The
first clue to the question of specificity was provided by
Johnson, et al (1972), who reported that preparations of
synaptic membranes from rat cerebrum contain only one protein whose phosphorylation is stimulated by cyclic AMP.
They have suggested that this protein acts specifically as
the mediator for cyclic AMP in its postulated role in the
generation of slow post-synaptic potentials. Subsequent
studies from the same laboratory (UEDA, et al, 1973) identified an additional protein whose phosphorylation and dephosphorylation was enhanced by cyclic AMP. These two proteins were designated by the Roman numerals I and II.
Protein II was later identified as the regulatory subunit
of membrane-bound, cyclic AMP-dependent protein kinase
(GREENGARD, 1976). Our initial studies (EHRLICH and
ROUTTENBERG, 1974; ROUTTENBERG and EHRLICH, 1975) confirmed
the report of Ueda, et al, (1973) and described two phosphoproteins which were designated D and E, tentatively
identified, respectively, as proteins I and II. In addition, we have described two other phosphoproteins characteristic of preparations enriched in synaptic membranes.

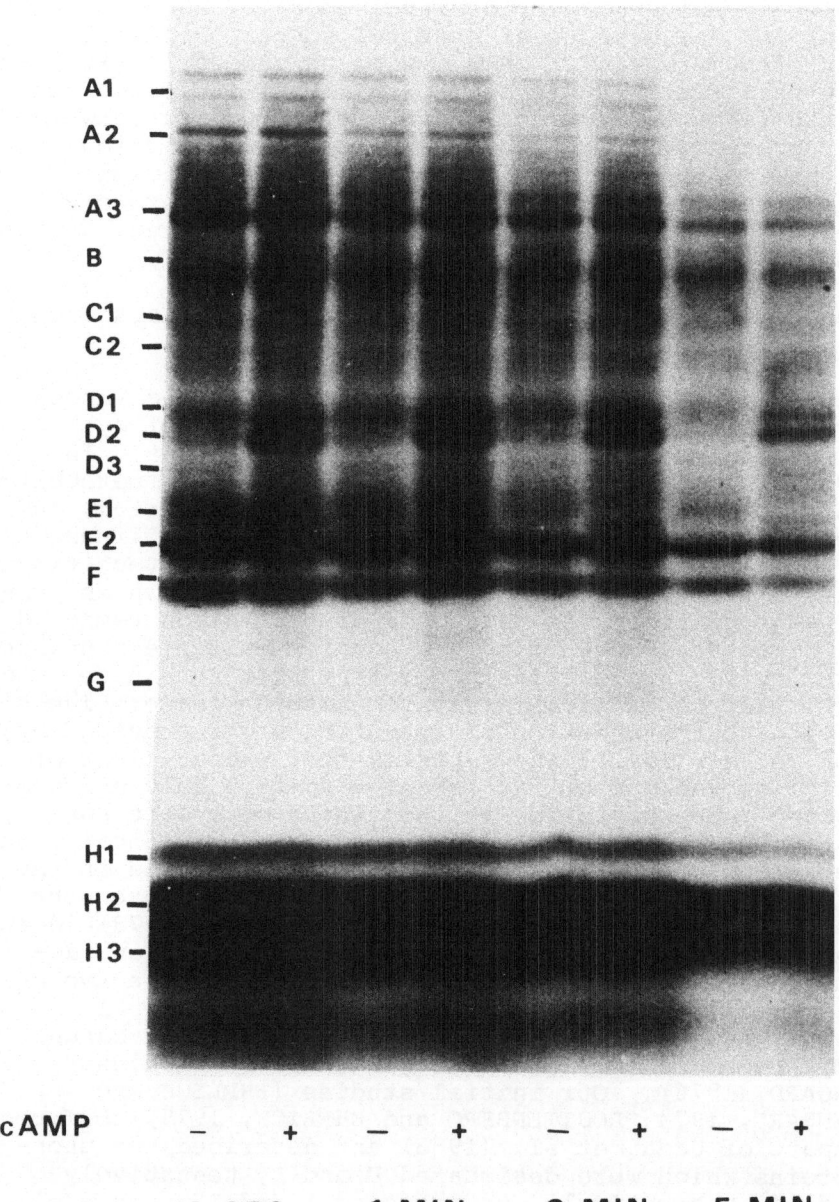

One, designated F, was shown to be a major phosphorylated component of synaptic membranes, cyclic AMP had only minimal effects on its phosphorylation, but its presence enhanced the rate of dephosphorylation of this protein (see Figure 3). The other, designated G, was a minor phosphoprotein, but its phosphorylation appeared to be virtually dependent on cyclic AMP. In vitro assays in which synaptic membranes are incubated with ATP labled with ^{32}P-phosphate, combined with the technique of polyacrylamide gel electrophoresis, have been used in these studies for identifying the endogenous substrates of protein kinase. Improved resolution obtained by using more advanced separation procedures has demonstrated that of about 70 protein components present in preparation enriched in synaptic membranes, over twenty can serve as substrates for endogenous phosphorylative activity (EHRLICH, et al, 1977c and Fig. 3, MAHLER, 1977; RODNIGHT, 1978). However, the heterogeneity of cells in brain tissue raised the possibility that different phosphoproteins are not located in the same cell. We have recently demonstrated that cytosol (EHRLICH, et al, 1977d) and membrane fractions (EHRLICH, et al, 1978d) prepared from a homogenous population of neuroblastoma cells grown in culture exhibit a heterogeneity of endogenously phosphorylated proteins which is as great as that found in preparations from brain. This finding indicates that individual nerve cells probably contain a multiplicity of phosphoproteins. Their presence in the same cell, and the differences in their properties could account for the specificity in the diverse physiological functions regulated by even a few protein kinases.

Characterization of the phosphorylative properties that different phosphoproteins exhibit in vitro may serve

Fig. 3: Time course of the endogenous phosphorylation of specific proteins in preparations containing synaptic membranes. Autoradiogram of reaction products from assay of the endogenous phosphorylation of cortical membranes using the methods described by Ehrlich, et al, (1977). Note that specific bands respond differently over time to the addition of 5 micromolar cyclic AMP to the reaction. Such effects can be seen for certain bands (e.g. band D1) during net phosphorylation (10 sec), and for others (e.g. band F) during net dephosphorylation (5 min). For more details see text.

as an indicator of potential differences in their physiological functions (RUBIN and ROSEN, 1975). Indeed, we have
demonstrated that different phosphoproteins in synaptic
membranes respond differently to various factors that affect
phosphorylative activity. These factors included cyclic
AMP, cyclic GMP, the concentration of ATP and various divalent cations such as magnesium, calcium and zinc, reaction time and preincubation time (EHRLICH & ROUTTENBERG,
1974; ROUTTENBERG & EHRLICH, 1975; EHRLICH, et al, 1977c;
RAM & EHRLICH, 1978). Moreover, these differences are not
characteristic only of endogenous phosphorylation systems
in membranes, where they might be attributed to constrains
imposed by the membrane structure, but also of endogenous
phosphorylation systems of the cytosol (EHRLICH, et al,
1977c, 1978c). Thus, preparations of soluble proteins from
rat cerebral cortex revealed the presence of four types of
endogenous phosphorylative activity: a. Cyclic AMP-
independent, b. Cyclic AMP-dependent, c. Cyclic AMP-
stimulated, and d. Cyclic AMP-inhibited. This classification evolved from analysis of phosphate incorporation
into specific phosphoproteins as a function of reaction
time, in the presence and absence of 5 μM cyclic AMP. A
chart illustrating the basis for these definitions is
depicted in Figure 4. Since these activities occur simultaneously in the same preparation, the results support the
notion that the specificity in the action of protein kinase
is determined, at least in part, by the nature of the substrates (EHRLICH, et al, 1977c). Furthermore, the presence
of numerous phosphate acceptor protein species in the cytosol that respond differently to increased levels of cyclic
AMP may account for the diversity of intracellular metabolic
activities known to be affected by neurohumoral stimulation.
Fractionation of this soluble preparation by gel filtration revealed that its different endogenous phosphorylation
systems are unequally distributed among large protein complexes (EHRLICH, et al, 1979). Such complexes may contain
also the metabolic machineries (e.g. tyrosine hydroxylase)
that are regulated by the specific phosphoproteins characteristic of each complex.

     The specificity in the physiological function subserved
by a phosphoprotein can be determined also by its subcellular location. Indeed, different phosphoproteins were
found to be characteristic of various subcellular fractions
from rat cerebral cortex, such as nuclei, microsomes, myelin,
synaptosomes and cytosol (EHRLICH, et al, 1977c). Recent
reports from the laboratory of H. Mahler (DE BLAS, et al,
1979) demonstrated that synaptic membranes contain endogenous phosphorylative activity directed towards 23

separate protein components, in the M.W. range of 40K-
300K. The phosphorylation of six of these proteins was
stimulated by cyclic AMP and that of all the other was
cyclic AMP-independent. Among the latter, five showed
stimulation by calcium ions and their phosphorylation required the presence of a heat stable synaptoplasmic factor,
identified previously by Schulman & Greengard (1978) as
CDR. Moreover, De Blas, et al, (1979) identified different
phosphoproteins characteristic of various subfractions
prepared from synaptic plasma membrane preparations.
Namely, synaptic junctional complexes, presynaptic structures, post-synaptic membranes and post-synaptic densities
(PSD). One of the phosphoproteins in PSDs with a M.W. of
47K, which showed cyclic AMP-independent phosphorylation,
was tentatively identified as the main polypeptide component of this structure. Two specific proteins which were
described first by De Lorenzo (1976), the phosphorylation
of which is stimulated by calcium ions and inhibited by the
anticonvalsant drug diphenylhydantoin, were found by De
Blas, et al, to be located presynaptically and assigned a
M.W. of 59K and 47K. De Lorenzo & Freedman (1978) found
that these phosphoproteins are associated with synaptic
vesicles and have suggested that they play a role in neurotransmitter release. A specific synaptosomal phosphoprotein whose phosphorylation is stimulated by calcium
ionophores and inhibited by magnesium ions was described
by Hershkowitz (1978) and its involvement in neurotransmitter release was postulated.

     In this context it should be emphasized that the
determination of the subcellular site of origin of specific phosphoproteins that is based on endogenous phosphorylation activity measured after the fractionation should be
regarded with a certain degree of caution; particularly
when membrane-bound proteins are involved. Endogenous
phosphorylative activity in membranes depends on the spacial
relationships between its various components (kinases,
substrates, phosphatases) and would therefore be highly
susceptible to structural changes induced by hyperosmotic
sucrose solutions (EHRLICH and ROUTTENBERG, 1974). Thus,
when synaptic plasma membranes were prepared using conventional sucrose density gradients, the specific activity of
endogenous phosphorylation increased by 40-70 percent only.
On the other hand, synaptic membranes prepared in a medium
that provides high density at iso-osmotic pressures (sodium
diatrizoate, TAMIR, et al, 1974) demonstrated a tenfold
enrichment in this activity (EHRLICH & ROUTTENBERG, 1975).
Under our standard assay conditions, the myelin fraction
obtained from sodium diatrizoate gradients (10% over 18%)

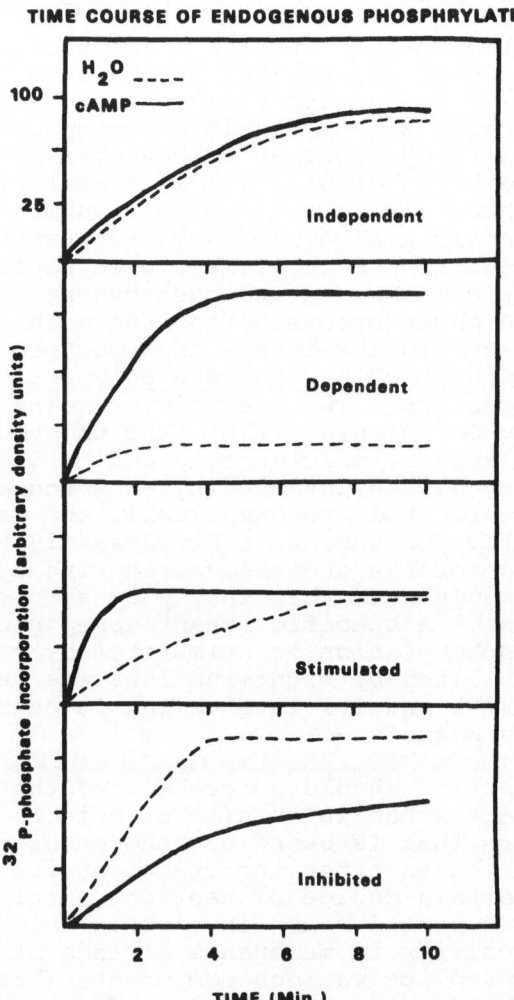

TIME COURSE OF ENDOGENOUS PHOSPHRYLATION

Fig. 4: Four types of endogenous phosphorylation activity
in brain cytosol. A fraction of the soluble proteins
from brain cytosol (designated S4; EHRLICH, et al,
1977c) was incubated with gamma-^{32}P-ATP in the presence
(———) and absence (---) of 5 micromolar cyclic AMP.
Phosphate incorporation into specific proteins was
determined at various reaction times, (as demonstrated
for the membrane fraction in Fig. 3). Addition of
cyclic AMP had different effects on the phosphorylation of different specific proteins. These activities
were classified according to the effects of cyclic AMP
on the time course of phosphorylation, as illustrated
in this figure.

incorporated 13.3 $\pm$ 1.7 pmole phosphate per mg protein
(mean $\pm$ SEM, n=3), for the mitochondrial fraction this
value was 9.76 $\pm$ 0.73. On the other hand, a fraction containing synaptic membrane fragments, designated TT (top
trizoate) and collected from the interphase of 10% sodium
diatrizoate to buffer layers, incorporated 227.3 $\pm$ 1.98
pmole phosphate per mg protein; compared to 23.1 $\pm$ 1.98
of the osmotically shocked crude synaptosomal fraction
which served as starting material in these studies (EHRLICH
& ROUTTENBERG, 1975, and unpublished observations). The
subcellular site of origin of the membranes in the fraction
exhibiting this tenfold increase has not been fully defined
yet. A similar fraction was prepared from synaptosomes
by Tamir, et al, (1976). De Blas, et al, (1979) describe
a fraction (P_3B_2) with a high content of synaptic membranes
and devoid of post-synaptic densities; the pattern of endogenously phosphorylated proteins observed in our TT fraction was very similar to that of the P_3B_2 fraction. The
use of membrane fractions isolated in a procedure using
iso-osmotic conditions throughout would undoubtedly prove
beneficial in future studies attempting to elucidate the
role of specific phosphoproteins in neuronal function.

A few of the numerous phosphoproteins which have been
identified in our studies are likely to be identical to
phosphorylated proteins studied elsewhere, for which a
role in brain function has already been postulated. Phosphoproteins in the A region of our gel (see Fig. 3) may
correspond to one of the microtubule-associated proteins
(SLOBODA, et al, 1975). The proteins which we have designated D_1 and D_2 are most probably identical to proteins
Ia and Ib, isolated recently by Ueda and Greengard, (1977).
Among the phosphoproteins in the E group, one may correspond to the phosphorylated subunit of tubulin and another
to that of the acetylcholine receptor (see GORDON & DIAMOND,
this volume). E_1 is probably the regulatory subunit of
membrane-bound cyclic AMP-dependent protein kinase (WALTER,
et al, 1978). In the soluble fraction, "E bands" could
correspond to the phosphorylated subunit of tyrosine hydroxylase (JOH, et al, 1978) and to the regulatory subunit of
the cytoplasmic cyclic AMP-dependent protein kinase
(MALKINSON, et al, 1975). Our phosphoprotein F may be the
protein whose phosphorylation was shown to be regulated
by ACTH fragments (ZWIERS, et al, 1976, 1977; see also
GISPEN, et al, this volume). Some of the phosphoproteins
in the H zone of our gel are, in all likelihood, myelin
basic proteins. However, their presence in membranes
from the cerebral cortex of newborn rats and from neuroblastoma cells (EHRLICH, et al, 1977c) indicates that

phosphoproteins of other subcellular origin co-migrate
with myelin basic-proteins. These inferences are not based
on electrophoretic mobility of a specific phosphoprotein
(M.W. estimates) only, but also on the relative phosphorylation state achieved by a specific protein band under
given set of conditions, on its subcellular location, and
on comparisons of various phosphorylative properties, such
as time course of phosphorylation, sensitivity to cyclic
nucleotides, effects of divalent cations and of neuropeptides.

        The picture which emerges from these data, then, is
of a multiplicity of phosphoproteins, the phosphorylation
of each is influenced by various physiological affectors
in a dissimilar fashion. The significance of these differential effects may be related to data concerning the
subcellular location of these various phosphoprotein components. In addition, the information available to date
indicates potential functional differences among various
phosphorylation systems, and suggests that the nature of
the substrate proteins may account for a great share of
the specificity exerted by phosphorylative activity in its
effects on neuronal function. Although most of the specific phosphoproteins whose function is known or postulated
have been implicated in the mediation of transient changes
in neuronal activity. It is conceivable that, in addition,
certain specific proteins may play a role in adaptive
processes underlying long-lasting alterations in neuronal
function.

Protein phosphorylation in neuronal adaptation.

        Various environmental, hormonal and pharmacological
influences are capable of producing long-lasting alterations in brain function. Such alterations, which in many
cases are manifested in persistent behavioral changes,
are believed to be the result of adaptive processes that
take place in cells of the central nervous system. Since
most all of the biological substrates with potential for
involvement in such processes are regulated by protein
phosphorylation (see Figure 2), it is very likely that
phosphorylative activity plays an important role in neuronal
adaptation.

        The reserpine-elicited increase in tyrosine hydroxy-
lase (TH) activity has been studied in greater detail than
most other adaptive processes that are induced by drugs
in the central nervous system. Reis, et al, (1975) have
demonstrated that a single injection (10 mg/kg) of reserpine caused a three fold increase in TH activity in

the nucleus locus ceruleus of rat brain, and this elevation remained significant for at least two weeks. The
same study demonstrated that this elevation was due entirely to accumulation of newly synthesized enzyme molecules. A series of studies at the laboratory of E. Costa
has provided evidence that induction of TH synthesis by
reserpine or cold-stress involves an increase in RNA synthesis, in a process that is triggered by neurotransmitters,
mediated by cyclic AMP, and specified by a cyclic AMP-
dependent protein kinase. According to these studies
(COSTA, et al, 1976, 1978), increased intracellular levels
of cyclic AMP causes a dissociation of the cytoplasmic
cyclic AMP-dependent protein kinase to its regulatory and
catalytic subunits. The free catalytic subunit trans-
locates to the cell nucleus. There it catalyzes the phosphorylation of chromosomal proteins, which in turn promotes
the synthesis of messenger RNA. Thus, by acting via this
phosphorylation process, neurotransmitters that activate
adenylate cyclases can cause a persistent alteration in
metabolic processes within the affected cell. It should
be mentioned, however, that these studies have used cells
of peripheral tissue, the adrenal medulla, as a target for
the triggering stimulus.

     Indication that similar processes may be involved
also in the differentiation and maturation of neurons came
from studies of neuroblastoma cells grown in culture. When
these cells are exposed to experimental conditions that
produce an increase in the intracellular levels of cyclic
AMP their growth is arrested. Moreover, the cells extend
long neurites and exhibit many physiological and biochemical
properties characteristic of mature neurons (PRASAD, 1975).
We have demonstrated that selective alterations occur in
the phosphorylation of specific proteins in the cytosol
from differentiating neuroblastoma cells (EHRLICH, et al,
1977b). The phosphorylation of a protein with apparent
molecular weight of 97,000 daltons (band 97, Fig. 5) decreases to about 50% of that measured in untreated control
cells. The phosphorylation of another protein (band 59),
tentatively identified as the regulatory subunit of cyclic
AMP dependent protein kinase, increased about twofold.
The phosphorylation of band 59 was found to be cyclic AMP
dependent, and that of band 97, independent. Yet, the
extent of the changes in the activities which support the
phosphorylation of either protein were not dependent on
whether cyclic AMP was added to the assay mixture used for
the enzymatic analysis (see Fig. 5). Recently, we have
extended these studies and have demonstrated that the
selective alterations in cytosolic phosphorylative

activities are accompanied by a 50% increase in the phosphorylation of nuclear proteins from differentiated neuroblastoma cells (EHRLICH, et al, 1978b). Further studies
would determine whether such events play a role also in
mechanisms underlying brain development and in their sensitivity to environmental influences.

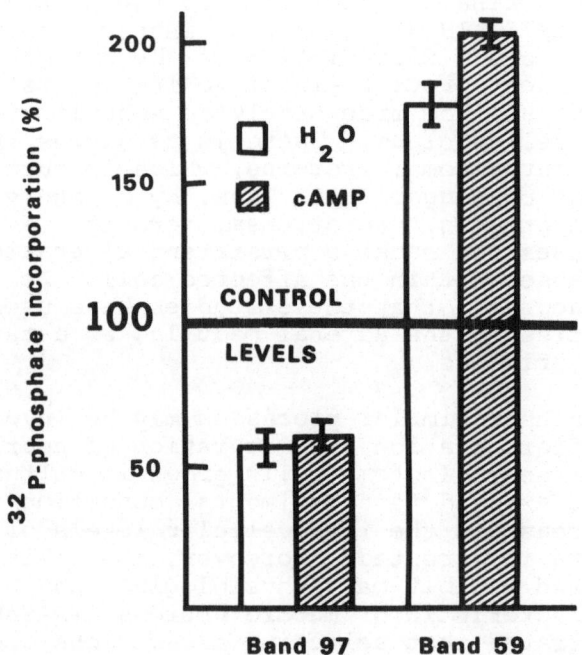

Fig. 5: Phosphate incorporation into two specific protein
         components of the cytosol from differentiating neuro-
         blastoma cells grown in culture. The treatment that
         induced the differentiation caused bidirectional
         changes. The phosphorylation of one protein (M.W.=
         59,000 daltons) increased, while that of another
         (M.W.=97,000), decreased. The phosphorylation of
         band 59 is stimulated by cyclic AMP and that of band
         97 is not. However, it can be seen that the magnitude
         of the changes occuring during differentiation does
         not depend on whether cyclic AMP is added ▨ , or
         ommitted ▭ from the reaction medium. Band 59 was
         tentatively identified as the phosphorylated regula-
         tory subunit of cytosolic cyclic AMP-dependent protein
         kinase. Data calculated from Ehrlich, et al, 1978b.

We have postulated that in addition to processes where phosphorylative mechanisms participate only in the initiation of adaptive responses, as described above, a persistent modification induced in a particular phosphorylation system may be sufficient for producing long-lasting alterations in neuronal activity. This required a demonstration that input to the brain can indeed cause selective alterations in the phosphorylation of specific proteins. We have initiated this line of investigation (EHRLICH, JOLLY & ROUTTENBERG, 1974; unpublished observations) by experiments that were repeated in later studies, (EHRLICH, et al, 1978d; CONWAY & ROUTTENBERG, 1978) in which the triggering input consisted of massive stimulation of brain tissue, induced by decapitation or electroconvulsive shock. For control in these studies we used cortical tissues from animals killed by cyrogenic techniques (immersion in liquid N_2). Compared to preparations from the controls, specific proteins in membranes from the experimental groups evidenced a significant and selective alteration in endogenous phosphorylation. Moreover, changes in certain specific proteins occurred in opposing directions. For example, ^{32}P-phosphate incorporation into protein in band F of Figure 3 decreased about twofold after decapitation, while that in band E_2 increased with about the same magnitude (EHRLICH, et al, 1978d). These results served to demonstrate the selectivity in the phosphorylative responses of different proteins to input. In addition, they pointed out the possibility of using in vitro assays of endogenous phosphorylation as means for detecting alterations in phosphorylation of specific proteins that may be affected in brain tissue by environmental influences.

These results have encouraged us to use the endogenous phosphorylation assays in attempts to find correlations between enduring effects of experimental input on animal behavior and on phosphorylation of specific proteins. In these studies, rats were trained in a single trial-passive avoidance conditioning task. Rats were trained to escape from a mild footshock and killed by a cyrogenic method 24 hours later. These rats were compared to controls (handled only) and to an additional experimental group consisting of rats that were foot-shocked but prevented from escaping the shock (stressed). Both, and each of the experimental groups evidenced a significant elevation in phosphate incorporation in vitro into proteins of membranes from their neostriata, as compared to controls. Moreover, the increase in phosphorylation of proteins in the bands designated F and H was significantly greater than that of the other bands (EHRLICH, et al, 1977a).

In another study in this line of investigation, we
have tested the effects of a pharmacological input on the
phosphorylation of synaptic membrane proteins. The treatment schedule was based on studies of Bonnet, Branchey
and Friedhoff at the Psychiatry Department of the New York
University Medical Center. They have developed an animal
model that mimics the state of protracted narcotic dependence characteristic of human addicts. Accordingly, rats
were implanted subcutaneously with pellets containing morphine twice a week for a period of three weeks. During
this period the dosage of morphine was gradually increased
from 18.5 to 75.0 mg morphine (BONNET, et al, 1978).
Seventy-two to 75 hours after the last implantation, animals
were killed by head immersion in liquid nitrogen, and blocks
of frozen neostriatal tissue were removed for biochemical
analyses.

Compared to preparations from placebo treated controls, samples from long-term morphinized rats evidenced
significant decrease in striatal cyclic AMP levels, tyrosine hydroxylase and protein kinase activities (BONNET,
et al, 1978). We then proceeded to test whether the
decreased phosphorylative activity was differential, or
directed equally toward all the endogenously phosphorylated
proteins in neostriatal membranes (EHRLICH, et al, 1978a).
We have found that the decreased phosphorylative activity
in neostriatal membranes from chronically morphinized rats
is selective, and directed specifically toward the phosphoproteins designated F and H. These were the same
phosphoprotein bands which demonstrated increased phosphorylation in neostriatal membrane from rats sacrificed
24 hours after learning experience (Table 1). The results
have thus supported our suggestion (EHRLICH, et al, 1977d)
that specific phosphoproteins in bands F and H of neostriatal membranes may be involved in the adaptive responses of this brain area to environmental influences.

Phosphoproteins and synaptic plasticity.

The mechanism whereby phosphoproteins induce adaptive
changes in neuronal function may involve alterations in
receptor-function (see the 2nd section of this volume).
Numerous studies have indicated that narcotic dependence
involves the development of latent hyperactivity of neurons
in the central nervous system, which is manifested behaviorally upon withdrawal of the individual from the addictive
drug (LAL, 1976; see also BONNET, this volume). Physiologically, such hyperactivity would be determined by
increased sensitivity of neurons to stimulation by

neurotransmitters. Biochemically, a state of latent receptor supersensitivity would be underlied by increased sensitivity of post-synaptic adenylate cyclase to stimulation
by neurotransmitters, accompanied by decreased supply of
pre-synaptically originating neurotransmitter molecules.
In the neostriatum such a state of supersensitivity most
likely involves the neurotransmitter dopamine. Alterations
in the sensitivity of neostriatal adenylate cyclase to
dopamine were demonstrated in morphine dependent rats
(IWATSUBO & CLUET, 1975). However, the same study has
also shown that morphine had no direct effects on adenylate
cyclase activity. The mechanism that underlies such

TABLE 1. Effects of Environmental Inputs on Phosphorylation of Specific Proteins in Neostriatal membranes.

Input / Protein Band	Learning Experience	Long-term Morphine
	Ratio of ^{32}P incorporation (Experimental / Control)	
D	1.40	0.71
E	1.16	0.75
F*	3.86	0.47
H*	2.11	0.35

Rats were subjected to the treatment and sacrificed by
cyrogenic methods. Neostriatal membrane were prepared
and assayed for endogenous phosphorylation as described
in the text. Values presented for the learning experience are the ratio of mean phosphate incorporation
into each band in membranes from rats subjected to one
trial passive avoidance conditioning, over that obtained
for handled (non shocked) controls (data from EHRLICH,
et al, 1977a). Values for long-term morphine are the
ratio of mean incorporation into bands of morphine-
dependent rats over that obtained for placebo treated
controls (data from EHRLICH, et al, 1978a). * The
differences for bands F and H in each study were statistically significant.

alterations may be the process of protein phosphorylation
in neostriatal membranes. According to the studies of
Gnegy, et al (1976a,b) cited above, a decrease in the phosphorylation of neostriatal membrane proteins could result
in increased association of calcium-binding protein (CDR)
with the membrane. Under these conditions the accumulation
of cyclic AMP, triggered by the interaction of dopamine
with its receptors, will be enhanced. Neostriatal membranes
from morphine dependent rats are capable of incorporating
only half the amount of phosphate (84 $\pm$ 2 picomoles per
milligram protein) than corresponding preparations from
control, placebo treated rats (151 $\pm$ 6 pmoles/mg, BONNET,
et al, 1978). They would therefore contain potentially
more than normal amounts of CDR and thus exhibit greater
sensitivity of adenylate cyclase to dopamine. The occur-
rence of such a mechanism can be tested experimentally,
and ongoing studies in this direction suggest that such
indeed may be the case. Thus, the decreased phosphorylation of proteins F and H in neostriatal membranes of rats
in a state of protracted narcotic dependence may represent
the underlying mechanism which accounts for the development
of latent receptor supersensitivity. Changes of this nature,
as suggested by Dismukes and Daly (1976), may constitute
a basis for synaptic plasticity.

Neuronal Modulation and Protein Phosphorylation.

        During our studies on the effects of narcotic exposure
on rat neostriatal biochemical processes we have noted
that a short term treatment was not sufficient to bring
about the decrease in protein phosphorylation described
above. We have therefore suggested that these phosphorylative modifications constitute part of the adaptive responses of nerve cells to persistent stimulation by narcotic
drugs. Long-term exposure to morphine was indeed shown
to result in compensatory biochemical changes in a direction
opposite that induced by the initial encounter with the
drug (BONNET, et al, 1978). The mechanisms that trigger
such adaptive processes most likely involve interactions
between neuroactive peptides and the opiate receptor.

        The laboratory of W. Gispen in the Netherlands has
been investigating extensively the role of various neuroactive peptides in mediating behavior related phenomena
(see GISPEN, et al, this volume). Using our conditions
for the analysis of endogenous protein phosphorylation in
brain membranes, they have studied the effects of ACTH-
like peptides on this activity. They have identified
specific phosphoprotein components whose phosphorylation

is selectively inhibited by ACTH 1-24 (ZWIERS, et al, 1976). These phosphoproteins have molecular weights and phosphorylative properties very similar to the phospho-proteins designated by us F and H (Fig. 3). Recent ob-servations (see DAVIS & EHRLICH, this volume) have indi-cated that the opioid peptide enkephalin has similar ef-fects on the phosphorylation of these proteins. ACTH fragments can influence the expressions of learned behavior and opioid peptides are believed to play a role in the development of narcotic dependence. It is of particular interest, therefore, that ACTH fragments and enkephalins affect the phosphorylation of presumably the same proteins which demonstrated phosphorylative alterations after a training experience and after long-term morphine treatment. These results suggest that phosphorylative modifications such as summerized in Table 1 may involve interactions between naturally occuring neuropeptides and enzymatic systems that phosphorylate the membrane-bound proteins F and/or H.

All these findings helped us to formulate a working hypothesis, summarized schematically in Figure 6. Accord-ing to this model, the phosphorylation state of membrane-bound proteins, specifically F and H, may determine the sensitivity of neurons to stimulation by neurotransmitters. This may be carried out via a mechanism involving altered sensitivity of post-synaptic adenylate cyclase to dopamine. A decrease in the capacity of the membrane to phosphory-late these proteins, such as evidenced in morphine depen-dent rats, would result in dopamine receptor supersensi-tivity. A decreased supply of pre-synaptically originating dopamine would cause this state to remain latent during morphine treatment. Changes in tyrosine hydroxylase activity and in calcium-dependent neurotransmitter re-lease that would underlie such decreased "supply", are known to be regulated by presynaptic phosphoproteins. Thus, adaptive processes may involve also phosphorylative modifications in presynaptic locations. Opioid peptides, such as the enkephalins, may modulate neuronal sensitivity by exerting direct effects on the phosphorylation of proteins F and/or H. ACTH-like peptides, by affecting the phosphorylation of these same proteins could mediate neuronal responses to stress and/or learning experiences. An input that causes persistent changes in the balance of any of these neuropeptides in certain brain areas could lead to adaptive changes in the phosphorylation of proteins F and H. These changes would result in alterations in the efficacy with which a stimulus activates or inhibits neurons in the affected brain area. Such modifications may

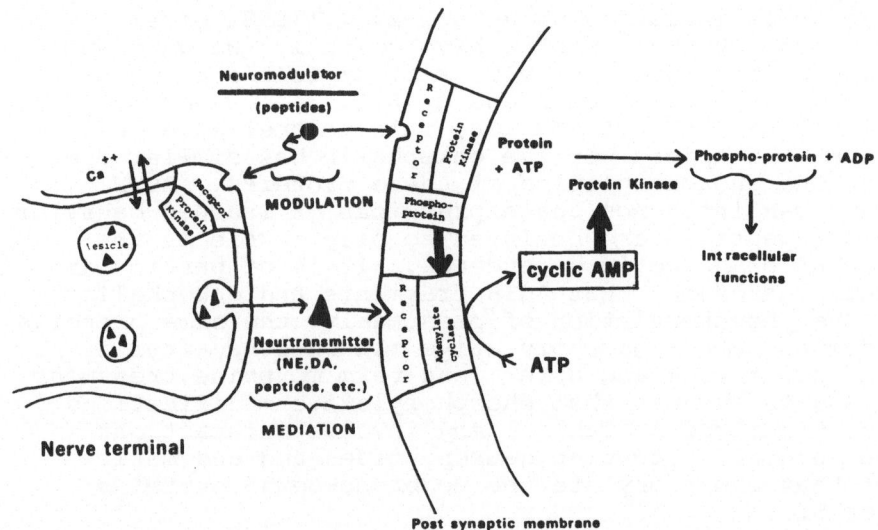

Fig. 6:   Role of protein phosphorylation in the mediation
          and modulation of synaptic transmission: hypothetical
          mechanisms of the mode of action.

A. MEDIATION. Neurotransmitter molecules are released
from the pre-synaptic terminal of either catecholaminergic-
(norepinephrine, NE; dopamine, DA), peptidergic-(endorphins,
ACTH-fragments) or amino acid (GABA, glutamate) neurons,
and interact with specific receptors on adjacent post-
synaptic membranes. The receptor is coupled with adenylate (or guanylate) cyclase. The newly formed cyclic
nucleotide accumulates intracellulary and acts as the
second messenger in the process of synaptic transmission.
By regulating the phosphorylation of specific proteins,
the cyclic nucleotides affect various cellular functions
(see Fig. 2).

B. MODULATION. Neuromodulators, such as endorphins and
other neuroactive peptides, interact with pre-synaptic
and/or post-synaptic receptors. In either case, the receptor action may involve membrane-bound protein kinase
activity. In the nerve terminal phosphorylative activity
would regulate calcium-dependent release of neurotransmitter molecules. In the post-synaptic-membrane, peptide-
regulated protein phosphorylation could affect the sensitivity of receptors to stimulation by neurotransmitters.
A specific example for such a case would be the interaction
of neuropeptides with the opiate receptor that inhibits
the phosphorylation of specific membrane-bound proteins
(F,H), and thus may increase the sensitivity of adenylate
cyclase to dopamine.

be involved in altered receptor function associated with
Schizophrenia and Tardive dyskinesia (VOLAVKA, DAVIS
and EHRLICH, 1979).

CONCLUSIONS

The phosphorylation of membrane-bound proteins has
been implicated in the regulation of a multiplicity of
neurochemical events that occur both pre- and post-synapti-
cally. Thus, irrespective of the nature of the environmental inputs that induce modifications in the phosphorylation of these proteins, they will always result in alterations in the functional state of the affected synapses.
Specificity in the behavioral manifestation of these alterations would be determined, in all likelihood, by the location in the brain of the modified synapses. However, the
proteins whose phosphorylation is modified may represent
a general site of neuronal adaptation. We have shown that
environmental inputs can induce long-lasting alterations
in the phosphorylation of specific proteins, designated
F and H (M.W. 47,000 and 15-20,000, respectively). The
capacity of the membranes to phosphorylate these proteins
can increase or decrease, depending on the input. Isolation of these proteins and characterization of the mechanisms
that regulate their phosphorylation and dephosphorylation
may result in a better understanding of the processes
whereby environmental inputs induce long-lasting alterations in brain function. Elucidation of the biochemical
processes which are regulated and modulated by these
phosphoproteins may reveal the identity of the molecular
mechanisms which enable neuronal circuits to store ex-
periential information. These experiments are in progress.

On the basis of our present findings we have formulated
a working hypothesis (see legend to Fig. 6) that links the
phosphorylation of proteins F and/or H with the opiate
receptor. Following this model, we have tested the effects
of a detergent reported to extract opiate receptor from
brain membranes (SIMON, et al, 1975) on our membrane preparations. We have found that this treatment detaches from
the membranes a complex containing protein F, its kinase
and phosphatase (EHRLICH, 1979). Thus, this hypothesis
has already helped in advancing our experimental work.
We hope that the presentation of this model will stimulate
further investigation into the possible role or specific
phosphoproteins in mediating and modulating mechanisms
associated with receptor functions in the central nervous
system.

ACKNOWLEDGEMENTS

The author is indebted to Drs. E. G. Brunngraber and L. G. Davis for helpful discussions. Some of the studies reported here were supported, in part, by NSF grant BMS19481, by a grant from the Epilepsy Foundation of America and by intramural funds from the Missouri Institute of Psychiatry.

REFERENCES

BONNET, K. A., BRANCHEY, L. G., FRIEDHOFF, A. J. and EHRLICH, Y. H. (1978) Life Sciences, 22: 2003-2008.

BROWNING, M., DUNWIDDLE, T., BENNETT, W., GISPEN, W. H. & LYNCH, G. (1979) Science, 203, 60-62.

CHEUNG, W. Y., BRADHAM, L. S., LYNCH, T. G., YING, M. N. and TALLANT, E. A. (1975) Biochem. Biophys. Res. Comm., 66, 1055-1063.

CONWAY, C. G. and ROUTTENBERG, A. (1978) Brain Res. 139, 366-373.

COSTA, E., CUROSAWA, A. and GUIDOTTI, A. (1976) Proc. Natl. Acad. Sci. US, 73, 3887-3891.

COSTA, E. (1978) Presented at the 4th International Catecholamine Symposium, California.

DAVIS, L. G. and EHRLICH, Y. H. (1978) Trans. Am. Soc. Neurochem. 9: 192.

DE BLAS, A. L., WANG, Y. J., SORENSEN, R. and MAHLER, H. Manuscripts submitted for publication, (1979).

DE LORENZO, R. J. (1976) Biochem. Biophys. Res. Comm. 71, 590-597.

DE LORENZO, R. J. and FREEDMAN, S. D. (1977) Epilepsia, 18: 357-365.

DISMUKES, R. K. and DALY, J. W. (1976) Experientia, 32: 730-732.

EHRLICH, Y. H. (1979) Proceedings of the 7th Meeting of the International Society for Neurochem. in-press.

EHRLICH, Y. H. and ROUTTENBERG, A. (1974) FEBS Lett. 45, 237-243.

EHRLICH, Y. H. and ROUTTENBERG, A. (1975) Trans. Am. Soc. Neurochem. 6, 83.

EHRLICH, Y. H., RABJOHNS, R. R. and ROUTTENBERG, A. (1977a) Pharmac. Biochem. Behav. 6, 169-174.

EHRLICH, Y. H., BRUNNGRABER, E. G., SINHA, P. K. and PRASAD, K. N. (1977b) Nature (Lond.) 265, 238-240.

EHRLICH, Y. H., DAVIS, L. G., GILFOIL, T. and BRUNNGRABER, E. G. (1977c) Neurochemical Research 2, 533-548.

EHRLICH, Y. H., BONNET, K. A., DAVIS, L. G. and BRUNNGRABER, E. G. (1977d) In: Mechanisms, Regulation and Special Functions of Protein Synthesis in Brain (Eds. Roberts, S., Lajtha, A., and Gispen, W. H.) Elsevier, Amst. 271-278.

EHRLICH, Y. H., BONNET, K. A., DAVIS, L. G. and BRUNNGRABER, E. G. (1978a) Life Sci., 23, 127-136.

EHRLICH, Y. H., PRASAD, K. N., DAVIS, L. G., SINHA, P. K. and BRUNNGRABER, E. G. (1978b) Neurochem. Res. 3, 803-813.

EHRLICH, Y. H., DAVIS, L. G., and BRUNNGRABER, E. G. (1978c) Trans. Am. Soc. Neurochem., 9, 77.

EHRLICH, Y. H., DAVIS, L. G. and BRUNNGRABER, E. G. (1978d) Brain Res. Bull. 3, 251-256.

EHRLICH, Y. H., KEEN, P. B., DAVIS, L. G. and BRUNNGRABER, E. G. (1979) Trans. Am. Soc. Neurochem., 10, in-press.

GAZIT, Y., OHAD, I. and LOYTER, A. (1976) Bioch. Biophys. Acta. 436, 1-14.

GNEGY, M. E., COSTA, E. and UZUNOV, P. (1976a) Proc. Natl. Acad. US, 73, 352-355.

GNEGY, M. E., UZUNOV, P. and COSTA, E. (1976b) Proc. Natl. Acad. US, 73, 3887-3890.

GOLDSTEIN, M., BRONAUG, B., EBSTEIN, B. AND ROBERGE, C. (1976) Brain Res. 109, 563-574.

GREENGARD, P. (1976) Nature (London), 260, 101-108.

GREENGARD, P. (1978a) Science, 199, 146-152.

GREENGARD, P. (1978b) Cyclic Nucleotides, Phosphorylated Proteins, and Neural Function. Raven Press, N.Y. 124, pp.

GREENGARD, P. and KEBABIAN, J. W. (1974) Fed. Proc., 33, 1059-1067.

HEALD, P. J. (1957) Biochem. J., 66, 659-663.

HERSHKOWITZ, M. (1978) Biochim. Biophys. Acta., 542, 274-283.

IWATSUBO, K. and CLOUET, D. H. (1975) Biochem. Pharm. 24, 1499-1503.

JOH, T. H., PARK, D. H., BRODSKY, M. J. and REIS, D. J. (1978) Proc. Soc. Neuroscience, 8th annual meeting.

JOHNSON, E. M., MAENO, H. and GREENGARD, P. (1971) J. Biol. Chem. 246, 7731-7739.

JOHNSON, E. M., UEDA, T., MAENO, H. and GREENGARD, P. (1972) J. Biol. Chem. 247, 5650-5652.

KUO, J.-F., and GREENGARD, P. (1969) Proc. Natl. Acad. Sci. US, 64, 1349-1355.

LAL, H. (1976) Life Sci., 17, 483-496.

LIBET, B. and TOSAKA, T. (1970) Proc. Natl. Acad. Sci. US 67, 667-673.

LOVENBERG, W., BRUCKWICH, E. A. and HANBAUER, I. (1975) Proc. Natl. Acad. Sci. US, 72, 2955-2958.

MAHLER, H. R.   (1977) Neurochem. Res., 2, 119-148.
MALKINSON, A. M., KRUEGER, B. K., RUDOLPH, S. A., CASNELLIE,
    J. E., HALEY, B. E. and GREENGARD, P. (1975) Metabolism
    24, 333-341.
MIYAMOTO, E., KUO, J.-F. and GREENGARD, P. (1969) Science,
    165, 63-65.
PRASAD, K. N. (1975) Biological Rev., 50, 129-165.
RAM, J. L. and EHRLICH, Y. H. (1978) J. Neurochem. 30,
    487-491.
REIS, D. J., JOH, T. H. and ROSS, R. D. (1975) J. Pharm.
    Exp. Therap., 193, 775-784.
RODNIGHT, R. and LAVIN, B. E. (1966) Biochem. J., 101,
    495-501.
RODNIGHT, R. (1975) In: Metabolic Compartmentalization
    and Neurotransmission (Eds. Berl, Clark, and Schneider)
    Plenum Press, N.Y., 205-228.
RODNIGHT, R. (1977) In: Mechanisms, Regulation and
    Special Function of Prot. Synth. in Brain. (Ed:
    Roberts, S., Lajtha, A., and Gispen, W. H.) Elsevier,
    Amsterdam, p. 255-266.
ROUTTENBERG, A. and EHRLICH, Y. H. (1975) Brain Research,
    92, 415-430.
RUBIN, C. S. and ROSEN, O. N. (1975) Ann. Rev. Biochem.
    44, 831-887.
SCHULMAN, H. & GREENGARD, P. (1978) Nature, 271, 478-479.
SIMON, E. J., HILLER, J. M. and EDELMAN, I. (1975) Science,
    190, 389-390.
SLOBODA, R. D., RUDOLPH, S. A., ROSENBAUM, J. L. and
    GREENGARD, P. (1975) Proc. Natl. Acad. Sci. US.,
    72, 177-181.
TAMIR, H., MAHDIK, S. P. and RAPPORT, M. M. (1976) Anal.
    Biochem, 76, 634-647.
TAMIR, H., RAPPORT, M. M. and ROZIN, L. (1974) J. Neuro-
    chem., 23, 943-949.
UEDA, T., MAENO, H. and GREENGARD, P. (1973) J. Biol.
    Chem., 248, 8295-8325.
UEDA, T. and GREENGARD, P. (1977) J. Biol. Chem., 252,
    5155-5163.
UNO, I., UEDA, T. and GREENGARD, P. (1976) J. Biol. Chem.
    251, 2192-2195.
VOLAVKA, J., DAVIS, L. G. and EHRLICH, Y. H. (1979)
    Schizophrenia Bull. in-press.
WALTER, U., KANOF, P., SCHULMAN, H. and GREENGARD, P.
    (1978) J. Biol. Chem., 253, 6275-6280.
WELLER, M. and RODNIGHT, R. (1970) Nature (Lond.), 225,
    187-188.
WELLER, M. and RODNIGHT, R. (1973) Biochem. J., 132,
    483-492.

WELLER, M. and MORGAN, I. (1976)   Biochim. Biophys. Acta.
     433, 223-228.
WELLER, M. and MORGAN, I. (1977)   Biochim. Biophys. Acta.
     465, 527-534.
WILLIAMS, M., PAVLIK, A. and RODNIGHT, R. (1974)   J.
     Neurochem., 22, 373-376.
WILLIAMS, M. and RODNIGHT, R. (1977)   Prog. in Neurobiol.
     8, 183-250.
ZWIERS, H., VELDHUIS, D., SCHOTMAN, P. and GISPEN, W. H.
     (1976)   Neurochem. Res., 1, 669-677.
ZWIERS, H., WIEGANT, V. M., SCHOTMAN, P. & GISPEN, W. H.
     (1977)   In: Mech. Regulation and Special Function
     of Protein Synthesis in the Brain. (Roberts, S.,
     Lajtha, A. & Gispen, W. H. eds.). Elsevier-North
     Holland Biomed. Press, Amsterdam, p. 267-272.

PHOSPHOPROTEINS AS PROPOSED MODULATORS OF VISUAL FUNCTION

D.B. FARBER AND R.N. LOLLEY

Jules Stein Eye Institute and Department of
Anatomy,
UCLA School of Medicine, Los Angeles, Ca. 90024
and Veterans Administration Medical Center,
Sepulveda, Ca. 91343

The retina is a region of the central nervous system
that is highly specialized for the reception of light and
the transmission of encoded visual information to the
brain. It develops embryonically from the primitive forebrain and, in the adult stage, it has a layered structure
with a limited number of cellular classes. The outermost
layer of cells, lying closest to the pigment epithelium,
corresponds to the photoreceptor cells. The innermost
layer, which is closest to the vitreous, is formed by the
retinal ganglion cells; between the two layers, there is
a discrete stratum of nuclei that belongs to the bipolar,
horizontal, amacrine and the glial-like Müller cells.

There are two general classes of photoreceptor cells,
i.e. rods and cones, with subclasses having been identified in some species (STELL, 1972). For the purposes of
this chapter, we will restrict our attention to rod photoreceptors.

The outer segment of the rod photoreceptor cell contains tightly packed membranous disks which are encased
by, but not attached to, the plasmalemma of the cell.
Most of the visual pigment rhodopsin is concentrated in
the disk membranes but the plasmalemma also has rhodopsin
(JAN & REVEL, 1974; DEWEY et al., 1969). The rest of the
photoreceptor cell is constituted by an inner segment that
is rich in mitochondria, a nucleus located in the cell
soma and a synaptic ending joined to the cell soma by a

short axon (YOUNG, 1969). When the photoreceptor cell
is in the dark, Na+ is actively extruded from the cell
body and returns across the outer segment membrane. This
Na+ flux is referred to as the "dark current". When the
photoreceptor cell is exposed to illumination, light is
absorbed by rhodopsin causing the photochemical isomeriza-
tion of its chromophore 11-cis retinal to the all-trans
configuration. The photoisomerization reaction is followed
by a sequence of dark reactions which yield, at least in
vitro, the products all-trans retinal and opsin (WALD,
1968). This process is known as bleaching of rhodopsin
since ROS suspensions change from a deep purple to a pale
yellow color. The bleaching of rhodopsin leads to amplification of the initial light signal and, through a mechanism as yet unresolved, to a reduction in the Na+ current.
In photoreceptor cells of vertebrates, the response to
light is reflected as a hyperpolarization of the membrane
(WERBLIN, 1974) and, in invertebrates, as a depolarization (HAGINS, 1965; KNIGHT et al., 1970).

     The link between photon capture that occurs in the
disk membrane and the change in potential that takes place
in the plasmalemma would be easy to visualize if one supposed that light causes the disks to release a substance
which diffuses to the plasma membrane and reduces its conductance for Na+. This is the basis of the model proposed
by HAGINS & YOSHIKAMI (1974), which suggests that Ca++
is the intracellular transmitter of the light signal.

     Cyclic GMP may also be an intermediate in the visual
process, but its role is not clearly established as yet.
Cyclic GMP content and metabolism in photoreceptor cells
are influenced by light. Rod outer segments of dark-adapted retinas from several species possess high levels of
cyclic GMP (about 100 times higher than those of cyclic
AMP; KRISHNA et al., 1976) and high levels of the enzyme
required for cyclic GMP synthesis and degradation (KRISHNA
et al., 1976; PANNBACKER, 1973a; GORIDIS & VIRMAUX, 1974;
MIKI et al., 1973). Upon bleaching of rhodopsin by light,
cyclic GMP phosphodiesterase activity is enhanced (MIKI
et al., 1973) causing the fall in cyclic GMP concentration (FLETCHER & CHADER, 1976) throughout the photoreceptor cell, but without affecting the cyclic GMP content of
the inner retinal layers (ORR et al., 1976). The dark-
adapted levels of cyclic GMP may be associated with the
depolarized state of the photoreceptor cell in darkness,
since cellular depolarization has been shown to elevate
cyclic GMP content in other tissues of the central nervous
system (FERRENDELLI et al., 1973). About one-half of the

cyclic GMP in frog or bovine retina is hydrolyzed within
3-10 sec of exposure to laboratory illumination (GORIDIS
et al., 1974). In isolated rod outer segments, bleaching
one molecule of rhodopsin leads to the hydrolysis of 1,000-
2,000 molecules of cyclic GMP, and this occurs rapidly
within 100-300 milliseconds (WOODRUFF et al., 1977).

Cyclic nucleotides are known to express their actions
through the activation of protein kinases which phosphorylate proteins selectively (GREENGARD, 1976). These phosphoproteins, in turn, may be specifically involved in the
regulation of functions that are fundamental to the cell,
e.g. in neurons they have been implicated in the control
of membrane permeability to certain ions (NATHANSON, 1977).

Rod outer segments possess the necessary biochemical
components for translating light-induced changes in cyclic
GMP concentration into a physiiologically significant message, i.e. a rod outer segment-specific phosphoprotein
(LOLLEY et al., 1977). In addition to the cyclic nucleotide-dependent phosphorylating system, rod outer segments
have also a cyclic nucleotide-independent protein kinase
which phosphorylates rhodopsin in a reaction that is
strongly stimulated by light (BOWNDS et al., 1972; FRANK
et al., 1973; KÜHN & DREYER, 1972; KÜHN et al., 1973;
WELLER et al., 1975a). For clarity, we will discuss these
two phosphorylating systems separately, indicating how
the respective phosphoproteins (products of the reactions)
may be involved in different control mechanisms related
to visual excitation and dark/light adaptation. Finally,
we will integrate the biochemical information and propose
a functional model of a rod outer segment.

CYCLIC NUCLEOTIDE-INDEPENDENT PHOSPHORYLATION
OF ROD OUTER SEGMENT MEMBRANES

Rhodopsin is phosphorylated subsequent to light absorption by a protein kinase of rod outer segment membranes in a slow, dark process which has been reported to
take from only a few minutes up to more than an hour (KÜHN
& DREYER, 1972; BOWNDS et al., 1972; FRANK et al., 1973;
KÜHN & BADER, 1976). The reaction occurs only after rhodopsin is bleached by light. In fact, light does not activate protein kinase but converts rhodopsin in the accept-
able substrate (WELLER et al., 1975; FRANK & BUZNEY, 1975).
The action spectrum of the light effect coincides with
the absorption spectrum of rhodopsin (BOWNDS et al.,
1972). ATP and GTP act as phosphate donors (CHADER et
al., 1976); their terminal phosphate group is transferred

and bound covalently to serine and threonine residue(s) in the protein moiety of rhodopsin (KÜHN & DREYER, 1972). That this phosphorylation is independent of cyclic AMP and cyclic GMP has been shown by several investigators (KÜHN & DREYER, 1972; FRANK et al., 1973; WELLER et al., 1975a). The specificity of the protein kinase of rod outer segment membranes for bleached rhodopsin is still an open question. Some authors have reported that bleached rhodopsin is the only substrate (WELLER et al., 1975a; SHICHI & SOMERS, 1978) while others have been able to phosphorylate in addition proteins such as histone, protamine or phosvitin (CHADER et al., 1976; PANNBACKER, 1973b; FRANK & BUZNEY, 1975; FRANK & BENSINGER, 1974; FARBER et al., 1978b). Still, the only kinase that seems to phosphorylate bleached rhodopsin is that of the rod outer segments; protein kinases from other tissues are ineffective (FRANK & BUZNEY, 1975).

Fully bleached suspensions of crude frog rod outer segments incorporate up to 4 moles of phosphate per mole of rhodopsin (MILLER & PAULSEN, 1975; KÜHN et al., 1977), whereas purified cattle rod outer segments give lower phosphorylation yields of 0.3-2 phosphate per rhodopsin. Isoelectric focusing of phosphorylated rhodopsin has shown that some rhodopsin molecules are highly phosphorylated, whereas a large fraction of the total rhodopsin is not phosphorylated at all (KÜHN et al., 1977). SHICHI & SOMERS (1978) have proposed that only newly synthesized rhodopsin, localized in the basal membranes of the rod outer segment, is phosphorylated after illumination. When less than 1% of rhodopsin from frog rod outer segments is bleached, the incorporation of phosphate is increased extraordinarily: up to 50 phosphates are incorporated per mole of rhodopsin (BOWNDS et al., 1972; MILLER et al., 1977). This effect is not observed for cattle rod outer segments (FRANK et al., 1973). The possibility of an amplification mechanism in the phosphorylation reaction is as yet an unsettled question.

If the phosphorylation of rhodopsin has a physiological role, the rod outer segment must have not only the enzyme to catalyze the incorporation of phosphate but also that which will catalyze the dephosphorylation; otherwise, no turnover of rhodopsin phosphate could occur in vivo. Several laboratories have measured the activity of phosphoprotein phosphatase in rod outer segments (WELLER et al., 1975a; KÜHN, 1974; KÜHN & BADER, 1976). There is agreement that the reaction is very slow and it is not affected by cyclic AMP or cyclic GMP (GORIDIS & WELLER,

1976). The reaction rates of phosphorylation (which has a half-time of about 2 min at 21°C) and dephosphorylation in the dark (with a half-time of about 13 min [KÜHN et al., 1977]) are very slow as compared to the reactions involved in visual excitation. But the time required for dephosphorylation in the dark is similar to the time required for dark adaptation after strong bleaching illumination. In other words, the data of KÜHN et al. (1977) could suggest that the state of phosphorylation of rhodopsin may be involved in controlling the light sensitivity of the rods.

The state of phosphorylation of rhodopsin has been implicated also in the control of the Ca++ permeability of rod outer segments (GORIDIS & WELLER, 1976). Several laboratories have shown that light exposure causes an increase in the rate of entry of Ca++ into the disks and in the rate of efflux of Ca++ from preloaded disks when compared to values of material kept in the dark (LIEBMAN, 1974; MASON et al., 1974; WELLER et al., 1975b). However, phosphorylation of the bleached disks lowers the rates of Ca++ entry and efflux to the values found for material kept in the dark (WELLER et al., 1975b). Both light exposure and phosphorylation have no effect on the binding capacity of rod outer segments for Ca++ (WELLER et al., 1975b). Therefore, it seems that exposure to light increases the Ca++ permeability of rod outer segment disks, whereas phosphorylation of the bleached rhodopsin lowers the permeability. On the basis of these data, WELLER et al (1975c) have proposed a hypothetical model for the possible involvement of rhodopsin phosphorylation in light and dark adaptation, since the reaction is fast enough to be involved in the regulation of the light sensitivity of the rod. According to this model Ca++ is released from the disks after bleaching of rhodopsin. But, by exposure to light, rhodopsin becomes phosphorylated, preventing now the release of Ca++; that is, Ca++ pores will close and Ca++ will be loaded back into the disks, switching them "off". If the light-exposed phosphorylated disks are returned to the dark, rhodopsin will be regenerated and dephosphorylated. By these means, the disks will be turned "on" again. Such a process, in which the sensitivity of photoreceptors to light is changed by the background illumination, would have an important role in light and dark adaptation.

CYCLIC NUCLEOTIDE-DEPENDENT PHOSPHORYLATION
OF ROD OUTER SEGMENT CYTOSOL

The first report indicating the presence of a cyclic
nucleotide-dependent phosphorylation reaction in rod outer
segment preparations was that of PANNBACKER (1972). However, most of the work done on protein kinases of rod
outer segments failed to corroborate any involvement of
cyclic nucleotides in the incorporation of phosphate
molecules into protein. The reason for this lack of
effect of cyclic nucleotides on the protein kinases
studied became evident when we started to investigate the
phosphorylation of soluble components of rod outer segments. Cyclic nucleotides stimulated a protein kinase
of the rod outer segments only when it was in soluble
form, but did not activate the enzyme when it was associated with the disk membranes (LOLLEY et al., 1977).

When isolated bovine rod outer segments are extracted with Tris buffer, pH 7.6, protein kinase activity is
partitioned, after centrifugation at 100,000 x g, into
a soluble and a membrane-bound form, and the membrane
kinase can be solubilized by the detergent Lubrol PX.
The soluble and membrane-solubilized activities are both
unaffected by light and stimulated by cyclic nucleotides,
and they phosphorylate exogenous histones used as substrates (LOLLEY et al., 1977). Half-maximal stimulation
is observed at 1 x 10^{-7} M cyclic AMP and 4-5 x 10^{-6} M
cyclic GMP (FARBER et al., 1978b). The Lubrol PX-solu-
bilized kinase is activated 5-6 times by cyclic AMP and
4-5 times by cyclic GMP. This degree of activation is
higher than that observed with the soluble, Tris-extracted
enzyme, suggesting that detergents extract predominantly
the holoenzyme from the rod outer segment membranes. The
soluble enzyme may be already dissociated in situ, since
cyclic GMP levels are high in dark-adapted rod outer
segments.

DEAE-cellulose chromatography of the Tris-extract
of rod outer segment resolved three peaks which showed
kinase activity; two of them correspond to the Type I and
Type II cyclic AMP-dependent protein kinases described
in most tissues (CORBIN et al., 1975). The third cyclic
nucleotide-independent peak eluted with the flow-through
volume and is probably free catalytic subunit of the
cyclic nucleotide-dependent enzyme. These data, together
with the observations that the soluble enzyme is responsive to lower concentrations of cyclic AMP than cyclic
GMP, in vitro, and the apparent level of activation is

greater with cyclic AMP, suggest that the soluble protein
kinase of rod outer segments has the characteristics of
a cyclic AMP-dependent enzyme (FARBER et al., 1978b).
However, cyclic GMP is the major cyclic nucleotide of rod
photoreceptors (KRISHNA et al., 1976), and its concentration is regulated by light in vitro and in vivo (WOODRUFF
et al., 1977). In contrast, the content of cyclic AMP
in rod outer segments is not changed by illumination.
Taking into account these considerations, we have suggested that rod outer segments possess a protein kinase
that has cyclic AMP-dependent characteristics but is modulated directly by light-induced changes in cyclic GMP concentrations in vivo. The soluble fraction of rod outer
segments contains also phosphoprotein phosphatase activity,
indispensable for the turnover of phosphorylated proteins
in vivo (FARBER et al., 1978b).

The physiological action of cyclic GMP will be expressed through the mediation of the soluble, cyclic nucleotide-dependent protein kinase and specific phosphoproteins. We have found that from the several proteins
of the soluble fraction of rod outer segments that are
phosphorylated by the kinase, only one protein is phosphorylated in a cyclic nucleotide-dependent manner (LOLLEY
et al., 1977; FARBER et al., 1978b). This soluble protein,
MW 30,000 on SDS gels, is present in the Tris-extracts
of rod outer segments of bovine, mouse and rat. PANNBACKER
(1974) also observed a similar protein in human rod outer
segments, with an estimated MW of 25,000. In rod outer
segments of frog (POLANS et al., 1978) and toad (unpublished observations), the 30,000-dalton protein is not
detected but, instead, two proteins of MW 12,000 and
13,000 are phosphorylated in a cyclic nucleotide-dependent manner. The content of all of these phosphoproteins
seems to be reduced upon exposure of the retina to light
(unpublished observations).

When we incubated the soluble, Tris-extracted kinase
with rod outer segment membranes that had been depleted
of kinase activity but enriched in purified rhodopsin
(WELLER et al., 1975a), we found that an apparent reassoci-
ation of the enzyme with the membranes had occurred (FAR-
BER et al., 1978b). Once on the membrane, the kinase
phosphorylated bleached rhodopsin independent of cyclic
nucleotides, just as we observed when investigating the
characteristics of the phosphorylation reaction using rod
outer segment membranes. Our findings suggest that the
soluble and membrane-associated protein kinases are interchangeable. What it is that controls their being "on"

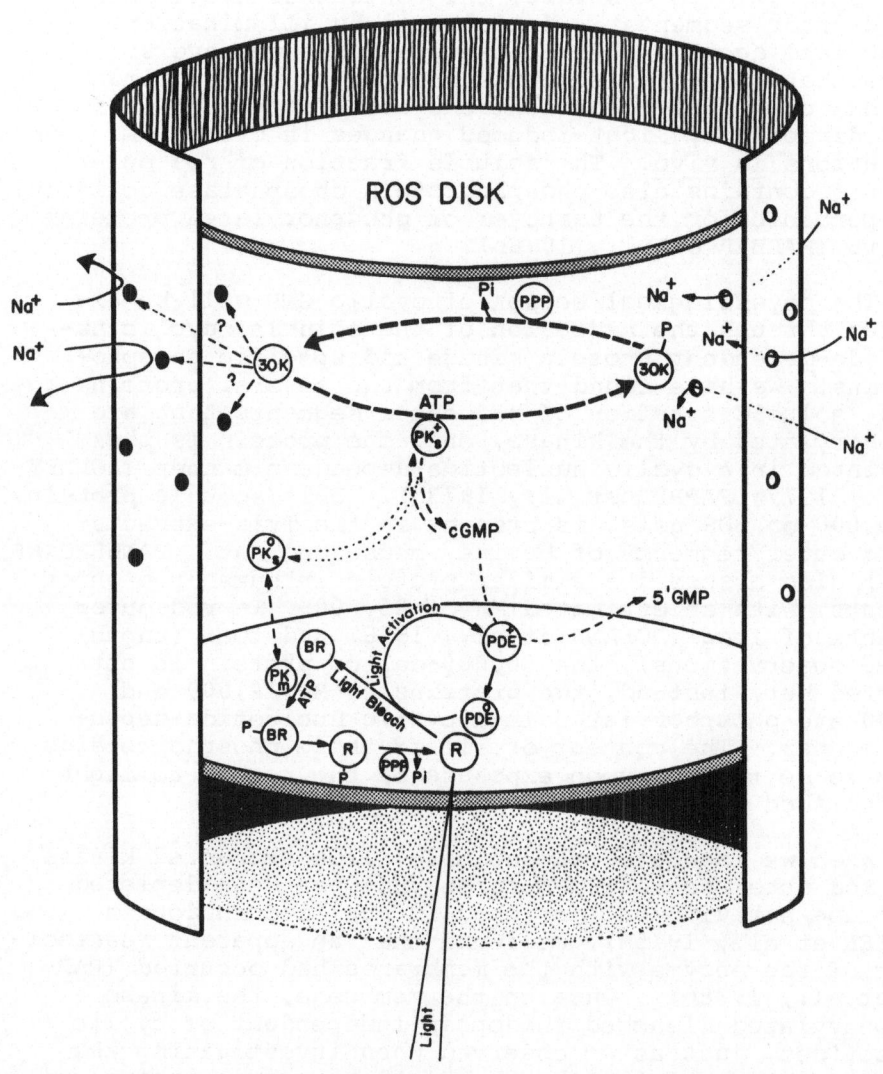

or "off" the membrane still remains an open question.

Another piece of information was obtained from an experiment similar to the one described above. After combining the soluble fraction of rod outer segments with the purified rhodopsin membranes, aliquots of the mixture were phosphorylated under different conditions, with or without cyclic AMP, in the dark or in the light. As expected, we found that in the dark, the 30,000-dalton soluble protein was phosphorylated and that this reaction was stimulated by cyclic AMP; no phosphorylation of rhodopsin was observed. Upon bleaching the membranes however, rhodopsin and the 30,000-dalton protein both incorporated phosphate, but only the soluble protein showed higher incorporation in the presence of cyclic AMP. This indicated that the proteins of rod outer segments are phosphorylated selectively in response to biological signals, which might correspond to those that act in vivo. That is to say, light promotes the phosphorylation of rhodopsin, independent of cyclic nucleotides, and cyclic nucleotides stimulate the phosphorylation of a 30,000-dalton soluble protein, independent of light.

Fig. 1:   The phosphorylation of specific proteins in rod outer segments (ROS) could regulate separately the sensitivity of the visual pigment to light and the amplitude of the visual response. In ROS disks, rhodopsin (R) is bleached (BR), phosphorylated by membrane protein kinase (PK_m), regenerated (P-Br→P-R) and dephosphorylated by phosphoprotein phosphatase (PPP). These reactions could regulate light/dark adaptation (see text). At the same time, the bleaching of rhodopsin activates cyclic GMP phosphodiesterase (PDE°→PDE+) which hydrolyzes cytoplasmic cyclic GMP to 5'-GMP. The reduced levels of cyclic GMP are not sufficient to activate the soluble protein kinase ($PK_s^o \nrightarrow PK_s^+$), and the content of 30 K-P falls as a result of PPP activity. The Na+ channels, that were kept open by 30 K-P (○), close as the concentration of 30 K increases (●). Therefore, the ratio 30K/30 K-P could perhaps control the amplitude of the visual response. The model also shows that PK_m and PK_s could interchange and be regulated by different biological signals: cyclic nucleotides would stimulate PK_s whereas light would make rhodopsin available to PK_m.

On the basis of all of the data that we had available
we proposed a model of the rod outer segment (Figure 1)
which shows the relationship between rhodopsin, cyclic
GMP modulation of protein kinase activity/phosphoprotein
content and ion permeability of the plasmalemma (FARBER
et al., 1978a). In essence, it suggests that the phosphorylation of specific proteins in the rod outer segment
cytosol or membrane compartments could regulate separately
the amplitude of the visual response and the sensitivity
of the visual pigment to light.

The key point of the cytoplasmic events is how much
of the 30,000-dalton protein is phosphorylated or unphos-
phorylated in the rod outer segment. We propose that the
phosphorylated protein interacts with the rod outer segment plasmalemma alone or in conjunction with calcium
facilitating in this way the free movement of ions across
the open channels of the membrane and providing a mechanism for the sustained state of depolarization that characterizes dark-adapted rod outer segments. When light
bleaches rhodopsin, the levels of cyclic GMP are rapidly
reduced (WOODRUFF et al., 1977) as a consequence of the
activation of cyclic GMP phosphodiesterase (BITENSKY et
al., 1975). This fall in cyclic GMP concentration will
reduce the activity of the soluble protein kinase and slow
the rate of phosphorylation of the 30,000-dalton protein.
This, together with the continued activity of the phosphoprotein phosphatase, will decrease the concentration of
the 30,000-dalton phosphoprotein. The net result will
be a reverse of the dark state, e.g. closure of the ion
channels and hyperpolarization of the visual cell.

The model shows also that bleached rhodopsin is
phosphorylated by the membrane-associated protein kinase,
which is interchangeable with the soluble protein kinase.
As has been mentioned before, the time-course of the phosphorylation-dephosphorylation reactions suggests that
these events may be related to the mechanisms of dark/light
adaptation. The phosphorylation of rhodopsin could be
a means for reducing the sensitivity of the visual pigment
to light.

In summary, specific phosphoproteins of rod photoreceptor cell outer segments may modulate, and perhaps
regulate, the physiological response of visual cells.

ACKNOWLEDGEMENTS

We wish to acknowledge the thoughtful assistance of Louise V. Eaton, Andrea Williams and Leana Sussman in the preparation of this chapter. Special thanks to Elisabeth Racz, Bruce Brown, Bernice Lieberman and Dennis Souza for their technical assistance in the laboratory. This work was supported by the National Eye Institute grants # EY 2651 and EY 331 and by the Medical Research Service of the Veterans Administration.

REFERENCES

BITENSKY, M. W., MIKI, N., KEIRNS, J. J., KEIRNS, M., BARA- BAN, J. M., FREEMAN, J., WHEELER, M. A., LACY, J. and MARCUS, F. R. (1975) in Advances in Cyclic Nucleotide Research (Drummond, G. I., Greengard, P. and Robison, G. A., eds.) Vol. 5, pp. 213-240. Raven Press, New York.

BOWNDS, D., DAWES, J., MILLER, J. and STAHLMAN, M. (1972) Nature, New Biol. 237, 125-137.

CHADER, G. J., FLETCHER, R. T., O'BRIEN, P. J. and KRISHNA, G. (1976) Biochemistry 15, 1615-1620.

CORBIN, J. D., KEELY, S. and PARK, C. R. (1975) J. Biol. Chem. 250, 218-225.

DEWEY, M. M., DAVIS, P. K., BLASIE, J. K. and BARR, L. (1969) J. Molec. Biol. 39, 395-405.

FARBER, D. B., BROWN, B. M. and LOLLEY, R. N. (1978a) Vision Res. 18, 497-499.

FARBER, D. B., BROWN, B. M. and LOLLEY, R. N. (1978b) Bio- chemistry. In press.

FERRENDELLI, J. A., KINSCHERF, D. A. and CHANG, M. M. (1973) Molec. Pharmac. 9, 445-454.

FLETCHER, R. T. and CHADER, G. J. (1976) Biochem. Biophys. Res. Commun. 70, 1297-1302.

FRANK, R. N. and BENSINGER, R. E. (1974) Expl. Eye Res. 18, 271-280.

FRANK, R. N. and BUZNEY, S. M. (1975) Biochemistry 14, 5110-5117.

FRANK, R. N., CAVANAGH, H. D. and KENYON, K. R. (1973) J. Biol. Chem. 248, 596-609.

GORIDIS, C. and VIRMAUX, N. (1974) Nature 248, 57-58.

GORIDIS, C. and WELLER, M. (1976) in Advances in Biochemi- cal Psychopharmacology (Costa, E. and Greengard, P., eds.) Vol. 15, pp. 391-412. Raven Press, New York.

GORIDIS, C., VIRMAUX, N., CAILLA, H. L. and DELAAGE, M. A. (1974) FEBS Lett. 49, 167-169.

GREENGARD, P. (1976) Nature 260, 101-108.

HAGINS, W. A. (1965) Cold Spring Harbor Symp. Quant. Biol.
    30, 403-417.
HAGINS, W. A. and YOSHIKAMI, S. (1974) Expl. Eye Res. 18,
    299-305.
JAN, L. Y. and REVEL, J.-P. (1974) J. Cell Biol. 62, 257-
    273.
KNIGHT, B. W., TOYODA, J.-I. and DODGE, F. A. Jr. (1970)
    J. Gen. Physiol. 56, 421-437.
KRISHNA, G., KRISHNAN, N., FLETCHER, R. T. and CHADER, G.
    (1976) J. Neurochem. 27, 717-722.
KÜHN, H. (1974) Nature 250, 588-590.
KÜHN, H. and BADER, S. (1976) Biochim. Biophys. Acta 428,
    13-18.
KÜHN, H. and DREYER, W. J. (1972) FEBS Lett. 20, 1-6.
KÜHN, H., COOK, J. H. and DREYER, W. J. (1973) Biochem-
    istry 12, 2495-2502.
KÜHN, H., McDOWELL, J. H., LESER, K.-H. and BADER, S. (1977)
    Biophys. Struct. Mechan. 3, 175-180.
LIEBMAN, P. A. (1974) Invest. Ophthalmol. 13, 700-701.
LOLLEY, R. N., BROWN, B. M. and FARBER, D. B. (1977) Bio-
    chem Biophys. Res. Commun. 78, 572-578.
MASON, W. T., FAGER, R. W. and ABRAHAMSON, E. W. (1974)
    Nature 247, 562-563.
MIKI, N., KEIRNS, J. J., MARCUS, F. R., FREEMAN, J. and
    BITENSKY, M. W. (1973) Proc. Natl. Acad. Sci. U.S.A.
    70, 3820-3824.
MILLER, J. A. and PAULSEN, R. (1975) J. Biol. Chem. 250,
    4427-4432.
MILLER, J. A., PAULSEN, R. and BOWNDS, M. D. (1977) Bio-
    chemistry 16, 2633-2639.
NATHANSON, J. A. (1977) Physiol. Rev. 57, 157-256.
ORR, H. T., LOWRY, O. H., COHEN, A. I. and FERRENDELLI,J.A.
    (1976). Proc. Natl. Acad. Sci. U.S.A. 73, 4442-4445.
PANNBACKER, R. G. (1973a) Science, N.Y. 183, 1138-1140.
PANNBACKER, R. G. (1973b) in Prostaglandins and Cyclic AMP
    (Kahn, R. H. and Lands, W. E., eds.) pp. 251-252.
    Academic Press, New York.
PANNBACKER, R. G. (1974) Invest. Ophthalmol. 13, 535-538.
POLANS, A., WOODRUFF, M., HERMOLIN, J. and BOWNDS, D.
    (1978) Invest. Ophthalmol. Visual Sci. 17 (Suppl.),256.
SHICHI, H. and SOMERS, R.L. (1978) J. Biol. Chem. In
    press.
STELL, W. K. (1972) in Handbook of Sensory Physiology
    (Fuortes, M. F. G., ed.) Vol. 7, Part 2, pp. 111-213.
    Springer-Verlag, New York.
WALD, G. (1968) Science, N.Y. 162, 230-239.
WELLER, M., VIRMAUX, N. and MANDEL, P. (1975a) Proc. Natl.
    Acad. Sci. U.S.A. 72, 381-385.
WELLER, M., VIRMAUX, N. and MANDEL, P. (1975b) Nature 256,

        68-70.
WELLER, M., GORIDIS, C., VIRMAUX, N. and MANDEL, P. (1975c)
        Expl. Eye Res. 21, 405-408.
WERBLIN, F. S. (1974) in The Eye (Davson, H. and Graham,
        L. T. Jr., eds.) Vol. 6, pp. 257-281. Academic Press,
        New York.
WOODRUFF, M. L., BOWNDS, D., GREEN, S. H., MORRISEY, J. L.
        and SHEDLOVSKY, A. (1977) J. Gen. Physiol. 69, 667-
        669.
YOUNG, R. W. (1969) in The Retina: Morphology, Function
        and Clinical Characteristics (Straatsma, B. R., Hall,
        M. O., Allen, R. A. and Crescitelli, F., eds.) pp.
        177-210. University of California Press, Los Angeles.

EVIDENCE FOR THE PRESENCE OF SUBSTRATES FOR cGMP DEPENDENT PROTEIN PHOSPHORYLATION IN HUMAN SYNAPTOSOMAL MEMBRANES

D.H. BOEHME*, R. KOSECKI* AND N. MARKS**

*VA Medical Center, East Orange, New Jersey 07019
**Center for Neurochemistry, Rockland Research
Institute, Wards Island, New York 10035

Substantial evidence exists that cyclic nucleotide effects are mediated by specific (intracellular) protein kinases. Since biological regulatory agents can affect intracellular levels of cyclic nucleotides, it has been proposed that protein phosphorylation is a central mechanism mediating their actions and thereby involved in many aspects of cellular function (see WALSH, 1978). Relevance to studies on changes in protein phosphorylation with age arise from findings that neurotransmitters have age-related effects on levels of cyclic nucleotides (SCHMIDT & THORN-BERRY, 1978; SCHMIDT et al., 1978) coupled with observations that there are changes in synaptic function and morphology during senescence (WALKER & WALKER, 1973; CRAGG, 1975; MCGEER & MCGEER, 1975). At the present time very few studies exist on alterations in protein phosphorylation in synaptic membranes--the major target sites of neurotransmitter action (DAVIS, 1977; TRUEX et al., 1978; BOEHME et al., 1978).

In studies on mouse brain cytosol, MALKINSON (1977) observed a postnatal increase in phosphate incorporation into a 49K protein with a greater response to added cAMP in neonates. KUNUNGO & THAKUR (1977) found a differential response to addition of Ca^{2+} in phosphorylation of histone and non-histone proteins extracted from rat brain slices in rats aged up to 84 weeks.

Man himself is the best model for the study of aging owing to his longevity and the morphological changes that

occur in the CNS which are absent in animal models. In
the present study, synaptic membranes prepared from different anatomical areas were found to incorporate phosphate and
were exceptional in showing a high sensitivity to stimulation by cGMP but not cAMP. The finding of cGMP stimulation
for a synaptosomal fraction may have implications for functions, since cGMP levels in tissues are sensitive to change
on addition of a large number of biological regulatory
agents. Brain and especially cerebellum contain (soluble)
cGMP stimulated protein kinases and one such enzyme has
been purified from bovine cerebellum (TAKAI et al., 1975).

                        MATERIALS AND METHODS

     Synaptic membrane fractions were prepared from male
Wistar rats of known age, and from human brain removed
within five hours of death. Human brains were dissected
to yield the following areas: frontal cortex, white matter
striatum, caudate nucleus and putamen. Brains were routinely
subjected to morphological examination, and if abnormal,
they were not included in this study. Purified synaptic
membranes were prepared according to the procedures of
COTMAN & MATTHEWS (1971) and HERNANDEZ (1974). This involved purification of the P_2 fraction containing crude
synaptosomes on a sucrose gradient 0.8-1.2M after centrifugation at 27,000 rpm for 2 h (Beckman SW-27 Head); the
synaptic membranes present at the interfaces of 0.9 and
1.0M sucrose were harvested by puncturing the base of the
tube with a 20 gauge needle coupled with tubing to a fraction collector. Membranes were pelleted by further centrifugation at 18,000 g for 20 minutes and then suspended in
0.32M sucrose and stored at -20°C before use. The purity
of fractions was determined by electron microscopy.

     In assays for protein kinase activity, the reaction
mixture of 0.2 ml 50mM sodium acetate buffer pH 6.5 contained 0.5 mg protein and 2 μmoles of $MgCl_2$. It was pre-
incubated for three minutes at 37°C, after addition of 20μl
of 50mM tris-HCl buffer containing γ-^{32}P-ATP (final con-
centration--7.5 μM), it was incubated for an additional 30
seconds. The reaction was terminated with 0.1 ml of a mixture containing 9% w/v SDS in a 3mM Tris buffer pH 8.0,
6% 2-mercaptoethanol, 3mM EDTA, and 15% sucrose according
to the procedure of UEDA et al. (1973). cAMP or cGMP was
added to some experiments as required.

     Tracking dye (Bromophenol Blue) was added to incubated
samples or standards and then run on acrylamide gels

(12.5% with a 5% spacer gel) in the presence of 0.1% w/v
SDS. Gels were fixed in a methanol-acetic acid mixture,
stained by immersion in Coomassie blue, destained, and
then dried, prior to densitometry or radioautography (Kodak
NS-2T X-ray film 2-3 days). Bands found to be labelled
on slab gels were quantified using a Transidyne General
RFT scanning densitometer fitted with an integrator. Slab

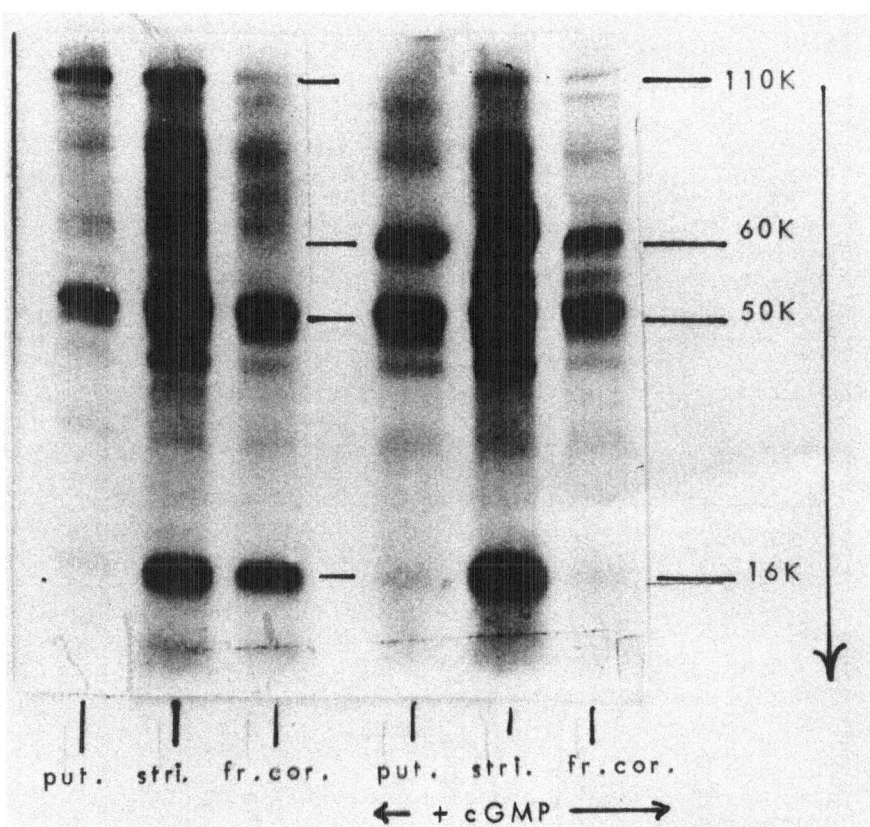

Fig. 1: Slab-gel electrophoresis of purified synaptosomal
    membranes prepared from human postmortem brain. The
    figure represents the pattern of ^{32}P incorporation fol-
    lowing incubation with labelled ATP and radioautography.
    Note presence of an intense band at approximately 110K
    present in putamen (Put), and striatum (stri), but ab-
    sent in frontal cortex (fr.cor). The concentration of
    cGMP and Mg^{2+} (three wells on the left as indicated) was
    10^{-6} M and 10 mM, respectively.

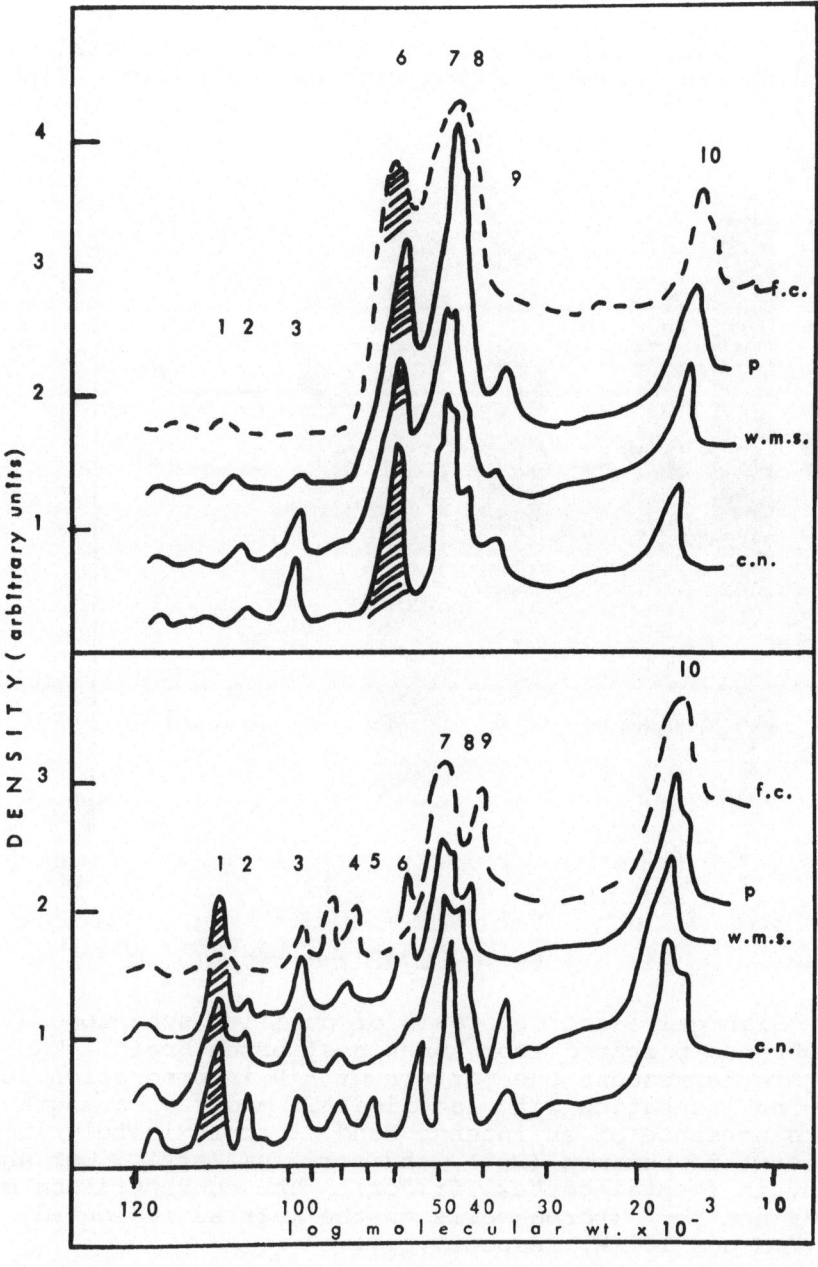

gels were calibrated for M.W. using a protein kit containing materials of 21,500-165,000 daltons (soybean trypsin inhibitor, bovine serum albumin and RNA polymerase subunits).

RESULTS AND DISCUSSION

Synaptic membranes were purified from caudate nucleus, white matter striatum and putamen (areas known to be prone to pathological changes in aging in presenile dementia or Alzheimer's disease) and compared to those of the frontal cortex. It was shown by means of slab gel electrophoresis that all anatomical regions indicated, contained 30-40 proteins ranging in M.W. from above 110K to 16K when compared to reference proteins. Striatal regions from senile brains were characterized by a high content of a 110K component which was found to be largely absent from the frontal cortex. Approximately one third of these bands incorporated radioactivity, especially those with approximate M.W.s of 110K, 38K and 16K; the band at 50K was a doublet and corresponded with a similar band(s) stained by dye (Figs. 1, 2). According to the terminology of WILLIAMS & RODNIGHT (1977), the 110K protein was present in region "α" or according to ROUTENBERG & EHRLICH (1975) to A-C; the 50K component appears to be analogous to the 49K component present in rat cytosol termed protein "II" according to GREENGARD (1976). The 16K component may represent some myelin contamination, since myelin elements were detectable in electron micrographs of "purified" membranes prepared from human and also from adult rat brains. It is

Fig. 2: Comparison of ^{32}P incorporation into synaptosomal membranes of four different anatomical areas prepared from a 60 year old male subject in absence of added cGMP (lower half) and presence of cGMP (upper half). Areas compared were the caudate nucleus (c.n), white matter striatum (w.m.s), putamen (p) and the frontal cortex (f.c),(also indicated by the dotted line). The hatched area in the lower figure indicates position of the high M.W. striatal component and in the upper half the position of major cGMP dependent protein kinase substrate found after addition of cGMP (10^{-6}M in presence of 10 mM Mg^{2+}). Proteins are numbered 1-10 in descending order of M.W. Band densities were read in a Transidyne General RFT scanning densitometer fitted with an integrator.

known that the 16K region of the gel contains phospho-
peptides derived from breakdown of protein and radioactivity entrapped in SDS micelles and lipids (WILLIAMS &
RODNIGHT, 1977). The 16K component could be removed
following delipidization by the method of RODNIGHT et al.
(1975) but this generally led to a loss of resolution of
the other bands.

Studies on age-related changes in protein phosphorylation were limited to cases where the cause of death could

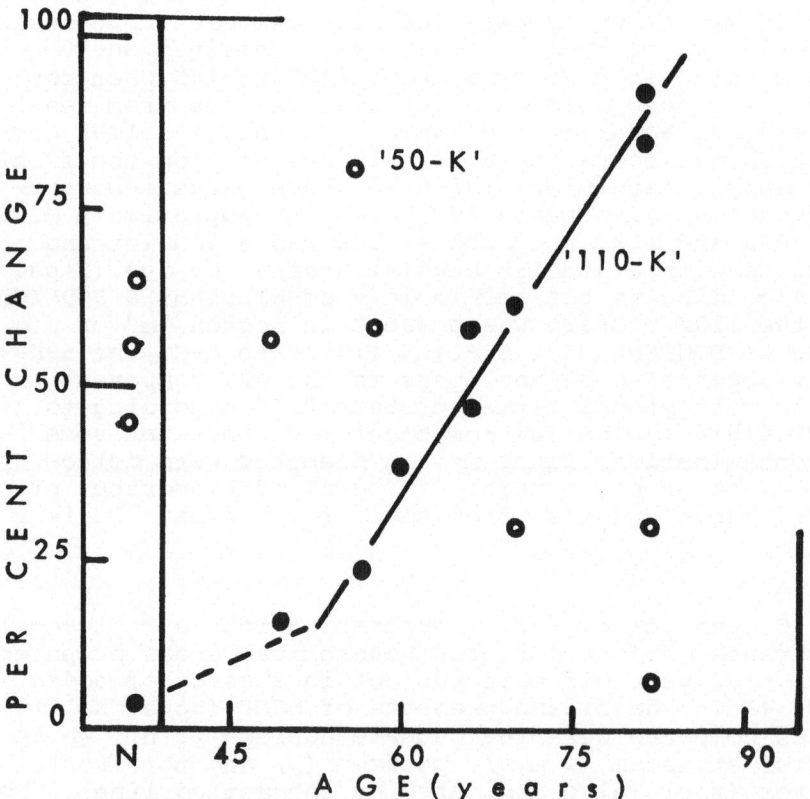

Fig. 3: Change with age in the incorporation of ^{32}P into
two protein components present in striatal areas (bands
1 and 7-8, Fig. 2). N denotes data for three brains
from a foetus, a 1 and 5 year old. Incorporation into
the 110K component (full circles) appeared to increase
with advancing age but no distinct pattern can be dis-
cerned for the 50K component (open circles).

not be attributed to CNS pathology. The comparison of
densitometric values for the major bands in such cases
indicated that only two were subject to significant changes
with advancing age. Incorporation into the 110K (striatal)
component increased from age 45-85, in contrast to the 50K,
where incorporation appeared to decrease only at extreme
ages (Fig. 3). In studies now in progress on post-develop-
mental changes in protein phosphorylation, we have observed
rather good incorporation into the 50K component of synaptic membranes prepared from a foetus, and infants ranging
in age from 1-5 years comparable to that observed in young
adults. Additional studies are required to determine the
overall significance of these changes in post-natal periods.
The high M.W. 110K component is of interest in terms of
age-related pigments since these contain neurofibrils and
neurotubulin which are known to be associated with high
M.W. (accessory) proteins and which can act as substrates
for protein kinases (TERRY & WISNIEWSKI, 1972; SOIFER,
1975). Present findings suggest, therefore, that striatal
regions may reflect an increase in membrane bound accessory
proteins. This is supported by the observation that the
110K component was absent in membranes prepared from a
foetal brain and ones aged 1 and 5 years (Fig. 3).

     No stimulation was observed for ^{32}P incorporation
upon addition of cAMP to human membranes from adult brains
incubated in vitro under optimal conditions. Comparable
experiments were performed with membranes from adult rats,
where cAMP stimulation was observed similar to that of
published studies. To decide if the lack of response was
attributable to postmortem handling of tissue, rats were
killed by rapid asphyxiation to maintain their body temperature then held at room temperature or at 0°C for 5 h
and then were stored at 4°C (morgue conditions) for periods
of up to 20 hours prior to removal of the brains and preparation of membranes. In most such experiments, the
membranes were found to respond to added cAMP (Fig. 4).
To decide if the levels of cAMP were optimal with respect
to human brain preparations, the levels of cyclic nucleotide were varied from 10^{-5} to 10^{-7}M without evidence of
stimulation, although at higher concentrations there was a
marked inhibition of ^{32}P incorporation into several bands.
It was, therefore, concluded that the failure to demonstrate
cAMP stimulation ^{32}P incorporation into human synaptic
proteins of adults was attributable not to postmortem
deterioration per se, or to the levels of cyclic nucleotide
in the incubation mixture, but to other factors. This was
reinforced by the finding that human membranes were extremely sensitive to addition of cGMP at 10^{-6}M, which led

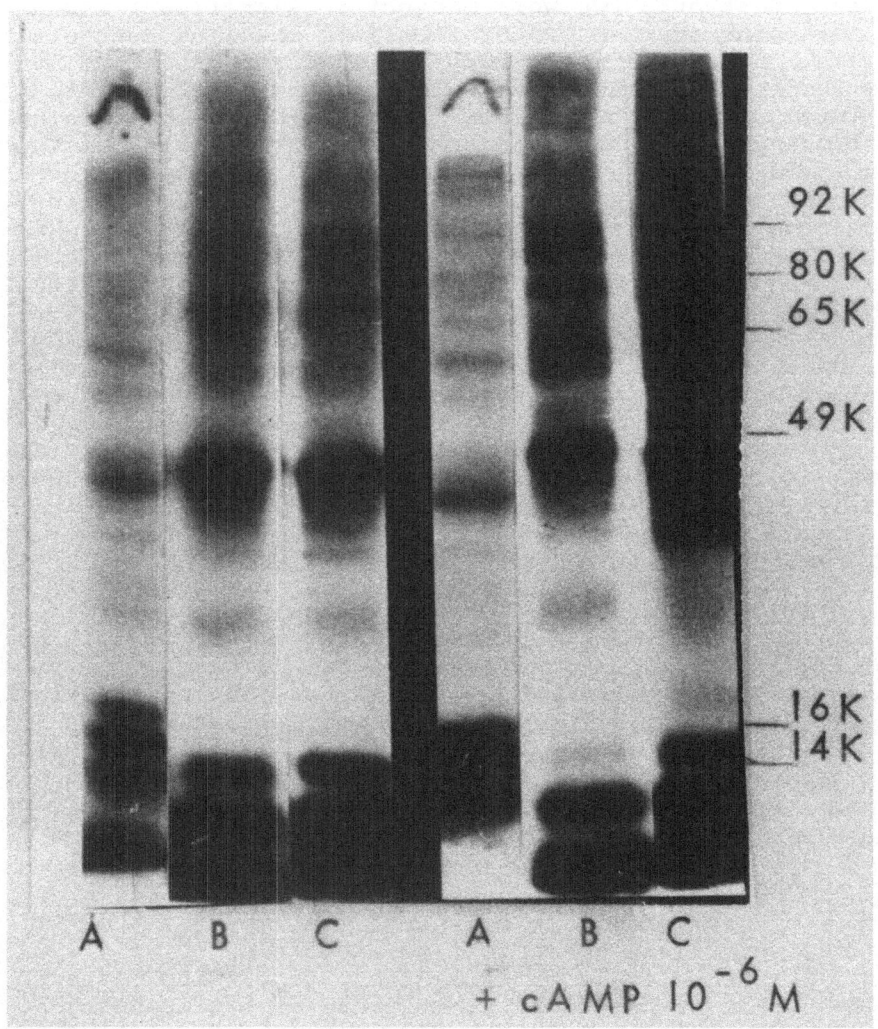

Fig. 4: Comparison of the incorporation of ^{32}P into synaptosomal membranes of rat brain prepared after storage of the brain at 0°C for 20 h (A), or processed after leaving the brain <u>in situ</u> for 4 h at room temperature or 0°C followed by storage of the body at 4° for 20 h (B), or processed immediately following the death of the animal (C). Note that the incorporation into proteins was lower in condition A and that the best cAMP effects were obtained in conditions B and C.

to an intensification of several existing bands along with
the apparent appearance of a new band at 60K (Fig. 1).
To determine if the lack of cAMP response was evident at
younger ages, we studied phosphorylation in one human foetus
and two infant brains. In the case of the five year old
but not younger ages, we did find a small but measurable
increase in several protein bands with addition of 10^{-6}M
cAMP although the changes were less than observed in comparable experiments using rat membranes. These preliminary
results suggest a differential change in sensitivity in
cyclic nucleotides with a switch from cAMP to cGMP with advancing age. It is of interest that KUO (1975) found an
increased level of cGMP dependent protein kinases in guinea
pig cytosol fractions; the ratio of cGMP to cAMP dependent
enzymes increased in brain, but were decreased for heart
and lung. This observation suggests that there may be a
reciprocal relationship between these two enzymes with differences between brain and peripheral tissues. It thus
will be of interest with respect to aging to determine if
a similar relationship such as this applies to membrane
bound protein kinases.

The question of cAMP stimulated phosphorylation in
synaptic membranes of postmortem human brain may require
additional studies to determine if assay conditions were
optimal in all respects. We are aware that methods of
sacrifice may affect cAMP sensitivity in experimental
animals, and that rates of phosphorylation for individual
proteins may be dependent on ATP/membrane ratios in view
of the rapid hydrolysis of the phosphate donor by phosphatases (WIEGANT et al., 1978). As a corollary, it may be
relevant to study age-related changes in phosphatases
present in membranes, or presence of inhibitory proteins
affecting cAMP dependent kinases, or factors reported to
modulate cGMP dependent protein kinases (SHOJI et al., 1977).

The existence in mammalian brains of cGMP-dependent
protein kinases in synaptosomal membranes has not been
reported previously. Such enzymes exist in other tissues
(WILLIAMS & RODNIGHT, 1977) and were recently reported by
RAM & EHRLICH (1978) to be present in nerve roots of Aplysia.
As such, human brain appears to be an excellent source of
cGMP dependent protein kinase substrates, since it could be
observed in all the selected anatomical regions of senile
brains. The presence of cGMP sensitive protein kinases may
have a bearing on functional processes in view of the known
sensitivity of tissue levels of cGMP to alteration by a
large number of biological regulatory agents. These include
neurotransmitter substances such as acetylcholine (muscar-

inic receptors), catecholamines (α-adrenergic), histamine
(H$_1$-receptors), glutamic acid, and drugs such as morphine
(GREENGARD, 1976; WILLIAMS & RODNIGHT, 1977). Moreover,
it has been established that addition of cGMP to cervical
sympathetic ganglia and cerebral cortex can, in some instances, mimic the action of neurotransmitters (HASHIGUHI
et al., 1978; DUNN et al., 1977). By way of comparison
the agents elevating cAMP levels include norepinephrine
(acting on β-receptors), histamine (H$_2$-receptors), prostaglandin E, dopamine, steroids, Ca^{2+} and peptidyl hormones.
The manner in which these agents regulate phosphorylation
is not completely understood, but a number of mechanisms
have been proposed. These include direct effects of agents
on the enzyme or effects on the de novo synthesis of substrates (as proposed for steroids), or metabolism of the
substrate (phosphatases, proteolytic enzymes) or of the
nucleotides (phosphodiesterases).

The distribution and properties of cGMP dependent
kinases are less well known than those of the cAMP dependent protein kinases. cGMP dependent protein kinases have
been demonstrated in several vertebrate and invertebrate
tissues (GREENGARD, 1976); enzyme purified from the lung
and vascular smooth muscle has been shown to consist of
a dimer composed of two identical subunits each of which
appears to contain cGMP binding and catalytic activity
of M.W. 75-81K (GILL et al., 1977; CASANELLI et al., 1978).
It is hoped that the availability of a system utilizing
human (post mortem) tissue will present opportunities to
extend our knowledge of the linkage between phosphorylative
processes and altered functional states and provide scope
for clinical applications.

ACKNOWLEDGEMENT

This study was supported by the Medical Research
Service of the Veterans Administration.

REFERENCES

BOEHME, D.H., DOSECKI, R. & MARKS, N. (1978) Protein phos-
     phorylation in human synaptosomal membranes: evidence
     for the presence of cGMP-dependent protein kinases.
     Brain Res. Bull. 3, in press.
CASANELLI, J.R., SCHLICHTER, D.J., WALTER, U. & GREENGARD,
     P. (1978) Photoaffinity labelling of cGMP-dependent
     protein kinase from vascular smooth muscle. J. Biol.

Chem. 253, 4771-4776.
COTMAN, C.E. & MATTHEWS, D.W. (1971) Synaptic plasma mem-
    branes from rat brain synaptosomes: isolation and
    partial characterizations. Biochem. Biophys. Acta
    249, 380-403.
CRAGG, B.G. (1975) The density of synapses and neurons in
    normal mentally defective and aging human brains.
    Brain 98, 81-90.
DAVIS, L.G. (1977) The subcellular distribution of endo-
    genously phosphorylated proteins from cerebral cortex
    of newborn rats. Proc. Soc. Neurosci. III, 103.
DUNN, N.J., KAIBARA, K. & KARCZMAR, A.G. (1977) Direct post-
    synaptic membrane effect of dibutryl cyclic GMP on
    mammalian sympathetic neurons. Neur. Pharmacology
    16, 715-717.
GILL, G.N., WALTON, G.M. & SPERRY, P.J. (1977) cGMP-depen-
    dent protein kinase from bovine lung. J. Biol. Chem.
    252, 6443-6449.
GREENGARD, P. (1976) Possible role for cyclic nucleotides
    and phosphorylated membrane proteins in postsynaptic
    action of neurotransmitters. Nature 260, 101-107.
HASHIGUCHI, T., USHIYAMA, N.S., KOBAYASHI, H. & LIBET, B.
    (1978) Does cyclic GMP mediate the slow excitatory
    synaptic potentials in sympathetic ganglia? Nature
    271, 267-268.
HERNANDEZ, A.G. (1974) Protein synthesis by synaptosomes
    from rat brain. Biochem. J. 142, 7-17.
KATZ, J.B., CRATRAVAS, G.N., VALASES, C. & WRIGHT, S.J.
    (1978) Morphine reduces cerebellar cGMP content and
    elevates CSF cGMP content in rhesus monkey. Life
    Sci. 22, 467-472.
KUNUNGO, M.S. & THAKUR, M.K. (1977) Phosphorylation of
    chromosomal proteins as a function of age and its
    modulation by Ca^{2+}. Biochem. Biophys. Res. Commun.
    79, 1031-1035.
KUO, J.F. (1975) Changes in relative levels of cGMP and
    cAMP-dependent protein kinase in lung, heart and brain
    of developing guinea pigs. Proc. Natl. Acad. Sci.
    72, 2256-2259.
MALKINSON, A.M. (1977) Developmental changes in the cAMP
    dependent phosphorylation and dephosphorylation of
    a protein endogenous to murine brain and liver.
    Biochem. Biophys. Res. Commun. 78, 91-98.
MCGEER, E.G. & MCGEER, P.L. (1976) Neurotransmitter
    metabolism in the aging brain. In Aging, Vol. 3.
    Neurobiology of Aging, pp. 398-403. Raven Press,
    New York.
RAM, J.L. & EHRLICH, Y.H. (1978) Cyclic GMP-stimulated
    phosphorylation on membrane bound proteins from nerve

roots of Aplysia California. J. Neurochem. 30, 487-491.

RODNIGHT, R., REDDINGTON, M. & GORDON, J. (1975) Methods for studying protein phosphorylation in cerebral tissues. Res. Meth. Neurochem. 3, 325-367.

ROUTENBERG, A. & EHRLICH. Y.H. (1975) Endogenous phosphorylation of four cerebral cortical membrane proteins: role of cyclic nucleotides, ATP and divalent cations. Brain Res. 92, 415-430.

SCHMIDT, M.J., PALMER, G.C. & ROBISON, G.A. (1978) The cyclic nucleotide system in brain during development and aging. In Psychopharmacology and Aging (Eisdorfer, C. & Fann, W.E., eds.). Spectrum Press, New York.

SCHMIDT, M.J. & THORNBERRY, J.F. (1977) Cyclic AMP and cyclic GMP accumulation in vitro in brain regions of young, old and aged rats. Brain Res. 139, 169-177.

SHOJI, M., PATRICK, J.G., TSE, J. & KUO, J.F. (1977) Studies on cGMP-dependent protein kinases from bovine aorta. J. Biol. Chem. 252, 4347-4353.

SOIFER, D. (1975) Enzymatic activity in tubulin preparations: cAMP dependent protein kinase activity of brain microtubule protein. J. Neurochem. 24, 21-33.

TAKAI, Y., NISHIYAMI, K., JAMAMURA, H. & NISHIZUKA, Y. (1975) cGMP-dependent protein kinase from bovine cerebellum. J. Biol. Chem. 250, 4690-4695.

TERRY, R.D. & WISNIEWSKI, H.M. (1970) The ultrastructure of the neurofibrillary tangle and the senile plaque. In Ciba Fdn. Symposium on Alzheimer's Disease and Related Conditions (Wolstenholme, G.E.W. & O'Connor, M., eds.), pp. 145-168. J & A Churchill, London.

TRUEX, L., CONWAY, A., ROUTENBERG, A. & SCHMIDT, M.J. (1978) cAMP-dependent protein kinase and protein phosphorylation in human brain during aging. Proc. Soc. Neurosci. IV, 129.

UEDA, T., MAENO, H. & GREENGARD, P. (1973) Regulation of endogenous phosphorylation of specific proteins in synaptic membrane preparations from rat brain by cAMP. J. Biol. Chem. 248, 8295-8305.

WALKER, J.B. & WALKER, J.P. (1973) Properties of adenylate cyclase from senescent rat brain. Brain Res. 54, 391-396.

WALSH, D.A. (1978) The role of the cAMP-dependent protein kinase as the transducer of cAMP action. Biochem. Pharmacol. 27, 1801-1804.

WIEGANT, V.M., ZWIERS, H., SCHOTMAN, P., GISPEN, W.H. (1978) Endogenous phosphorylation of rat brain synaptosomal plasma membranes in vitro; some methodological aspects. Neurochem. Res. 3, 443-454.

WILLIAMS, M. & RODNIGHT, R. (1977) Protein phosphorylation
    in nervous tissue: possible involvement in nervous
    tissue functions and relationship to cyclic nucleo-
    tide metabolism. Progress in Neurobiology 8, 183-250.

SECTION II

Interaction of Neuropeptides, Cyclic Nucleotides and
Phosphoproteins in Mechanisms Underlying Receptor Function

INTRODUCTION

This section integrates recent findings in the study
of neuroactive peptides with those in the fields of cyclic
nucleotides and protein phosphorylation. The chapters in
this section demonstrate functional interactions between
the three title systems. Thus, it has been demonstrated
that neuropeptides interact with systems that generate
cyclic nucleotides as well as with enzyme systems that
phosphorylate membrane-bound proteins. These systems can
act in concert to establish long-term adaptive changes in
the brain. Such adaptive processes involve modification
in receptor function. Therefore, the section begins with
chapters concerned with the regulation of adrenergic,
(Bylund), cholinergic, (Gordon and Diamond), and opiate,
(Blume, Boone and Lichtshtein) receptors. These chapters
are followed by descriptions of the effects of ACTH (Gispen,
Zwiers, Wiegant, Schotman, and Wilson) and of opioid peptides on cyclic nucleotide generation (Klee) and on protein
phosphorylation (Davis and Ehrlich). The overall picture
which emerges from this section indicates a dual role of
cyclic nucleotides and phosphoproteins in receptor function.
On the one hand, cyclic nucleotides and phosphoproteins
play the role of "second messenger" in synaptic transmission.
On the other hand, the phosphorylation of proteins may
modulate the sensitivity of receptors to stimulation by
neurotransmitters, possibly through its regulation by
neuropeptides.

REGULATION OF CENTRAL ADRENERGIC RECEPTORS

DAVID B. BYLUND

Department of Pharmacology, University of
Missouri, School of Medicine
Columbia, Missouri 65212

INTRODUCTION

The mechanisms through which the catecholamines
produce their effects in the central nervous system are
not yet understood. Norepinephrine, which could be
acting either as a neurotransmitter or neuromodulator,
initially binds to its receptor site on the external
surface of the cell membrane. This binding interaction
eventually results in alterations of the electrical properties of the membrane and perhaps alterations in other
cellular functions. Many of these effects appear to be
mediated through the activation of the membrane bound
enzyme, adenylate cyclase. It is well established that
the responsiveness of the receptor systems which mediate
the effects of norepinephrine is subject to regulation.
Thus, under certain conditions, the response to a given
concentration of norepinephrine may be either greater
than or less than normal, i.e., supersensitive or subsensitive. There appear to be several biochemical events
which are regulated, one or more of which might be important under a given set of conditions: the receptor binding site for norepinephrine; enzymes such as adenylate
cyclase, phosphodiesterase, protein kinase, phosphoprotein
phosphatase; the functional coupling between the receptor
binding site and adenylate cyclase; and membrane properties such as fluidity, asymmetry, and lipid composition.
It is the intent of this chapter to review the evidence
suggesting that changes in the number and properties of
receptor binding sites is one important regulatory mechanism.

     Until recently, neurotransmitter receptors were de-
fined solely in a functional sense, and their existence
as actual entities was only inferred. In the early 1970's
techniques were developed in several laboratories to specifically and reversibly label opiate receptors using radioactive ligands. This ability to radiochemically label
the binding site of the receptor has been the basis for
the labeling of many other putative neurotransmitter
receptors, including α- and β-adrenergic receptors (YAMA-
MURA, ENNA & KUHAR, 1978). Using these techniques the
number (B_{max}) and affinity (K_D) of the receptor binding
sites can be determined. For the purposes of this review,
the term receptor will be used to signify the macromole-
cule on the outer surface of the plasma membrane which
binds norepinephrine and initiates the biochemical events
leading to a physiological response. Under appropriate
conditions, the receptor can be experimentally defined by
the binding of certain radioactive compounds to membrane
suspensions. This review is limited to α- and β-receptor
binding studies in mammalian CNS dealing with receptor
regulation. Other reviews, dealing with various aspects
of adrenergic receptors, are available (MAGUIRE, ROSS &
GILMAN, 1977; WOLFE, HARDEN & MOLINOFF, 1977; WILLIAMS &
LEFKOWITZ, 1978; SNYDER, U'PRICHARD & GREENBERG, 1978;
U'PRICHARD & SNYDER, 1978d). For purposes of organization, a rather arbitrary subdivision into biochemical
regulation, physiological regulation, pharmacological
regulation and pathological alterations has been made.
This will be preceded by a few comments on the receptor
binding technique.

                        Radioreceptor Assay

     Although the receptor binding techniques, i.e.,
radioreceptor assay, are conceptually simple and relatively easy technically, it must be strongly emphasized
that it is not a trivial assay, and great care should be
taken in the experimental design and in the interpretation of the results. It is important to carefully select
the total assay volume, the membrane (receptor) concentration and the radioactive ligand concentration with respect
to the K_D for the labelled ligand and receptor. For example, if the receptor concentration is greater than 0.1
K_D, the K_D determined from Rosenthal analysis (often mis-
takenly called Scatchard analysis; Rosenthal, 1967) is
an apparent K_D which is dependent on the receptor concentration (CAUTRECASAS & HOLLENBERG, 1976; FIELDS et al.,
1978). This concept could be of practical importance in
experiments involving a change of receptor properties.

Assume, for instance, that treatment X resulted in a
doubling of the B_{max} and a 50% decrease in K_D. If both
experimentals and controls are assayed at the same tissue
concentration such that the receptor concentration is significantly greater than 0.1 K_D, the $K_D$app might change
very little, thus hiding the actual K_D change.

The definition of non-specific binding may also be
very important. For example, data were recently reported
which suggested an absence of β-adrenergic receptors in
the guinea pig cerebral cortex (CLARK et al., 1977). However, a more detailed study showed that β-receptors were
indeed present (BYLUND, 1978). The results of the latter
study indicated the presence of a population of moderate
affinity, non-specific binding sites for ^{3}H-DHA (dihy-
droalprenolol). Both (-)- and (+)-propranolol inhibited
the binding of H-DHA to these sites, but agonists, such
as (-)-isoproterenol and (-)-epinephrine did not. The
previous failure to observe β-receptor binding appears
to have been due to a combination of two factors: first,
the use of a high concentration of [^{3}H]DHA (7 nM) which
labeled a large fraction of the nonspecific binding sites;
and, secondly, the inappropriate use of an antagonist to
determine non-specific binding, thus including a considerable amount of non-specific binding in the presumed
specific binding, which, in turn, obscured the true β-
adrenergic binding sites.

An important question which has been only partially
answered is whether adrenergic receptors labeled in CNS
binding studies are located on neurons, glia or both.
There is evidence to suggest that dopamine receptors are
partially located on glia (HENN et al., 1977) and that
benzodiazapine may be primarily localized to glia (CHANG
et al., 1978). In preliminary studies, we have obtained
evidence indicating the presence of β-adrenergic, serotonin and muscarinic cholinergic receptors on both neuronal
and glial elements of rat cortex (BYLUND, PODUSLO & SNYDER,
unpublished). Recent results from lesion studies in
Molinoff's laboratory appear to suggest that $β_1$-adrener-
gic receptors in the rat cerebral cortex are neuronal,
while $β_2$-receptors are non-neuronal (MOLINOFF, personal
communication). The applicability of these results to
other species and other brain regions is not known.

Although the radioreceptor assay is rapidly increasing
our understanding of the mechanisms of action of neurotransmitters, the technique is far from perfect. It is
of utmost importance that, for each new tissue or species

studied, the binding characteristics be investigated in
detail to assure that physiologically relevant receptors
are being labeled.

BIOCHEMICAL REGULATION

Certain biochemical changes, such as the addition
of cations or other small molecules (guanyl nucleotides),
or the alteration of membrane properties can significantly alter the binding characteristics in an in vitro
adrenergic receptor assay system. In most binding experiments the standard solution against which comparisons are
made is 50 mM TRIS/HCl (pH 7.2-8.0). The in vivo ionic
composition of the microenvironment of the receptor is
not known, but it is certainly not 50 mM TRIS/HCl, and
thus the relevance of these in vitro experiments to the
in vivo situation must remain an open question. However,
these experiments do play a useful role in our attempt to
understand receptor function and do suggest possible
regulatory mechanisms.

β-Adrenergic Receptors

The monovalent cations have a dramatic effect on the
β-adrenergic receptor binding of ^{3}H-DHA and ^{3}H-epinephrine
to calf cerebellar homogenates which is specific for the
smaller cations and for agonist ligands (U'PRICHARD,
BYLUND & SNYDER, 1978). Lithium ion lowered the affinity
of agonists at the site labeled by ^{3}H-DHA by a factor of
12 but had only a minimal effect on the affinity of antagonists. This decrease in agonist affinity was a concentration dependent effect with a EC_{50} in the range of 27-75
mM Li+. Sodium and ammonium ions mimicked lithium although
the effect was not as great (6 to 8 fold), while, potassium, rubidium, cesium and choline were without effect.
This order of selectivity corresponds to the Eisenman
XI series (DIAMOND & WRIGHT, 1969). The effect of these
ions on the binding of a labeled agonist, ^{3}H-epinephrine
was also studied. Lithium and sodium markedly reduced
the binding of ^{3}H-epinephrine, half maximally at concentrations of 10 and 20 mM, respectively, while potassium,
rubidium and cesium were considerably weaker. Rosenthal
analysis of saturation experiments in the presence of
increasing concentrations suggested that incubation with
Na+ resulted in a conversion of a portion of the receptors
to sites having a 5 fold lower affinity for ^{3}H-epine-
phrine. An increase in the Na+ concentration resulted in
an increase in the proportion of sites in the lower affinity state, although the total number of β-receptors

did not change.

The magnitude of this effect of monovalent cations
appears to be related to their size. For example, if the
ratio of the IC_{50} values for epinephrine (against ^{3}H-DHA)
obtained in the presence and absence of the ion is plotted
against the log of the ionic radius, a nearly linear
relationship is obtained for the alkali metals. Furthermore, a line joining the points for the two nitrogenous
cations (ammonium and choline) has a slope similar to
that of alkali cations (Fig. 1). Although these relationships may well be strictly fortuitous, it will be of
interest to determine the effects of ions such as Cu+
(r=0.96) and methylammonium (r=1.81).

Although pronounced effects of guanyl nucleotides
on β-adrenergic receptor binding have been found in most
other systems, we were unable to identify any such effects
on ^{3}H-DHA binding (either in the presence or absence of
monovalent ions) in CNS tissues (U'PRICHARD & SNYDER,1978).

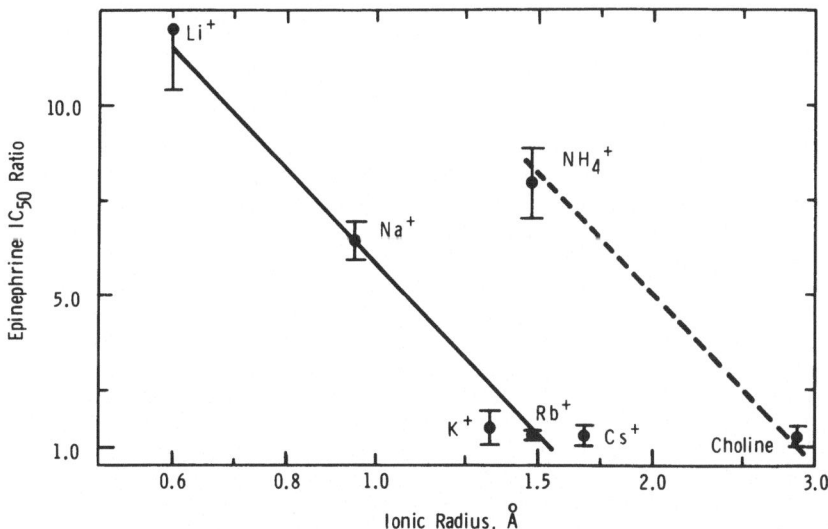

Fig. 1: The effect of ionic radius on the potency of
epinephrine at ^{3}H-DHA binding sites. The epinephrine IC_{50} ratio (from U'PRICHARD, BYLUND & SNYDER,
1978) for the various ions is the ratio of the IC_{50}
(for the inhibition of ^{3}H-DHA binding) in the
presence of 150 mM ion to the IC_{50} in its absence.
The Pauling ionic radii were taken from Table 5
of MORENO & DIAMOND (1975). For ammonium and
choline, the radius was calculated as r=(a+b)/4.

    Since receptors are physically and functionally
part of a membrane, the lipid matrix environment could
be important in the regulation of the receptor. For example, the insertion of specific free fatty acids into
the turkey erythrocyte membrane can cause a large increase in the isoproterenol stimulated adenylate cyclase
activity (ORLY & SCHRAMM, 1975). In mouse brain, a high
concentration of phospholipids inhibited the catechola-
mine-stimulated, but not the NaF-stimulated adenylate
cyclase activity (LEON et al., 1978). Recent evidence
from Axelrod's laboratory suggest that phospholipid methylation in the rat erythrocyte may be important in the
generation of phospholipid asymmetry, the regulation of
the number of β-receptors and the coupling between the
receptor and adenylate cyclase (HIRATA, AXELROD & STRITT-
MATTER, 1978). The increase in the number of β-adrenergic
receptors was correlated with the formation of phosphatidylcholine from phosphatidyl-N-monomethyl-ethanolamine
(STRITTMATTER, HIRATA & AXELROD, 1978). They also made
the intriguing observation that β-receptor stimulation
can increase the rate of phospholipid methylation by a
mechanism which is apparently independent of adenylate
cyclase activation (HIRATA, STRITTMATTER & AXELROD, 1978).
Although the brain also has phospholipid asymmetry (FON-
TAINE, HARRIS & SCHROEDER, 1979) there is no evidence
of central methylation of phosphatidylethanolamine to
give phosphatidylcholine (ANSELL, 1973). Thus phospholipid methylation would not appear to be important in the
regulation of central β-adrenergic receptors.

                    α-Adrenergic Receptors

    The effects of monovalent cations on α-adrenergic
receptor binding are somewhat similar to their effects
on the β-receptor. Sodium and lithium, but not potassium,
drastically reduced the binding of labeled agonists (^{3}H-
epinephrine, ^{3}H-norepinephrine and ^{3}H-clonidine) but had
only minor effects on labeled antagonist binding (^{3}H-WB-
4104 and ^{3}H-dihydroergocryptine, GREENBERG et al., 1978).
In contrast to the β-receptor, Li+ tended to be less
potent than Na+ (Eisenman X Series, DIAMOND & WRIGHT,
1969). The reduction in binding appeared to result from
a decrease in the number of binding sites without a change
in their affinity. We have observed that this reduction
in the number of sites may actually be due to an increased
K_D if a fraction of the α-receptors is converted to a
state which has such low affinity for the agonist ligand
that specific binding to that state cannot be measured

with the present experimental techniques (U'PRICHARD,
BYLUND & SNYDER, 1978). Sodium ions (100 mM) also appear
to reduce by 5-10 fold the affinities of agonists at the
component of CNS dihydroergokryptine binding corresponding
to the ^{3}H-agonist binding site (U'PRICHARD, unpublished
observations). A selective reduction in agonist affinities by sodium ions has also been observed at platelet
α-receptors (TSAI & LEFKOWITZ, 1978).

     The guanyl nucleotides appear to play an important
regulatory role at various receptor-adenylate cyclase
systems. At the α-adrenergic receptors labeled by agonists, GTP and Gpp(NH)p (a phosphohydrolase resistant
analogue) markedly reduced the binding of the labeled
ligands with IC_{50} values of approximately 2 μM (U'PRICH-
ARD & SNYDER, 1978a). The corresponding adenyl nucleotides
were about two orders of magnitude weaker. The decreases
in binding produced by 10 μM GTP (but not those produced
by Gpp[NH]p) were reversed by Ca++ and Mg++ (IC_{50} values
of 0.1 mM), although these ions had no effects in the
absence of guanyl nucleotides (U'PRICHARD, 1978e).
Saturation experiments suggested that the reduction in
binding was due to a decreased affinity of the labeled
agonist for the receptor with no change in B_{max}. Furthermore, 10 μM GTP reduced the potency of agonist inhibitors of ^{3}H-epinephrine binding by 4-to-5-fold, while
the potency of antagonist inhibitors was not affected
(U'PRICHARD & SNYDER, 1978a). In contrast to the effects
on α-receptors labeled by agonists, none of the binding
parameters at α-receptors sites labeled by the antagonist,
^{3}H-WB-4101 was altered by guanyl or adenyl nucleotides at
concentrations up to 0.1 mM. Although both GTP and sodium
decrease binding at α-receptor sites labeled by agonists,
they appear to have different effects, i.e., GTP increases
the K_D (decreases the affinity) while sodium decreases
the B_{max}. These effects appear to be mutually independent, since the effect of either is not altered by the
other (U'PRICHARD & SNYDER, 1978a, d).

     Thus, the predominant effect of cations and guanyl
nucleotides at adrenergic receptors appears to be the
regulation of the agonist affinity. In a synaptosomal
preparation, Li+ appeared to have a regulatory influence
on the norepinephrine stimulated adenylate cyclase system, but not on the dopamine cyclase system (RECHES,
EBSTEIN & BELMAKER, 1978; also see DALY, 1977). Although
one would like to believe that these regulatory effects
on central adrenergic receptors are of some importance
in the intact animal, the evidence which might support

such a hypothesis is not yet available.

PHYSIOLOGICAL REGULATION

It would be reasonable to assume that central adrenergic receptors might be regulated by normal physiological
processes, and that changes in receptor properties might
be detected by studying the brain removed from animals
which were in various physiological states or conditions.
The best example of this type of regulation of β-adrenergic receptors comes from the peripheral nervous system--
the pineal gland (reviewed by AXELROD & ZATS, 1977). In
the CNS, variations in adrenergic receptor characteristics
as a function of age, species and brain region have been
investigated.

Age

The effect of age on ^{3}H-HDA binding in the cerebellum
and cerebral cortex has been studied in both rat and
human brains. In the rat cerebral cortex, the density
of β-adrenergic receptors is very low at birth. The
number of receptors increases rapidly during the next
several weeks to adult levels and then remains constant
throughout the life of the animal (Tables 1 and 2). There
was no change in the K_D as a function of age. HARDEN et
al. (1977a) studied both β-adrenergic binding (using
^{125}I-iodohydroxybenzylpindolol) and adenylate cyclase
activity in rat cerebral cortex during the first few weeks
after birth. The fluoride stimulated adenylate cyclase
activity developed faster than catecholamine stimulated
activity, while the increase in β-adrenergic binding was
temporally correlated with the increase in catecholamine
cyclase activity. Thus, it would appear that the development of the β-adrenergic receptor and the adenylate cyclase
enzyme are independent of each other. This conclusion
is in agreement with previous findings in other systems
(CHARNESS et al., 1976; INSEL et al., 1976; ORLY & SCHRAMM,
1976), with results in the brain stem of aging rats (MAGGI
et al., 1978) and with the mobile receptor theory of
hormone action (BENNETT, O'KEEFE & CUATRECASAS, 1975).
Since the receptor and the cyclase are physically separable,
have different time courses of development, and can be
independently regulated they might well be considered
as autonomous entities, even though they are functionally
coupled under appropriate conditions.

TABLE 1

Influence of rat age on ^{3}H-DHA binding to cortical membranes.

Age, days	% Adult Binding	Age, months	% Adult Binding
4	15	1.5	100
9	33	6.5	95
13	63	15	103

Data from BYLUND et al. (1977) and BYLUND & SNYDER, unpublished.

In contrast to the approximately constant level of cerebral β-receptors in aged rat brains, ^{3}H-DHA binding was reduced in the cerebellum and corpus striatum (Tables 1 and 2). MAGGI et al. (1978) also studied β-receptors in human brains obtained at autopsy. No age-related change was found in binding in the cerebral cortex but after age 61 the binding decreased by about 50% in the cerebellum (Table 3). In agreement with the studies noted above, the B_{max} but not the K_D was altered. The significance of these observations is not yet clear,

TABLE 2

Influence of rat age on ^{3}H-DHA binding to brain membranes.

Receptor Binding

Age, Months	Cerebellum[a]	Cerebellum[b]	Cerebral Cortex[b]	Corpus Striatum[c]	Brain Stem[b]
	fmol/mg prot. (% of maximal binding)				
3	58±6* (78)	20±2 (100)	38±3 (100)	150± 9 (94)	11±1 (100)
6	74±5 (100)	-	-	160±18 (100)	-
12	43±9‡ (58)	20±2 (100)	32±3 (84)	132±22* (83)	7±2 (64)
12	29±9 (30)	14±2* (70)	30±4 (79)	107±17‡ (67)	5±1* (45)

*p<0.05; ‡p<0.001; [a]Data from GREENBERG & WEISS (1978) using 10 nM H-DHA; [b]Data from MAGGI et al. (1978) using 0.5 nM $_3$H-DHA; [c]Data from GREENBERG & WEISS (1978) using 20 nM ^{3}H-DHA.

although it has been suggested that the reduction in
β-adrenergic receptor number in the cerebellum may contribute to behavioral impairments which are associated
with old age (MAGGI et al., 1978).

Species and Brain Regions

Another approach to the study of physiological mechanism of adrenergic receptor regulation involves the differences in number and properties of β-adrenergic receptors
in various brain regions and in various mammalian species.
Peripheral β-adrenergic receptors have been divided into
β_1 and β_2 subtypes (LANDS et al., 1967). Briefly, epinephrine is more potent than norepinephrine at β_2-receptors,
while they have approximately equal potencies at β_1-recep-
tors. Similarly, for antagonists, practolol is more potent
at β_1 receptors, while butoxamine is more potent at
β_2 receptors. On this basis, the heart is considered to
contain predominantly β_1-adrenergic receptors and the lung
β_2-receptors. The possible presence of both β_1- and β_2-
receptors in a given tissue must also be given due consideration (BARNETT, RUGG & NAHORSKI, 1978).

The present evidence suggests that central β-adrenergic receptors can be subclassified according to the identical β_1, β_2 scheme. Using both agonist and antagonist
radioligands, we made a detailed comparison of the β-receptor binding in the calf cerebellum and cortex with the

TABLE 3

Influence of human age on ^{3}H-DHA binding to brain membranes.

Age Years	Receptor Binding fmol/mg prot. (% of 0-2 year level)	
	Cerebellum	Cerebral Cortex
0-2	18+3 (100)	21+3 (100)
40-60	14+3 (78)	18+4 (86)
61-80	9+1* (50)	18+3 (86)

*p<0.025. Data from MAGGI et al. (1978).

TABLE 4

Comparison of peripheral and central β-adrenergic
receptors.

Drugs	$\dfrac{\text{Cortex}[a]}{\text{Heart}}$	$\dfrac{\text{Cere-}}{\text{Lung}}$ bellum[a]	$\dfrac{\text{Cere-}}{\text{Lung}}$ bellum[b]
	Ratio of K_i Values[c]		
MJ-9184-1	1.05	0.55	0.63
Cc-25	5.47	0.90	0.79
Cc-34	2.84	0.87	1.00
Isoproterenol	7.00	0.51	1.14
Epinephrine	4.87	1.27	0.46
Norepinephrine	2.73	2.29	0.34
Salbutamol	3.89	0.91	0.27
Terbutaline	2.25	1.05	0.52
Hydroxybenzylpindolol	2.18	0.95	1.41
Propranolol	5.00	1.23	2.89
Alprenolol	3.92	1.08	1.25
Dihydroalprenolol	2.40	1.22	1.45
Dichloroisoproterenol	3.88	1.29	1.57
Soltalol	2.43	1.35	2.25
Butoxamine	4.31	1.08	2.33
Practolol	2.57	0.85	0.52
Mean	3.55	1.09	1.18
±SEM	0.38	0.10	0.19
	Linear Regression of K_i Values[c,d]		
Correlation coefficient	0.990	0.955	0.982

[a]Labeled ligand was ^{3}H-DHA. [b]Labeled ligand was ^{3}H-epine-
phrine. [c]Calculated from data in Tables 1 and 2 of
U'PRICHARD, BYLUND & SNYDER (1978). [d]For the regressions,
the K_i values in the peripheral tissue was the independent variable and the K_i values in the central tissue the
dependent variable.

binding of the rat lung and heart (U'PRICHARD, BYLUND &
SNYDER, 1978). Using the apparent K_i values determined
in that study, the K_i ratios, i.e., the ratio of the K_i
in cerebral cortex to the K_i in heart and the K_i for cere-

bellum to the K_i for lung, were determined for each drug
(Table 4). If pharmacologically similar receptors are
labeled in the CNS and the periphery, these ratios should
be similar for the 16 different drugs. Considering first
the cerebellum and lung, all the ratios are close to 1.0,
using both ^{3}H-DHA and ^{3}H-epinephrine as labeled ligands,
suggesting that an identical type of receptor was labeled
in both tissues. The cortex to heart ratios are also very
consistent, although the drugs were 3 to 4 fold more potent
in the heart. Linear regression analyses were also performed, and excellent correlations were obtained in all
three cases (Table 4). Similar analyses were done for
cortex vs lung, and as expected, poor correlations were
obtained: the K_i ratio varied from 0.20 to 29, and the
correlation coefficient was 0.324. Thus, it appears
reasonable to conclude that central and peripheral β-
receptors are similar pharmacologically and that the calf
cerebral cortex contains predominantly β_1-receptors while
the calf cerebellum is predominantly β_2.

TABLE 5

Comparison of K_i ratios in cerebellum and cortex of
various mammalian species.

Species	Norepinephrine / Epinephrine		Practolol / Butoxamine	
	Cerebellum	Cortex	Cerebellum	Cortex
Horse	15	0.9	20	1.1
Calf	17	1	5	0.2
Sheep	14	–	2.2	–
Goat	25	–	2.7	–
Hamster	25	9	8	2.6
Rat	--	0.7	–	0.44
Guinea Pig	13	0.6	6	0.09
	Lung	Heart	Lung	Heart
	10	2.2	6	0.27

Values are the ratio of K_i (for the inhibition of ^{3}H-DHA
binding) for the two indicated drugs (SUNN & BYLUND, 1978;
U'PRICHARD, BYLUND & SNYDER, 1978; BYLUND & SNYDER, 1976).

In order to see if this anatomical localization of
β-receptor subtypes in the calf CNS was a more general
phenomenon, we studied a number of mammalian species
(SUNN & BYLUND, 1978). Table 5 gives the ratios of the
K_i values of (-)-norepinephrine to (-)-epinephrine, and
(-)-practolol to (±)-butoxamine. High ratios are characteristic of β_2 receptors while low ratios would indicate
β_1 receptors. For all species examined, the cerebellum
is similar to rat lung (predominantly β_2) and the cortex
is similar to rat heart (predominantly β_1). An exception
may be the hamster cortex, which has fairly high ratios
for both norepinephrine to epinephrine and practolol to
butoxamine, suggesting the presence of comparable numbers
of β_1 and β_2 receptors.

A possibly important aspect of physiological regulation would be variations in the number and affinity of
β-adrenergic receptors among species. The affinity (K_D)
of [³H]-DHA for the receptor binding sites in the cortex
and cerebellum is relatively constant among the different
species although the cerebellum usually has a lower K_D.
By contrast, the number of [³H]DHA binding sites vary over
10 fold in the cerebellum (from 3 to 40 pmol/g tissue)

TABLE 6

Characteristics of ³H-DHA binding in various mammalian
species.

Species	(n)	B_{max}, pmol/g tissue		K_D, nM	
		Cere-bellum	Cortex	Cere-bellum	Cortex
Horse	(2)	40+2	4 +1	0.24+.01	0.40+.02
Calf	(4)	17∓1	3.4∓0.4	0.22∓.04	0.90∓.04
Sheep	(3)	14∓4	<2	0.59∓.03	–
Goat	(2)	13∓3	–	0.50∓.06	–
Hamster	(8)	12∓1	11 +1	0.28∓.03	0.26+.02
Dog	(1)	11	4	0.26	0.53
Rat	(9)	4+.2	7 +0.4	0.32+.03	0.60+.05
Guinea Pig	(6)	3∓.2	4.5∓0.6	0.49∓.04	0.45∓.05

Values for calf and guinea pig cortex were taken from
U'PRICHARD, BYLUND & SNYDER (1978) and BYLUND (1978),
respectively. Other values from SUNN & BYLUND (1978).
(In Table 3 of U'PRICHARD, BYLUND & SNYDER, 1978 the
values listed under lung and cortex are backwards and
should be interchanged.)

and from less than 2 to 11 pmol/g tissue in the cortex.
The implications of these differences for the physiological actions of the catecholamines in these brain regions is unclear.

PHARMACOLOGICAL REGULATION

The area of adrenergic receptor regulation which is
receiving the greatest emphasis at the present time is
the effect of various in vivo drug treatments on the number of adrenergic receptors which is subsequently determined in vitro by the radioreceptor assay. An increase
in receptor number is often called supersensitivity while
a decrease, is termed subsensitivity. It appears that
such changes may be important both in disease states and
in the mechanism of action of numerous drugs. The effects
of acute drug treatment will be considered first, followed
by chronic drug treatment and the effects of neurotoxins.

Acute Drug Treatment

Acute drug treatment does not appear to alter β-adrenergic receptors. A single dose of the following drugs
did not significantly change the amount of β-receptor
binding using the radioreceptor assay: d-amphetamine
(BANERJEE et al., 1978c); ethanol (BANERJEE et al., 1978b);
trazodone (CLEMENTS-JEWERY, 1978) phenobarbital (WADDING-
HAM et al., 1978); propranolol (BYLUND, unpublished);
desipramine (WOLFE et al., 1978); doxepin and iprindole
(BANERJEE et al., 1977). The only reported exception to
this generalization is cocaine. Twelve hours after a dose
of 10 mg/kg, there was a 52% increase in ^{3}H-DHA binding,
although no increase was observed at either 1 or 24 hr
after injection (BANERJEE et al., 1978a).

Chronic Drug Treatment

During the past year, it has rapidly become apparent
that a large number of chronically administered drugs can
alter β-adrenergic receptor binding (Table 7). In all reported studies where saturation experiments were performed,
the change in binding was found to be due to an alteration
in the B_{max} and not in K_D. There are two particularly
interesting issues with respect to these experiments.
First, the time course of the effects both during treat-

TABLE 7

Influence of Chronic Drug Treatments on
β-Adrenergic Receptor Binding to Brain Membranes

Drug	Treatment days	Dose, mg/kg	Doses per day	% change	Bmax or single[a]	Comment[b]	Ref.
Propranolol	12	10	2	+35	Bmax		Wolfe, et al. (1978)
Phenobarbital	7	380	diet	+63	Single	Mouse	Waddingham, et al. (1978)
Reserpine	21	0.25	1	+51	Bmax	Forebrain, 1 day WD	U'Prichard & Snyder (1978b)
Reserpine	7	0.25	1	+28	Bmax	5 days WD	Bergstrom, et al. (1978)
Ethanol	60	13 g	diet	-10	Single	WB-CBL	Banerjee, et al. (1978b)
Ethanol	60	13 g	diet	+43	Bmax	48 hr WD; WB-CBL	Banerjee, et al. (1978b)
Cocaine	42	10	1	+142	Bmax	12 hr WD; WB	Banerjee, et al. (1978a)
D-Amphetamine	42	10	1	+82	Single	12 hr WD; WB-CBL	Banerjee, et al. (1978c)
D-Amphetamine	42	10	1	+54	Bmax	12 hr WD; WB	Banerjee, et al. (1978a)
D-Amphetamine	8	0.5	2	-20	Single		Sellinger, et al. (1978)
Pargyline	13	25	1	-13	Single		Wolfe, et al. (1978)
Desipramine	10	10	2	-38	Bmax		Wolfe, et al. (1978)
Desipramine	42	10	1	-38	Bmax	WB-CBL	Banerjee, et al. (1977)
Desipramine	3	10	2	-35	Bmax		Sellinger, et al. (1978)
Desipramine	3	10	2	-34	Bmax		Sarai, et al. (1978)
Doxepin	42	10	1	-31	Single	WB-CBL	Banerjee et al. (1977)
Iprindole	42	10	1	-42	Single	WB-CBL	Banerjee, et al. (1977)
Iprindole	11	10	2	-16	Single		Wolfe, et al. (1978)
Trazodone	25	10	1	-79	Single	24 hr; WD	Clements-Jewery (1978)
Tranylcypromine	16	5	1	-15	Single		Sellinger, et al. (1978)

[a] B_{max} indicates that multiple concentrations of the radioligand were used, and the number of binding sites calculated. Single indicates that only one concentration of radioligand was used and thus differences in binding could result from either a change in K_D or B_{max}.

[b] Cerebral cortex from rat except as indicated; WD, withdrawal; WB, whole brain; WB-CBL, minus cerebellum.

[c] Data from control rats was somewhat variable and tended to emphasize differences. Therefore an average control value was used to calculate percent change.

ment and after withdrawal, and secondly, possible mechanisms which are responsible for sub- and supersensitivity.

A major impetus for the study of the effects of
chronic treatment with antidepressants (e.g., desipramine)
on β-receptors is the discrepancy between the immediate
effect of the presumed biochemical action--inhibition of
norepinephrine uptake--and the 2 to 3 weeks needed for
clinical improvement (VETULANI et al., 1976). BANERJEE
et al. (1978a) studied β-adrenergic receptor binding in
rats treated daily with desipramine (10 mg/kg, i.p.).
After 10 days of treatment, binding was still at control
levels, but was significantly decreased by two weeks.
Maximal subsensitivity was observed after five weeks of
treatment. In a similar series of experiments (desipramine
10 mg/kg, i.p., twice daily), MOLINOFF et al. (1978) observed a small, but significant decrease in the number of
β-adrenergic receptor binding sites after 1 day of treatment (i.e., 2 injections) with maximal reduction evident
after seven days. They also found that the isoproterenol
stimulated adenylate cyclase activity decreased with a
similar time course. Similar observations have been made
by SARAI et al. (1978). Although these results differ
somewhat (perhaps due to the frequency of injections and
other experimental variables) they suggest that time is
needed for the decrease in receptors to develop, and thus
β-receptor subsensitivity may be important in the actions
of antidepressant drugs.

The time course of β-receptor effects after withdrawal from chronic drug treatment has also been studied
as a function of time. Reserpine treatment resulted in a
β-receptor supersensitivity which persisted for at least
five days after the final injection (BERGSTROM, TREISER &
KELLAR, 1978). After termination of desipramine treatment,
the decreased number of β-receptors gradually returns to
control levels over a five day period (MOLINOFF et al.,
1978). Thus, for these two drugs, no large changes would
be apparent for at least several days after withdrawal.
By contrast, withdrawal from ethanol, amphetamine and cocaine produced significant effects within 1-24 hours.
During ethanol intoxication, a reduction in β-receptors was
found. Twenty-four hours after withdrawal, this reduction
was reversed and, the number of receptors was greater than
control with a maximal increase being observed at 72 hr
(BANERJEE et al., 1978b). With amphetamine and cocaine
the changes were more rapid, with maximal increases in
β-receptor number 12 hr after withdrawal (BANERJEE et al.,
1978b).

The mechanisms which mediate changes in β-receptor number as a result of chronic drug treatment are still an open question. An attractive hypothesis is that the sensitivity of the receptors is inversely proportional to the concentration of norepinephrine which interacts at the receptor sites (RAFF, 1976). Thus propranolol which prevents the interaction of norepinephrine with the β-receptor causes an increase in receptor number (WOLFE et al., 1978). Chronic treatment with reserpine, which depletes norepinephrine from the neurons, would similarly result in increased β-receptor number (U'PRICHARD & SNYDER, 1978). On the other hand, compounds which would increase the norepinephrine concentration in the synapse such as pargyline and desipramine, decrease the number of β-receptors (WOLFE et al., 1978). Studies from Molinoff's laboratory using multiple drug treatments also support this hypothesis (WOLFE et al., 1978). As would be expected, concomitant treatment with desipramine and propranolol blocks the effect of desipramine. The effect of desipramine can also be blocked by the destruction of noradrenergic nerve terminals with 6-hydroxydopamine. Although the antidepressant, iprindole, does not appear to significantly block norepinephrine reuptake, the subsensitivity observed after chronic treatment with iprindole is also reversed by 6-hydroxydopamine, indicating that the pre-synaptic terminal is necessary for its action and suggesting that it may increase the norepinephrine concentration by another mechanism. d-Amphetamine and cocaine (which stimulate release of norepinephrine and block its reuptake) should cause a decrease in β-receptor number, but this is not necessarily the case. BANERJEE et al. (1978a) did not find subsensitivity after treatment with a high dose (10 mg/kg daily), although at a 10 fold lower dose (0.5 mg/kg twice daily) subsensitivity was observed (SELLINGER et al., 1978). Thus, it is not clear whether the experiments of BANERJEE et al. (1978a) represent an exception to the hypothesis, or whether the effective norepinephrine concentration was decreased under their conditions. The observed effects of "non-adrenergic" drugs on β-receptor might also be explained by this hypothesis. The supersensitivity seen after phenobarbital treatment may result from a reduction in the availability of norepinephrine (SABAN & DEMORAES, 1977; WESTFALL & FEDAN, 1975). In addition, the effects of norepinephrine metabolism during ethanol intoxication and withdrawal appear to be consistent with the changes seen in receptor sensitivity (FRENCH, PALMER & WIGGERS, 1975; BANERJEE et al., 1978b).

Thus, the effective concentration of norepinephrine
at the receptors definitely seems to be an important mechanism for the control of β-adrenergic receptor sensitivity.
Although it may not be the only mechanism, it is consistent
with most of the presently available experimental data,
and compelling evidence to the contrary is yet to be reported.

Neurotoxins

The effects of 6-hydroxydopamine on the β-adrenergic
receptors in rat cerebral cortex have been studied in considerable detail by Molinoff's laboratory. After intraventricular administration to mature rats, resulting in the
destruction of noradrenergic nerve terminals, the norepinephrine content decreased rapidly to 10% of control levels.
The binding of ^{125}I-iodohydroxybenzylpendolol gradually increased and reached a maximum of 150% of control by 16
days (SPORN et al., 1977). This increase in binding was
due to a 50% increase in the number of binding sites without a change in their affinity. Similar results were obtained by SKOLNICK et al. (1978) using ^{3}H-DHA. The effects
of neonatal 6-hydroxydopamine treatment have also been
studied. In the cerebral cortex, there was a 40 to 75%
increase in the number of β-adrenergic receptors, but the
time course of development was not changed (HARDEN et al.,
1977b). In contrast to the denervation in the cerebral
cortex, there is a hyperinnervation of the cerebellum, accompanied by a two-fold increase in norepinephrine content,
and 20% decrease in the number of β-adrenergic receptors
(JONSSON & HALLMAN, 1978).

In Yamamura's laboratory, the effect of specific 6-
hydroxydopamine lesions on endogenous norepinephrine and
β-receptor binding in specific brain regions is being investigated (YAMAMURA, personal communication). Lesions in
the dorsal bundle of the locus ceruleus increased binding
in the hippocampus and cerebral cortex, and decreased
binding in the cerebellum, apparently as a result of opposite changes in norepinephrine levels (Table 8). The lack
of decreased binding in the hypothalamus may be due to the
relatively modest decrease in norepinephrine concentration
(66% as compared to 95% in the cortex). Lesions of the
nigrostriatal pathway increased ^{3}H-DHA binding in the cortex
by 44%, but decreased binding in the striatum by 31%. The
levels of norepinephrine under these conditions have not
yet been determined.

TABLE 8

The influence of dorsal bundle lesions on ^{3}H-DHA
binding in rat brain membranes.

	^{3}H-DHA Binding pmol/mg prot.			Norepinephrine Concentration µg/g	
	Control	Lesion	% Change	Control	Lesion
Frontal Cortex	32+2	48+2	+50#	305+8	15+2
Hippocampus	15+1	20+2	+33*	432+9	17+1
Cerebellum	17+1	14+1	-18*	247+22	302+24
Hypothalamus	13+1	13+1	0	2793+87	914+82

Lesions produced by 6-hydroxydopamine; from YAMAMURA
(personal communication). *p<0.05; #p<0.01

The results with 6-hydroxydopamine also appear to be
consistent with the hypothesis that the number of β-adrenergic receptor sites is inversely related to the availability of norepinephrine.

α-Adrenergic Receptors

The effects of in vivo treatment with drugs and neurotoxins on the in vitro characteristics of α-adrenergic
receptors has not yet been investigated as fully as the
β-adrenergic receptor. Reserpine treatment, as would be
expected, results in an increase in the number of α-adrenergic binding sites in the forebrain (Table 9). Amitriptyline, a tricyclic antidepressant, which might be expected to decrease binding (due to the inhibition of norepinephrine uptake), also increases ^{3}H-WB 4101 binding.
This may be due to the potent inhibition of ^{3}H-WB 4101 receptor binding by amitriptyline (in vitro K_i of 24 nM;
U'PRICHARD, GREENBERG & SNYDER, 1978) which reduces the
effective concentration of norepinephrine at α-receptor
sites(analogous to the reduction in β-receptors after
propranolol treatment). Amitriptyline is much weaker at
^{3}H-epinephrine binding sites (U'PRICHARD, personal communication) and did not significantly change the number of

TABLE 9

Influence of chronic drug treatments on α-adrenergic receptor binding to brain membranes.

Drug	Treatment	Interval[a]	Brain Region	B_{max} or Single	% Change in B_{max} using			Ref.
					3H-Epi	3H-WB401	3H-Clo	
Reserpine	0.25 mg/kg, daily, 3 wk	1 day	Forebrain	B_{max}	+25*	+45‡		1
Amitriptyline	30 mg/kg, daily, 3 wk	1 day	Forebrain	B_{max}	+21	+51‡		2
6-hydroxydopamine	2x250 μg, i.c.v.	4 weeks	Cortex	B_{max}	+47‡	+46‡	+86*	3,4
			Striatum	Single			+317*	3
			Cerebellum	Single			+256‡	3

*p<0.05, ‡ p<0.01. 3H-Epi, 3H-epinephrine; 3H-Clo, 3H-clonidine (high affinity site)
[a]Interval between last injection and assay
References: 1. U'PRICHARD & SNYDER (1978b); 2. U'PRI-
     CHARD, GREENBERG & SNYDER (1978); 3. U'PRICHARD &
     SNYDER (1978c); 4. U'PRICHARD et al. (1979)

receptors labeled by 3H-epinephrine. Snyder's laboratory has also found that 6-hydroxydopamine treatment increases the number of α-receptor binding sites labeled by 3 radioligands (Table 9). Although increases were not found in all brain regions, in none of the brain regions studied was a decrease in binding observed (U'PRICHARD & SNYDER, 1978c). By contrast, SKOLNICK et al. (1978) did not find an increase in 3H-WB 4101 binding in the cerebral cortex after 6 hydroxydopamine treatment (250 μg, intraventricular), although 3H-DHA binding increased about 25%.

Although the α-receptor data are not nearly as extensive as those for the β-receptor, the synaptic level of norepinephrine may well be an important mechanism for the regulation of the number of α-receptors.

PATHOLOGICAL ALTERATIONS

As has been described in the preceding sections, the responsiveness of adrenergic receptors appears to be subject to regulation at multiple levels. Thus, it is reasonable to ask if changes in receptor characteristics might be important in the etiology or symptoms of one or more disease states. For several receptors, such relationships are being established (MELNECHUK, 1978).

Several years ago we studied the β-adrenergic receptors in postmortem brains of patients with Huntington's disease (chorea). Binding was significantly reduced in the globus pallidus (p<0.01) and non-significantly reduced in the caudate and putamen, but was the same as control in the cerebral cortex (ENNA et al., 1976a, b). Similarly, Yamamura's laboratory recently found a nonsignificant decrease in the caudate, and a significant 50% decrease (p<0.05) in the putamen (personal communication). In brains from schizophrenic patients, we did not find any consistent changes in β-adrenergic receptor binding (BENNETT et al., 1978). In brain membranes from spontaneously hypertensive rats, the number of α-adrenergic binding sites labeled by ^{3}H-WB 4101 was also significantly increased as compared to normotensive rats of the same strain over the 4-19 week age period (U'PRICHARD, GREENBERG & SNYDER, 1978). The K_D for ^{3}H-WB 4101 was also significantly increased in the hypertensive animals. In contrast, there were no significant differences in the number of α-adrenergic sites labeled by ^{3}H-clonidine. The significance, if any, of these changes to the various disease states is not known.

A recent revision of the catecholamine hypothesis of affective disorders suggests a pathology involving a supersensitive postsynaptic norepinephrine receptor coupled adenylate cyclase system (SULSER, VETULANI & MOBLEY, 1978). This was based on the supersensitivity of norepinephrine-stimulated adenylate cyclase activity seen following reserpine treatment (a drug associated with depression in humans), and a subsensitive response after antidepressant drugs (tricyclic antidepressants and MAO inhibitors) and electroconvulsive shock. Parallel changes are seen in the number of β-adrenergic receptors (Table 7). Based on similarities in the magnitude and time course of the decreases in adenylate cyclase activity and in β-adrenergic receptor number following chronic treatment with desipramine, it was concluded that the decreased density of β-receptors is responsible for the decrease in cyclase activity (WOLFE et al., 1978). This model of depression would predict that the increase in β-receptors caused by reserpine treatment should be reversed--at least partially--by antidepressant drugs and by electroconvulsive shock. Although such experiments studying the effect of antidepressant drugs on reserpine induced supersensitivity have not yet been reported, the effect of the combined treatment of reserpine and electroconvulsive shock is as predicted. BERGSTROM, TREISER & KELLER (1978) have found a reduced β-receptor density after chronic, but not

acute, electroconvulsive shock. In addition, they treated
rats with reserpine (0.25 mg/kg) for 7 days followed by 5
days of shock (50-100 mA, 200 msec) or 5 day control treatment (electrodes placed, but no current passed). The increase (as compared to rats receiving neither shock nor
reserpine) in the number of β-adrenergic receptors was
significantly lower in the shocked animals than in the
reserpine only animals (Table 10). There was no change
in K_D under any of the conditions.

Although the specific involvement of altered adrenergic receptors in CNS disorders has not yet been established, this should prove to be a fruitful area of research.
It is probable that changes more subtle than simply an
increase or decrease in receptor number may be involved.
Interesting examples from other systems include the decreased ability of aged rats to produce adaptive changes
in pineal gland β-receptors (GREENBERG & WEISS, 1978) and
the prevention of chronic haloperidol-induced supersensitivity of dopamine receptors by lithium chloride (PERT et
al, 1978).

TABLE 10

Influence of chronic treatment with reserpine and electroconvulsive shock on β-adrenergic receptor binding.

Treatment	B_{max}	% Change from control	% Difference from reserpinized
Acute			
Control	10.9+0.9		
Shock	7.8+0.6	−28*	
Chronic			
Control	12.3+0.4		
Reserpine	15.7+0.5	+28#	
Reserpine + Shock	14.2+0.3	+15#	−13*

*$p<0.02$, #$p<0.01$, data from BERGSTROM, TREISER & KELLER
(1978).

CONCLUDING REMARKS

Until recently, it has been generally assumed that neurotransmitter receptors were relatively constant elements in the synaptic events related to neuronal transmission, and thus, alterations in receptor properties have not been considered in relation to drug treatment, disease, or physiological state. The data reviewed here is clear evidence that this is not the case, and suggests that adrenergic receptors are regulated entities, responding to several different stimuli.

The changes in central adrenergic receptor density seen following drug treatment or electroconvulsive shock tend to have a relatively slow time course (days) and changes are usually not seen following a single treatment. This is in contrast to the more rapid changes which have been observed in pineal β-adrenergic receptors (AXELROD & ZATS, 1977) and in central benzodiazepine receptors (PAUL & SKOLNICK, 1978). There are however, significant exceptions to the above generalization. A single cocaine treatment, or withdrawal from chronic cocaine or amphetamine produces maximal increases in β-receptors within 12 hr. Presumably, changes in receptor density could result from several different biochemical mechanisms which would be correlated with the different time courses.

In the rat brain, the corpus striatum is the region which has the highest density ^{3}H-DHA binding sites (BYLUND & SNYDER, 1976). This is a rather anomalous finding in view of the limited noradrenergic innervation (GRZANNA et al., 1977) and low norepinephrine content (VERSTEEG et al., 1976) of this region. Since the development of postsynaptic β-receptors is apparently not regulated by the presynaptic nerve terminals (HARDEN et al., 1977b), this high density of receptors may, in part, actually be the result of the low norepinephrine concentration: a physiological example of denervation supersensitivity.

Although changes in the effective norepinephrine concentration at the receptor sites may account for most, if not all of the changes in receptor number which have been noted in this review, the molecular mechanisms involved are unknown. Several possibilities can be considered, the most obvious of which, is a change in receptor turnover secondary to changes in synthesis, or degradation (perhaps preceded by some inactivation mechanism). Receptors might exist in "active" and "inactive" states

which are interconvertible by biochemical (phosphorylation?) or physical (steric factors? membrane fluidity?)
mechanisms. As noted previously, an experimentally observed decrease in receptor number may be actually due
to a large decrease in affinity, with the result that
the receptors can no longer be labeled by current techniques. Conversely, the affinity of the receptor for
norepinephrine could markedly increase, with the result
that norepinephrine remains tightly bound, thus preventing
the binding of the labeled ligand in the radioreceptor
assay. Some of these proposed mechanisms may be only
in vitro artifacts, and thus caution is needed in ascrib-
ing physiological significance to changes observed by
the radioreceptor assay. Hopefully, the molecular mechanisms responsible for the changes in adrenergic receptors
will be elucidated over the next several years.

                        ACKNOWLEDGEMENTS

     The author wishes to thank several colleagues for
providing manuscripts and data prior to publication and
DR. DAVID U'PRICHARD for helpful discussions. Supported
in part by USPH grant MH31428.

                        REFERENCES

ANSELL, G.B. (1973) Phospholipids and the nervous system,
     in Form and Function for Phospholipids (Ansell, G.B.,
     Hawthorne, J.N. & Dawson, R.M.C., eds.), pp. 377-
     422. Elsevier Science, New York.
AXELROD, J. & ZATZ, M. (1977) The β-adrenergic receptor
     and the regulation of circadian rhythms in the pineal
     gland, in Biochemical Actions of Hormones, Vol. IV,
     pp. 249-268. Academic Press, New York.
BANERJEE, S.P., KUNG, L.S., RIGGI, S.J., CHANDA, S.K.
     (1977) Development of β-adrenergic receptor sub-
     sensitivity by antidepressants. Nature 268, 455-
     456.
BANERJEE, S.P., SHARMA, V.K. & CHANDA, S.K. (1978a) Psycho-
     active drugs and central β-adrenoceptors, in Recent
     Advances in the Pharmacology of Adrenoceptors
     (Szabadi, E., ed.). In press.
BANERJEE, S.P., SHARMA, V.K. & KHANNA, J.M. (1978b) Al-
     terations in β-adrenergic receptor binding during
     ethanol withdrawal. Nature 276, 407-409.
BANERJEE, S.P., SHARMA, V.K., KUNG, L.S. & CHANDA, S.K.
     (1978c) Amphetamine induces β-adrenergic receptor

supersensitivity. Nature 271, 380-381.

BARNETT, D.B., RUGG, E.L. & HANORSKI, S.R. (1978) Direct
    evidence of two types of β-adrenoceptor binding site
    in lung tissue. Nature 273, 166-168.

BENNETT, J.P., ENNA, S.J., BYLUND, D.B., GILLIN, J.C.,
    WYATT, R.J. & SNYDER, S.H. (1978) Neurotransmitter
    receptor binding alterations in schizophrenic brain.
    Arch. Gen. Psychiat. In press.

BENNETT, V., O'KEEFE, E. & CUATRECASAS, P. (1975) Mech-
    anisms of action of cholera toxin and the mobile
    receptor theory of hormone receptor-adenylate cyclase
    interactions. Proc. Natn. Acad. Sci. U.S.A. 72,
    33-37.

BERGSTROM, D.A., TREISER, S. & KELLAR, K.J. (1978) Electro-
    convulsive shock and reserpine: Effects on monoaminer-
    gic receptor binding in rat brain, in Catecholamines:
    Basic and Clinical Frontiers (Usdin, E., ed.). In
    press.

BYLUND, D.B. (1978) β-Adrenergic receptor binding in
    guinea pig cerebral cortex. Brain Research 152, 391-
    395.

BYLUND, D.B. & SNYDER, S.H. (1976) Beta-adrenergic recep-
    tor binding in membrane preparations from mammalian
    brain. Molec. Pharmacol. 12, 568-580.

BYLUND, D.B., TELLEZ-INON, M.T. & HOLLENBERG, M.D. (1977)
    Age-related parallel decline in β-adrenergic receptors,
    adenylate cyclase and phosphodiesterase activity in
    rat erythrocyte membranes. Life Sci. 21, 403-410.

CHANG, R.S.L., TRAN, V.T., PODUSLO, S.E. & SNYDER, S.H.
    (1978) Proc. Natn. Acad. Sci. U.S.A. In press.

CHARNESS, M.E., BYLUND, D.B., BECKMAN, B.S., HOLLENBERG,
    M.D. & SNYDER, S.H. (1976) Independent variation of
    β-adrenergic receptor binding and catecholamine-
    stimulated adenylate cyclase activity in rat erythro-
    cytes. Life. Sci. 19, 243-250.

CLARK, C., HOYLER, E. & DAVIS, J.N. (1977) Failure to find
    β-adrenergic receptor binding in guinea pig cerebral
    cortex with [³H]dihydroalprenolol. Brain Res. 127,
    313-316.

CLEMENTS-JEWERY, S. (1978) The development of cortical
    β-adrenoceptor subsensitivity in the rat by chronic
    treatment with trazodone, doxepin and mianserine.
    Neuropharmacology 17, 779-781.

CUATRECASAS, P. & HOLLENBERG, M.D. (1976) Membrane recep-
    tors and hormone action. Adv. Protein Chem. 30,
    251-451.

DALY, J. (1977) Cyclic Nucleotides in the Nervous System,
    pp. 7-13. Plenum Press, New York.

DIAMOND, J.M. & WRIGHT, E.M. (1969) Biological membranes:

The physical basis of ion and nonelectrolyte selectivity. Ann. Rev. Physiol. 31, 581-646.

ENNA, S.J., BENNET, J.P., BYLUND, D.B., SNYDER, S.H.,
BIRD, E.D. & IVERSEN, L.L. (1976a) Alterations of
brain neurotransmitter receptor binding in Huntington's chorea. Brain Res. 116, 531-537.

ENNA, S.J., BIRD, E.D., BENNETT, J.P., BYLUND, D.B.,
YAMAMURA, H.I., IVERSEN, L.L. & SNYDER, S.H. (1976b)
Huntington's chorea: Changes in neurotransmitter
receptors in the brain. New Eng. J. Med. 294,
1305-1309.

FIELDS, J.Z., ROESKE, W.R., MORKIN, E. & YAMAMURA, H.I.
(1978) Cardiac muscarinic cholinergic receptors. J.
Biol. Chem. 253, 3251-3258.

FRENCH, S.W., PALMER, D.S. & WIGGERS, D. (1975) Changes
in receptor sensitivity of the cerebral cortex and
liver during ethanol ingestion and withdrawal. Adv.
Exp. Med. Biol. 85, 515-538.

FONTAINE, R.N., HARRIS, R.A. & SCHROEDER, F. (1979)
Neuronal membrane lipid asymmetry. Life. Sci. In
press.

GREENBERG, D.A., U'PRICHARD, D.C., SHEEHAN, P. & SNYDER,
S.H. (1978) α-Noradrenergic receptors in the brain:
Differential effects of sodium on binding of ^{3}H-
agonists and ^{3}H-antagonists. Brain Res. 140, 378-384.

GREENBERG, L.H. & WEISS, B. (1978) β-Adrenergic receptors
in aged rat brain: Reduced number and capacity of
pineal gland to develop supersensitivity. Science
201, 61-63.

GRZANNA, R., MORRISON, J.H., COYLE, J.T. & MOLLIVER, M.E.
(1977) The immunohistochemical demonstration of noradrenergic neurons in the rat brain: The use of ho-
mologous antiserum to dopamine-β-hydroxylase. Neurosci. Lett. 4, 127-134.

HARDEN, T.K., WOLFE, B.B., SPORN, J.R., PERKINS, J.P. &
MOLINOFF, P.B. (1977a) Ontogeny of β-adrenergic receptors in rat cerebral cortex. Brain Res. 125,
99-108.

HARDEN, T.K., WOLFE, B.B., SPORN, J.R., POULOS, B.K. &
MOLINOFF, P.B. (1977b) Effects of 6-hydroxydopamine
on the development of the β-adrenergic receptor/
adenylate cyclase system in rat cerebral cortex. J.
Pharmac. Exp. Ther. 203, 132-143.

HENN, F., ANDERSON, D. & SELLSTROM, A. (1977) Possible
relationship between glial cells, dopamine and the
effects of antipsychotic drugs. Nature 266, 637-638.

HIRATA, F., AXELROD, J. & STRITTMATTER, J. (1978) Methyla-
tion of membrane phospholipids, in Catecholamines:
Basic and Clinical Frontiers (Usdin, E., ed.).
In Press.

HIRATA, F., STRITTMATTER, W.J. & AXELROD, J. (1978) Proc. Natn. Acad. Sci., U.S.A. In press.

INSEL, P.A., MAGUIRE, M.E., GILMAN, A.G., BOENE, H.R., COFFINO, P. & MELMON, K.I. (1976) Beta-adrenergic receptors and adenylate cyclase: Products of separate genes? Mol. Pharmacol. 12, 1062-1069.

JONSSON, G. & HALLMAN, H. (1978) Changes in β-receptor binding sites in rat brain after neonatal 6-hydroxy-dopamine treatment. Neurosci. Lett. 9, 27-32.

LANDS, A.M., ARNOLD, A., MCAULIFF, J.P., LUDEUNA, F.P. & BROWN, T.G. (1967) Differentiation of receptor systems activated by sympathomimetic amines. Nature 214, 597-598.

LEON, A., BENVEGNU, D., TOFFANO, G., ORLANDO, P. & MASSARI, P. (1978) The effect of brain cortex phospholipids on adenylate cyclase activity of mouse brain. J. Neurochem. 30, 23-26.

MAGGI, A., SCHMIDT, M.J., GHETTI, B. & ENNA, S.J. (1978) Effect of aging on neurotransmitter receptor binding in rat and human brain. Life Sci. In press.

MAGUIRE, M.E., ROSS, E.M. & GILMAN, A.G. (1977) β-Adrener-gic receptor: Ligand binding properties and the interaction with adenylyl cyclase. Adv. Cyc. Nuc. Res. 8, 1-83.

MELNECHUK, T. (1978) Cell Receptor Disorders. Western Behavioral Sciences Intitute, La Jolla, California.

MOLINOFF, P.B., SPORN, J.R., WOLFE, B.B. & HARDEN, T.K. (1978) Regulation of β-adrenergic receptors in the cerebral cortex. Adv. Cyc. Nuc. Res. 9, 465-483.

MORENO, J.H. & DIAMOND, J.M. (1975) Nitrogenous cations as probes of permeation channels. J. Membrane Biol. 21, 197-259.

ORLY, J. & SCHRAMM, M. (1975) Fatty acids as modulators of membrane functions: Catecholamine-activated adenylate cyclase of the turkey erythrocyte. Proc. Natn. Acad. Sci., U.S.A. 72, 3433-3437.

ORLY, J. & SCHRAMM, M. (1976) Coupling of catecholamine receptor from one cell with adenylate cyclase from another cell by cell fusion. Proc. Natn. Acad. Sci., U.S.A. 73, 4410-4414.

PAUL, S.M. & SKOLNICK (1978) Rapid changes in brain benzo-diazepine receptors after experimental seizures. Science 202, 892-894.

PERT, A., ROSENBLATT, J.E., SIVIT, C., PERT, C.B. & BUNNEY, W.E., JR. (1978) Long-term treatment with lithium prevents the development of dopamine receptor sensi-tivity. Science 201, 171-173.

RAFF, M. (1976) Self regulation of membrane receptors. Nature 259, 265-266.

RECHES, A., EBSTEIN, R.P. & BELMAKER, R.H. (1978) The
    differential effect of lithium on noradrenaline- and
    dopamine-sensitive accumulation of cyclic AMP in
    guinea pig brain. Psychopharmacology 58, 213-216.
ROSENTHAL, H.E. (1967) Graphic method for the determina-
    tion and presentation of binding parameters in a
    complex system. Anal. Biochem. 20, 525-532.
SABAN, R. & DEMORAES, S. (1977) Noradrenergic supersensi-
    tivity in the rat vas deferens during barbital with-
    drawal. Eur. J. Pharm. 43, 195-198.
SARAI, K., FRAZER, A., BRUNSWICK, D. & MENDELS, J. (1978)
    Desmethylimipramine-induced decrease in β-adrenergic
    receptor binding in rat cerebral cortex. Biochem.
    Pharmacol. 27, 2179-2181.
SCHWARTZ, J.C., COSTENTIN, J., MARTRES, M.P., PROTAIS, P.
    & BAUDRY, M. (1978) Modulation of receptor mechanisms
    in the CNS: Hyper- and hypo-sensitivity to cate-
    cholamines. Neuropharmacology 17, 665-685.
SELLINGER, M., SARAI, K., FRAZER, A., MENDELS, J. & HESS,
    M.E. (1978) β-Adrenergic receptor binding in rat
    cerebral cortex after repeated administration of
    psychotropic drugs. Fed. Proc. 37, 309.
SKOLNICK, P., STALVEY, L.P., DALY, J.W., HOYLER, E. & DAVIS,
    J.N. (1978) Binding of α- and β-adrenergic ligands to
    cerebral cortical membranes: Effect of 6-hydroxydopa-
    mine treatment and relationship to the responsiveness
    of cyclic-AMP generating systems in two rat strains.
    Eur. J. Pharmacol. 47, 201-210.
SNYDER, S.H., U'PRICHARD, D.C. & GREENBERG, D.A. (1978)
    Neurotransmitter receptor binding in the brain, in
    Psychopharmacology: A Generation of Progress (Lipton,
    M.A. et al., eds.), pp. 361-370. Raven Press, New
    York.
SPORN, J.R., WOLFE, B.B., HARDEN, T.K. & MOLINOFF, P.B.
    (1977) Supersensitivity in rat cerebral cortex:
    Pre- and post-synaptic effects of 6-hydroxydopamine
    at noradrenergic synapses. Mol. Pharm. 13, 1170-
    1180.
STRITTMATTER, W.J., HIRATA, F. & AXELROD, J. (1978) Phos-
    pholipid methylation unmasks cryptic β-adrenergic
    receptors in rat reticulocytes. Science. In press.
SULSER, F., VETULANI, J. & MOBLEY, P.L. (1978) Mode of
    action of antidepressant drugs. Biochem. Pharmacol.
    27, 257-261.
SUNN, L. & BYLUND, D.B. (1978) Comparison of ^{3}H-dihydro-
    alprenolol binding to putative β_1- and β_2-adrenergic
    receptors in the CNS of several mammalian species,
    in Catecholamines: Basic and Clinical Frontiers
    (Usdin, E., ed.). In press.

TSAI, B.S. & LEFKOWITZ, R.J. (1978) Agonist-specific
    effects of nonvalent and divalent cations on adenylate
    cyclase-coupled α-adrenergic receptors in rabbit
    platelets. Mol. Pharmacol. 14, 540-548.
U'PRICHARD, D.C., BECHTEL, W.D., ROUCOT, B. & SNYDER, S.H.
    (1979) Multiple α-noradrenergic receptor binding
    sites in rat brain: Effect of 6-hydroxydopamine.
    Mol. Pharmacol. In press.
U'PRICHARD, D.C., BYLUND, D.B. & SNYDER, S.H. (1978) (+)-
    [H]-Epinephrine and (-)-[^{3}H]-dihydroalprenolol
    binding to β_1 and β_2 noradrenergic receptors in brain,
    heart and lung membranes. J. Biol. Chem. 253, 5090-
    5102.
U'PRICHARD, D.C., GREENBERG, D.A. & SNYDER, S.H. (1978)
    CNS α-adrenergic receptor binding: Studies with
    normotensive and spontaneously hypertensive rats, in
    Perspectives in Nephrology and Hypertension (Schmidt,
    H. & Meyer, P., eds.). Wiley in press.
U'PRICHARD, D.C. & SNYDER, S.H. (1978a) Guanyl nucleotide
    influences on ^{3}H-ligand binding to α-noradrenergic
    receptors in calf brain membranes. J. Biol. Chem.
    253, 3444-3452.
U'PRICHARD, D.C. & SNYDER, S.H. (1978b) ^{3}H-Catecholamine
    binding to α-receptors in rat brain: Enhancement by
    reserpine. Europ. J. Pharmacol. 51, 145-155.
U'PRICHARD, D.C. & SNYDER, S.H. (1978c) ^{3}H-Clonidine
    binding to two α-adrenergic sites in rat CNS: Dif-
    ferential effects of 6-hydroxydopamine, in Presynap-
    tic Receptors (Langer, S.Z. et al., eds.). Pergamon,
    in press.
U'PRICHARD, D.C. & SNYDER, S.H. (1978d) Catecholamine
    binding to CNS adrenergic receptors. J. Supramol.
    Struct. In press.
U'PRICHARD, D.C. & SNYDER, S.H. (1978e) Nucleotide and
    ion regulation of CNS adrenergic receptors, in
    Recent Advances in the Pharmacology of Adrenoceptors
    (Szabadi et al., eds.), pp. 153-162. Elsevier.
VERSTEEG, D.H.G., GUGTEN, J.V.D., DEJONG, W. & PALKOVITS,
    M. (1976) Regional concentrations of noradrenaline
    and dopamine in rat brain. Brain Res. 113, 563-574.
VETULANI, J., STAWARZ, R.J., DINGELL, J.V. & SULSER, F.
    (1976) A possible common mechanism of action of anti-
    depressant treatments. N.-S. Arch. Pharmacol 293,
    109-114.
WADDINGHAM, S., RIFFEE, W.R., BELKNAP, J.K. & SHEPPARD,
    J.R. (1978) Barbiturate dependence in mice: Evidence
    for β-adrenergic receptor proliferation in brain.
    Res. Comm. Chem. Path. Pharm. 20, 207-220.

WESTFALL, D.P. & FEDAN, J.S. (1975) The effect of pretreatment with 6-hydroxydopamine on the norepinephrine
concentration and sensitivity of the rat vas deferens.
Eur. J. Pharm. 33, 413-417.

WILLIAMS, L.T. & LEFKOWITZ, R.J. (1978) Receptor Binding
Studies in Adrenergic Pharmacology. Raven Press,
New York.

WOLFE, B.B., HARDEN, T.K. & MOLINOFF, P.B. (1977) In
vitro study of β-adrenergic receptors. Ann. Rev.
Pharmacol. Toxicol. 17, 575-604.

WOLFE, B.B., HARDEN, T.K., SPORN, J.R. & MOLINOFF, P.B.
(1978) Presynaptic modulation of β-adrenergic receptors in rat cerebral cortex following treatment with
antidepressants. J. Pharmac. Exp. Ther. 207, 446-
457.

YAMAMURA, H.I., ENNA, S.J. & KUHAR, M.J., eds. (1978)
Neurotransmitter Receptor Binding. Raven Press,
New York.

REGULATION OF THE NEUROBLASTOMA X GLIOMA HYBRID OPIATE

RECEPTORS BY Na$^+$ AND GUANINE NUCLEOTIDES

Arthur J. Blume, Gloria Boone and David Lichtshtein

Department of Physiological Chemistry and
Pharmacology, Roche Institute of Molecular Biology
Nutley, New Jersey   07110

INTRODUCTION

Neuroblastoma x glioma hybrid NG108-15 cells have
saturable, high affinity, sterospecific opiate receptors
(KLEE & NIRENBERG, 1974; BLUME, et al., 1977).   In viable
cells, opiates lower the intracellular concentration of
cAMP as well as prevent full elevation of cAMP by activators (i.e., PGE$_1$ and adenosine) of adenylate cyclase.
Inhibition of adenylate cyclase activity by opiates can be
directly demonstrated in vitro with NG108-15 cell membranes.
These actions of opiates appear to require agonist occupation of the opiate receptor and they are blocked by the
specific opiate antagonist naloxone (SHARMA, NIRENBERG &
KLEE, 1975; SHARMA, KLEE & NIRENBERG, 1975 & 1977).   However, the affinities of agonists under adenylate cyclase
assay conditions were generally 10-100 times poorer than
those observed with viable intact cells suspended in an
isotonic sucrose/Tris/HCl buffer.   No such discrepancies
were observed with the opiate antagonist naloxone.   A
priori, this specific alteration in opiate agonist affinity could be due either to (i) changes in the opiate
receptor per se after cell homogenization or (ii) the presence of high concentrations of cations or nucleotide
triphosphates in the adenylate cyclase assays.

Studies of ionic influences on opiate receptor binding (using rat brain membranes) have shown that certain
monovalent cations selectively decrease opiate agonist
binding.   This monovalent cation effect is best seen with

Na+, less with Li$^+$ and not with K$^+$ or choline$^+$. In contrast, the binding of a pure opiate antagonist is not decreased by cations (PERT, et al., 1973; SIMON & GROTH,
1975; PASTERNAK, et al., 1975; SIMANTOV & SNYDER, 1976).
Since the adenylate cyclase assays used by SHARMA, NIREN-
BERG & KLEE (1975) contain 20 mM creatine phosphate as a
di-Na$^+$ salt, sodium is a likely candidate for the cation
within the adenylate cyclase assays which selectively affects agonist binding.

The two nucleoside triphosphates which are present in
the adenylate cyclase assays in appreciable amounts are
ATP, which is used as enzyme substrate, and GTP, which is
a regulator of enzyme activity. Of these two, it appears
that the guanine nucleotide might very well regulate opiate
receptors in an agonist-specific manner. Previously published data has shown that GTP regulates a number of receptors (i.e., β-noradrenergic, and prostaglandin E_1)
which themselves regulate adenylate cyclase. The occupation of these receptors by their respective agonists causes
an activation of adenylate cyclase, and GTP uniformly decreases the affinity of the agonist but not of the antagonist at these receptors (BRUNTON, et al., 1976; MAGUIRE,
et al., 1976; LEFKOWITZ, et al., 1977; CHHABIRANI & LEF-
KOWITZ, 1976; LAD, et al., 1977).

This paper summarizes recent work in our laboratory
proving that Na$^+$ and guanine nucleotides are potent
agonist-specific effectors of the opiate receptors of
NG108-15 cells. The initial observation related to this
work has been published (BLUME, 1978a).

EXPERIMENTAL PROCEDURES: The mouse neuroblastoma x rat
glioma hybrid clone NG108-15 was grown as previously described (KLEE & NIRENBERG, 1974; BLUME, et al., 1977; BLUME,
1978a). For studies with intact cells, cells were shaken
off flasks and washed in 0.32 M sucrose/10 mM Tris/HCl
buffer, pH 8, 0.5 mM $CaCl_2$ and 1mM DTT. For preparation
of cell membranes, cells were washed twice with 50 mM
Tris/HCl buffer, pH 7.4, suspended in the same buffer and
homogenized at 4°. The ensuing membrane preparation was
washed once and finally suspended in the same buffer at
0.2-1.0 mg protein/ml. Binding assays with intact cells
were carried out at 37° as previously described (KLEE &

*ABBREVIATIONS USED: DTT, dithiothrietol; LeuENK, leu^5
enkaphalin; DHM, dihydromorphine; metENK, met^5 enkephalin;
DAMA, the met^5 enkephalin analogue Dala^2 met^5 NH_2 ;
Gpp(NH)p, guanyl-5'-ylimidodiphosphate.

NIRENBERG, 1974).   Assays with membranes were conducted at 32° or ·37° in 50 mM Tris/HCl buffer, pH 7.4 (other additions as noted in text), in a final volume of 0.5 ml and terminated by rapid filtration over Whatman GF/B filters (BLUME, et al., 1977; BLUME, 1978a).

In our initial survey of NG108-15 cells, we found that the opiate receptors in membrane preparations obtained from cell homogenates are quite similar to those in intact cells.   Some of the data obtained by competition binding studies in support of this conclusion are shown in Table 1.

Table 1.   Comparison of Opiate Affinity in Intact Cells and Membranes

		Non-radioactive ligand affinity (nM $K_d \pm$ S.D.)				
		$[^3H]$leuENK	$[^3H]$DAMA	$[^3H]$Etorphine	$[^3H]$DHM	$[^3H]$Naloxone
A.	Intact Cells					
	Morphine	110 + 10	N.D.	90 + 11	N.D.	160 + 20
	Naloxone	40 + 5	N.D.	40 + 5	N.D.	26 + 4
	leuENK	6 + 1	N.D.	N.D.	N.D.	N.D.
B.	Membranes					
	Morphine	N.D.	160 + 20	N.D.	115 + 11	115 + 10
	Naloxone	N.D.	N.D.	N.D.	N.D.	36 + 4
	DAMA	N.D.	N.D.	N.D.	108 + 1	1 + 1

Assays with both intact cells and membranes were carried out at 37° for 20 min.   S.D. = Standard deviation (n ≥ 3); N.D. not determined. K_d values were determined from the relationship:

$$K_d = IC_{50}/L + \frac{[^3H]\text{ligand}}{K_d\,[^3H]\text{ligand}}.$$

The opiate receptors in both intact cells suspended in
Tris buffer and sucrose and membranes suspended in Tris
buffer have a much higher affinity for the peptide agonist
leuENK and hydrolysis-resistant metENK analogue DAMA than
for the opiate alkaloid morphine. Furthermore, the affinity of a given opiate (agonist or antagonist) appears the
same regardless of whether the [^{3}H]ligand used in the competition binding assay is an agonist or an antagonist.
The above supports the notion that there is only one basic
type of opiate receptor in NG108-15. More importantly,
the affinity of morphine for these receptors is the same
whether affinity is monitored with intact cells or isolated membranes. Finally, there has been no significant
loss of receptors due to cell homogenization as the density of opiate receptors was found to be 1500-2400 fmol/
mg protein in the membrane preparations and 600-800 fmol/
mg protein of intact cells.

Since cell disruption does not apparently alter the
opiate receptor, we investigated the influences of Na$^+$
and nucleotides on opiate binding. As shown in Figure 1,
Na$^+$, as well as the hydrolysis-resistant GTP analogue
Gpp(NH)p, decreases opiate agonist binding in membrane
preparations, Na$^+$ inhibition, in the absence of Gpp(NH)p
and at physiological concentration (i.e., 150 mM), appears
to be 50%, and half maximum at 17 mM. Studies on the
cation specificity indicate that Li$^+$ and K$^+$ inhibit with
K$_i$s of 60-70 mM and Choline$^+$ with a K$_i$ of $\geq$ 170 mM. Na$^+$
also inhibits the binding of [^{3}H]DHM and [^{3}H]DAMA to about
50%, yet inhibits $\geq$ 10% of [^{3}H]naloxone or [^{3}H]naltrexone
binding (see later discussion). With regard to the effects of guanine nucleotides, maximum Gpp(NH)p inhibition
of agonist binding is also 50% (above 1 mM Gpp(NH)p the
increased inhibition observed is probably due to the fact
that each mole of nucleotide contains two moles of Na$^+$).
Previously published data (BLUME, 1978a,b) showed that
only GTP, GDP, ITP and Ipp(NH)p can effectively replace
Gpp(NH)p as an effector of opiate binding. Adenine nucleotides cannot replace Gpp(NH)p in this respect. This
specificity agrees with that observed for nucleotide control over the β-adrenergic and prostaglandin receptors
(BRUNTON, et al., 1976; MAGUIRE, et al., 1976; LEFKOWITZ,
et al., 1977; CHHABIRANI & LEFKOWITZ, 1978; LAD, et al.,
1977). The inhibition by Na$^+$ and Gpp(NH)p of opiate
agonist binding to NG108-15 membrane is clearly additive.

Previously published data on the effect of Na$^+$ on
rat brain opiate receptors and guanine nucleotides on
β-adrenergic receptors had shown that both agents

dramatically increase the dissociation rate of agonists.
We therefore investigated what are the effects of addition
of Na+ and Gpp(NH)p on pre-equilibrated [³H]opiate agonist-
receptor complexes. The data shown in Fig. 2 clearly indicate that both Na+ and Gpp(NH)p induce a large percentage (50-80%) of the bound agonist to dissociate rapidly.
The effects of both Na+ and nucleotide on dissociation
are also additive.

If these effects of Na+ and guanine nucleotides are
to account for the specific loss of agonist affinity observed under adenylate cyclase assay conditions, then they
must be agonist specific. This seems to be true for Na+,
as the cation does not significantly (i.e., ≤ 10%) inhibit
[³H]naloxone (unpublished data) or [³H]naltrexone binding
(Fig. 3). In contrast, Na+ clearly reduces the affinity

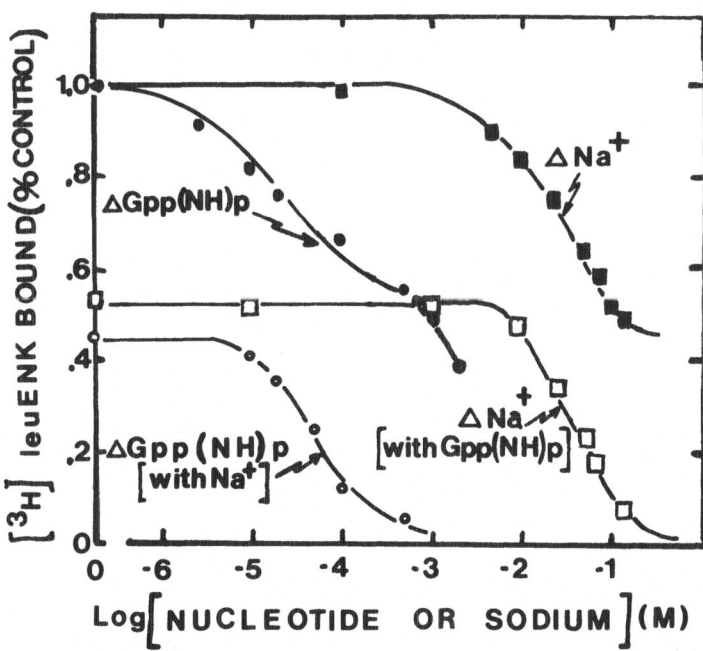

Fig. 1: Gpp(NH)p and Na+ Inhibition of [³H]leuENK Binding.
        Cell membranes were incubated in Tris/HCl buffer
        (pH 7.4) with 2 nM [³H]leuENK. 1 = 100% specific
        [³H]leuENK binding. Δ = increasing concentrations.
        [   ] = concentration fixed; Na+ at 135 mM and
        Gpp(NH)p at 0.5 mM.

of about half the receptors for the [3H]agonists DAMA or
leuENK. This loss in total agonist binding sites is an
artifact of the assay system. These sites now have a much
lower affinity for agonists, and can be shown to exist in
the presence of Na$^+$ by monitoring the ability of the
agonist to compete with [3H]naltrexone for binding (see
below).

The agonist specificity of the Gpp(NH)p effect is not
as readily apparent from inspection of Figure 3. On one
hand, in the presence of Na$^+$ and Gpp(NH)p, all of the receptors have reduced affinity for the agonist (i.e., K_d
for DAMA increases fivefold) and about half have increased
affinity for the antagonist (i.e., K_d for naltrexone decreases two-fold). On the other hand, about half of the

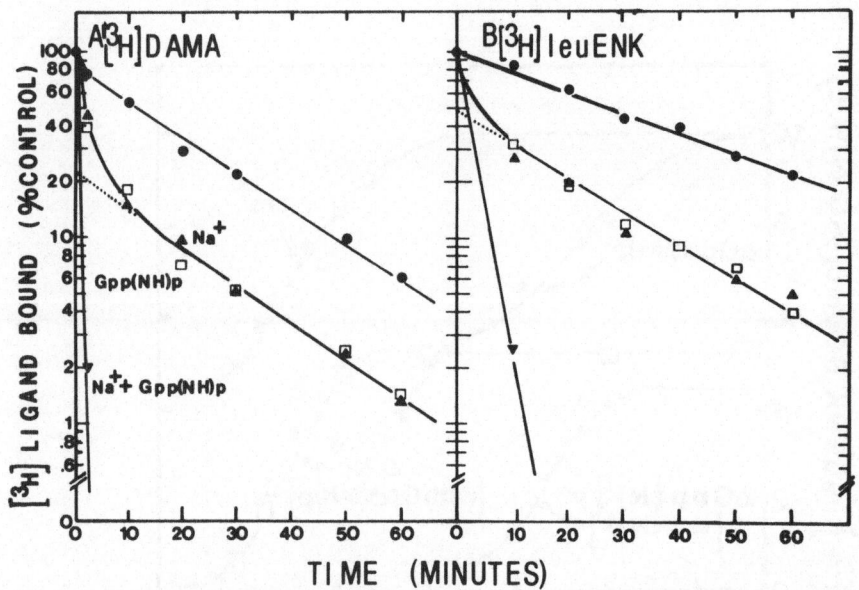

Fig. 2: Na$^+$ and Gpp(NH)p Induce Rapid Dissociation of
        [3H] Opiate Agonists.
[3H]DAMA (4 nM) or [3H]leuENK (2 nM) were allowed
to equilibrate for 10 min. at 32° in Tris/HCl at
pH 7.4 with cell membranes and then dissociation of
the specifically bound [3H]ligand was followed with
time (t = o) after addition of only 10 μM naloxone
or naloxone plus either Na$^+$ (135 mM) (▲); Gpp(NH)p
(0.5 mM) (▢) or both Na$^+$ + Gpp(NH)p (▼).

binding sites for [3H]naltrexone also disappear when
Gpp(NH)p is added in addition to Na+ .   This Gpp(NH)p-
induced decrease in [3H]naltrexone binding observed in the
presence of Na+ is also seen for [3H]naloxone binding (Fig.
4).   In fact, nucleotide inhibition of antagonist and agonist binding has the same IC_{50} value, and both appear to be
equally cooperative events (Fig. 4).   It appears, therefore, that nucleotides directly affect a fraction (about
half) of the NG108-15 opiate receptors so as to decrease

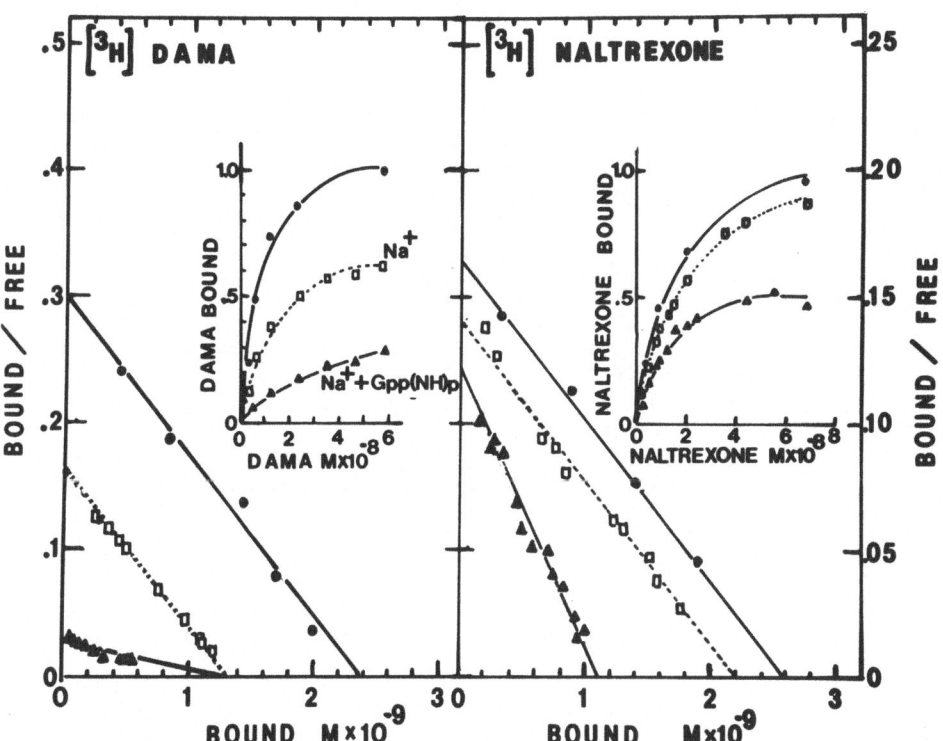

Fig. 3: Differences in the Effects of Na+ and Gpp(NH)p
on Opiate Agonist versus Antagonist Binding.
Specific binding assessed after 20 min. at increasing
concentrations of [3H]ligand in Tris buffer alone (o)
or with 135 mM Na+ (□) or Na+ and 0.5 mM Gpp(NH)p
(▲).   The large figures are Scatchard plots, and the
inserts are plots of specific binding vs. [Ligand],
the amount of ligand bound is given as fmol/0.5 mg
protein.

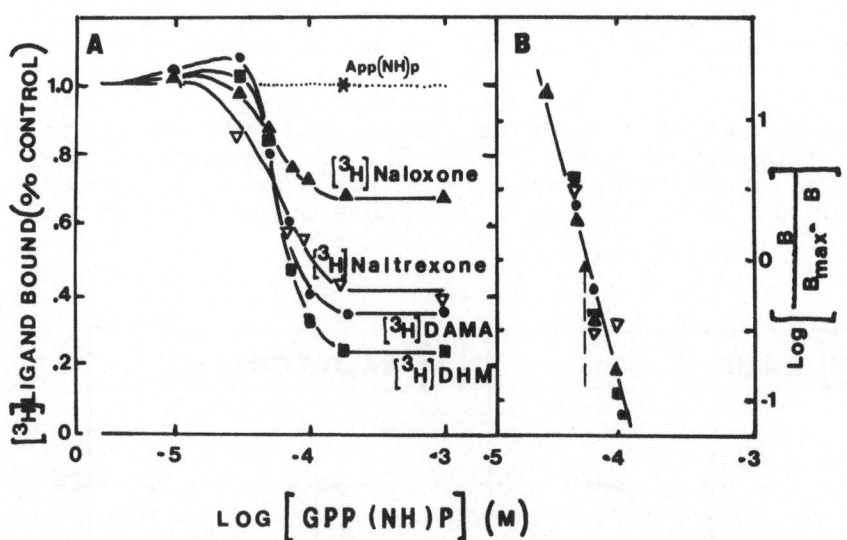

Fig. 4: Inhibition of Opiate Binding by Gpp(NH)p.
Specific binding of [3H]ligands (2-10 mM) in the
presence of 135 mM Na+ was monitored with cell membranes at 32°. Panel A: 1 = 100% specific binding
in the presence of 135 mM Na+ . Panel B: Hill plot
of nucleotide-directed inhibition; arrow indicated
the IC_{50} for Gpp(NH)p.

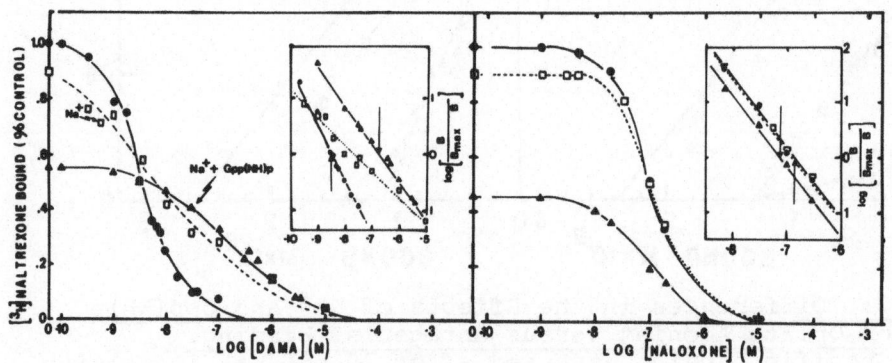

Fig. 5: Ligand Competition for [3H]Naltrexone Binding.
Specific binding of [3H]naltrexone (4 nM) was monitored in Tris buffer alone (●), with 135 mM Na+ (□),
or with 0.5 mM Gpp(NH)p and Na+ (▲) at increasing
concentrations of DAMA or naloxone. Insets are Hill
plots of data, and the arrows indicate the position
of the IC_{50} dose. 1 = 100% specific binding in Tris
buffer.

Table 2. Effects of Na$^+$ and Gpp(NH)p on Opiate Receptors

Ligand	nM IC$_{50}$ Additions:				Na$^+$Shift** $\frac{Na^+ + "G"}{"G"}$	"G"Shift** $\frac{"G"}{None}$	Na$^+$ + "G" Shift** $\frac{Na^+ + "G"}{None}$
	None	Na$^+$	Gpp(NH)p	Na$^+$ + Gpp(NHp)			
Antagonists							
Naltrexone	25(15.8*)	25(15.8*)	20(15.2*)	20(9.5*)	1.0	0.8	0.8
Naloxone	100	100	80	80	1.0	0.8	0.8
Mixed Antagonists: Agonists							
Levallorphan	5	Hetero	6.3	14	2.2	1.3	2.8
Cyclazocine	7.5	18	8.4	28	3.3	1.1	3.7
Agonists							
Fentenyl	200	Hetero	600	3700	6.2	3.0	18.5
Oximorphone	105	Hetero	167	1050	6.3	1.6	10.0
Morphine	200	Hetero	400	5000	12.5	2.0	25.0
Dala^{2}met^{5}NH$_2$	2.5	Hetero	3.8	200	53	1.5	80.0
Etorphine	1	Hetero	0.8	53	66	0.8	53.0

Assayed on cell membranes at 32° with 2-4 nM [^{3}H]naltrexone in 50 mM Tris/HCl buffer (pH 7.4). Where indicated, Na$^+$ (135 mM) and/or Gpp(NH)p (0.5 mM) was also present. G = Gpp(NH)p. *Apparent K$_d$ values. Hetero = Hill plots indicate heterogeneous competition (i.e. = n$_H$ << 1.0). **Shifts are the ratio of the observed IC$_{50}$ values noted in parentheses.

the binding of all opiate ligands at these sites. However, the other half of the receptor population appears to have had its agonist affinities decreased yet its antagonist affinities increased.

A detailed analysis of ligand competition against [3H]naltrexone binding was undertaken to better demonstrate the specificity of the nucleotide and Na^+ effects on opiate receptors. An example of the data obtained is given in Fig. 5, and a summary of all of the data is given in Table 2. The following are immediately apparent: (i) In the absence of Gpp(NH)p or Na^+ the competition by either agonist or antagonist results in a homogeneous displacement of the [3H]naltrexone (Hill slopes $[n_H] \cong 1.0$). (ii) In the presence of Na^+ the competition by the agonist only becomes heterogeneous (n<<1.0). This agrees with the earlier data (c.f. Figs. 1 and 3) which show that by itself Na^+ decreases agonist binding at only half of the receptor sites. (iii) The addition of Gpp(NH)p and Na^+ together depresses [3H]antagonist binding about 50% (this is due to the non-specific effect of Gpp(NH)p on half the receptor population). In addition, the competition with the remaining [3H]naltrexone binding is again homogeneous ($n_H \cong 1.0$) yet now indicates a much higher K_d for agonists and unaltered or even slightly lower K_d for antagonists. Since the displacements by the agonist in the presence of Gpp(NH)p and Na^+ show $n_H \cong 1$, a comparison of the two IC_{50} values obtained under these two conditions is appropriate, and an agonist-specific Na^+ shift (decrease in agonist affinity) becomes readily apparent. The Na^+ shifts (Table 2), although present, are very small for the mixed agonist-antagonists. Furthermore, the magnitude of the Na^+ shift differs dramatically even among the "pure" agonists. A similar observation was reported for the opiate receptors in rat brain (PERT & SNYDER, 1974).

What is also apparent from inspection of Table 2 is that there is a small but significant decrease in opiate affinity due to the presence of Gpp(NH)p, and this change is apparently agonist specific. If one then calculates the total shift (i.e., loss in agonist affinity) due to the combined presence of Na^+ and nucleotide, values as high as 80-fold are obtained. Interestingly, the affinity of morphine (i.e., $\sim$ 5 μM) and etorphine ($\sim$50 nM) under the latter conditions are remarkably close to the estimate of their receptor affinity based on their capacity to inhibit NG108-15 adenylate cyclase! Therefore, as guanine nucleotides and Na^+ are present in living cells, it is then possible that the low affinity state represents the true physiological state (i.e., receptor coupled to some responsive system) of the opiate receptor in these cells.

In summary, the opiate receptors of NG108-15 cells are regulated by Na^+ and guanine nucleotides. The Na^+ effect is agonist specific and matches that found with brain opiate receptors. There are two discernible effects of guanine nucleotides: (i) a nonspecific reduction of ligand binding at about one-half of the opiate receptors, and (ii) a specific reduction of agonist binding and increased binding of antagonists at the other half of the opiate receptor population. The reduced ability of opiate agonists to inhibit adenylate cyclase as compared to their ability to bind to membranes suspended in Tris buffer is most likely due to the presence in the cyclase assays of Na^+ and GTP. It will be extremely interesting to see whether the presence of Na^+ and/or guanine nucleotide is in fact a requirement for opiate inhibition of adenylate cyclase. Such an observation would support the notion that the lower affinity state of these receptors is the actual physiological state. Our initial observations of a similar regulation of brain opiate receptors by guanine nucleotides (BLUME, 1978b) have not only been confirmed by CHILDERS & SNYDER (1978) but extended so as to clearly indicate that nucleotides can cause a specific decrease in opiate agonist affinity. The similarity between the brain and NG108-15 cells' opiate receptors with regard to their regulation by ions and nucleotides is remarkable. However, it remains to be proven if the action of opiate on the brain is also mediated by cyclic nucleotides.

REFERENCES

BLUME, A. J. (1978a) Life Sci. 22, 1843-1852.
BLUME, A. J. (1978b) Proc. Nat. Acad. Sci. USA 75, 1713-1717.
BLUME, A. J., SHORR, J., FINBERG, J. P. M. and SPECTOR, S. (1977) Proc. Nat. Acad. Sci. USA 74, 4927-4931.
BRUNTON, L. L., WIKLUND, R. A., VAN ARSDALE, P. M. and GILMAN, A. G. (1976) J. Biol. Chem. 251, 3037-3044.
CHHABIRANI, M. and LEFKOWITZ, R. J. (1976) Proc. Nat. Acad. Sci. USA 73, 1494-1498.
CHILDERS, S. R. and SNYDER, S. H. (1978) Life Sci. In press.
GOLDSTEIN, A., COX, B. M., KLEE, W. A. and NIRENBERG, M. (1977) Nature 265, 362-363.
KLEE, W. A. and NIRENBERG, M. (1974) Proc. Nat. Acad. Sci. USA 71, 3474-3477.

LAD, P. M., WELTON, A. F. and RODBALL, M. (1977) J. Biol.
   Chem. 252, 5942-5946.
LEFKOWITZ, R. J., MULLIKIN, D., WOOD, C. L., GORE, T. B.,
   and CHHABIRANI, M. (1977) J. Biol. Chem. 252,
   5295-5303.
MAGUIRE, M. E., VAN ARSDALE, P. M. and GILMAN, A. J. (1976)
   Molec. Pharmacol. 12, 335-339.
PASTERNAK, G. W., SNOWMAN, A. M. and SNYDER, S. H. (1975)
   Molec. Pharmacol. 11, 735-744.
PERT, C. B., PASTERNAK, G. and SNYDER, S. H. (1973)
   Science, N.Y. 182, 1359-1361.
PERT, C. B. and SNYDER, S. H. (1974) Molec. Pharmacol.
   10, 868-879.
SHARMA, S. K., KLEE, W. A. and NIRENBERG, M. (1975)
   Proc. Nat. Acad. Sci. USA 72, 3992-3996.
SHARMA, S. K., KLEE, W. A. and NIRENBERG, M. (1977)   Proc.
   Nat. Acad. Sci. USA 72, 3365-3369.
SHARMA, S. K., NIRENBERG, M. and KLEE, W. A. (1975)   Proc.
   Nat. Acad. Sci. USA 72, 590-594.
SIMANTOV, R., SNOWMAN, A. M. and SNYDER, S. H. (1976)
   Molec. Pharmacol. 12, 977-986.
SIMON, E. J. and GROTH, J. (1975) Proc. Nat. Acad. Sci.
   USA 72, 2404-2407.

PHOSPHORYLATION OF THE ACETYLCHOLINE RECEPTOR

A.S. GORDON AND I. DIAMOND

Departments of Neurology and Pediatrics and the
Liver Center, University of California
San Francisco Medical Center
San Francisco, California 94143

There is increasing evidence that disorders of receptor function account for several major clinical diseases.
In the nervous system, two of the most striking examples
are myasthenia gravis and tardive dyskinesia. Each of
these conditions is associated with a change in the number
and/or sensitivity of a neurotransmitter receptor. In
myasthenia gravis, fatigue and weakness is best explained
by a decreased number of nicotinic acetylcholine receptors
(AChR) at the neuromuscular junction. On the other hand,
the movement disorder in tardive dyskinesia is best explained by an apparent increase in the number and sensitivity of dopamine receptors in the corpus striatum. It
seems likely therefore that changes in specific biochemical
mechanisms which regulate post-synaptic membrane receptor
proteins account for the pathophysiologic abnormalities
in each of these disorders. Such mechanisms might involve
covalent biochemical modification of the receptor such as
phosphorylation, methylation or acetylation, or conformational changes in the receptor as a result of its interaction with effectors. However, the biochemical regulatory mechanisms affecting receptor proteins at the synapse
have not yet been identified and we do not know how these
are altered in neurologic disease. Since phosphorylation
of specific proteins is a key cellular regulatory mechanism,
one of the best candidates for regulating receptor function
at the synapse is a reversible phosphorylation-dephosphorylation reaction.

In order to understand the role of membrane protein
phosphorylation at the synapse, we have chosen to study
acetylcholine receptor-enriched synaptic membranes purified from the electric organ of T. californica. This organ
is an ideal model system for such studies. In contrast
to brain, the electric organ is homogeneously innervated
by nicotinic cholinergic neurons and is a rich source of
the acetylcholine receptor. Membranes can be purified
from the electric organ which are enriched in the AChR
(DUGUID & RAFTERY, 1973) and which retain their ability
to show cholinergic agonist-dependent cation flux (POPOT,
SUGIYAMA & CHANGEUX, 1976). Moreover, the AChR from
T. californica has been purified (KARLIN et al., 1975),
biochemically characterized (MICHAELSON et al., 1974) and
used to generate specific antibodies (PATRICK & LINDSTROM,
1973; GORDON et al., 1977a). Receptor-enriched membranes
contain only a few other proteins which are closely associated with the receptor in the post-synaptic membrane.
Such associated proteins may play a critical role in regulating the function of the AChR in the post-synaptic membrane. We have taken advantage of these conditions to
study phosphorylation of the membrane-bound AChR in this
well-defined, homogeneous system.

CHARACTERIZATION OF AChR-ENRICHED MEMBRANES

When the electric organ of T. californica is homogenized and the resulting membranes further purified by
ultracentrifugation on a discontinuous sucrose gradient,
several membrane fractions are recovered (Fig. 1). Because
a post-synaptic membrane fraction should be enriched in
the AChR, we searched for a membrane fraction with high
specific activity for Naja naja siamensis toxin binding.
The distribution of toxin binding compared to other membrane markers in the various membrane fractions is shown
in Fig. 2. The AChR was enriched in fractions C and D
(Fig. 2a). Compared to the total membrane preparation
(H), fraction D had nearly a 15-fold increase in receptor
activity while fraction C was 5-fold enriched. In contrast to the distribution of AChR, acetylcholinesterase
was enriched in fraction A (Fig. 2b) and Na-K-ATPase activity was enriched in fractions B and E (Fig. 2c). Fraction
D, the fraction most enriched in AChR, contained very low
levels of NaKATPase and acetylcholinesterase activity.
This suggested that fraction D was enriched in membranes
of post-synaptic origin.

If phosphorylation regulates receptor function in

the post-synaptic membrane, then receptor-enriched membranes should contain endogenous protein kinase activity.
When fractions A-D were examined for endogenous membrane
protein phosphorylation, by incubating the membranes with
$(\gamma-^{32}P)$ATP in the presence of Mg2+, fraction D exhibited
maximal phosphorylation (Fig. 3). Several of the phosphorylated polypeptides have molecular weights corresponding to some of the subunits of the purified AChR. Significant phosphorylation of the same polypeptides was also
demonstrable in fraction C but markedly reduced in fraction
B and virtually absent in A. Thus, fraction D, a specialized membrane preparation enriched in the AChR, also
showed maximal endogenous membrane protein phosphorylation.

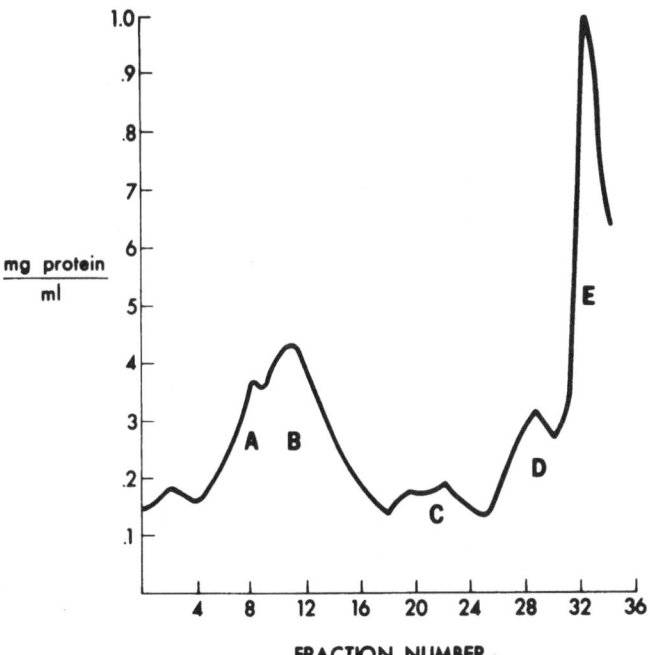

Fig. 1: Discontinuous sucrose gradient centrifugation
of a total membrane homogenate from Torpedo
californica.

PROPERTIES OF THE ENDOGENOUS PHOSPHORYLATION REACTION

If phosphorylation is important for synaptic function, then phosphorylation in synaptic membranes ought to
have special properties. Although protein phosphorylation in receptor-enriched membranes showed a typical dependence on divalent cations (Fig. 4), its response to monovalent cations was unique. At 0.1M, K+ produced a marked
stimulation of endogenous membrane protein phosphorylation while Na+ had no effect (Fig. 5). This differential
cation response has not been observed with other membrane

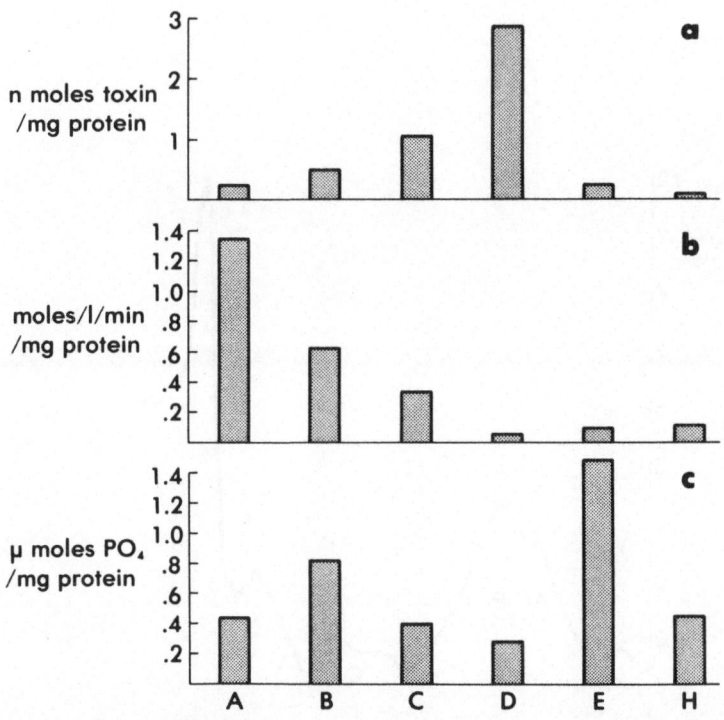

Fig. 2: Marker enzyme distribution in membranes purified
by sucrose gradient centrifugation: (a) I^{125}-
Naja naja siamensis toxin binding activity; (b)
acetylcholinesterase activity; (c) Na-K-ATPase
activity. Fractions A-E are recovered from the
sucrose gradient. Fraction H is the total membrane
homogenate applied to the sucrose gradient.

preparations and suggests we are studying a specialized
biochemical reaction at the synapse. In order to confirm
this possibility, we searched for evidence that cholinergic ligands have a specific effect on phosphorylation in
AChR-enriched membranes. As shown in Fig. 6, carbachol,
a nicotinic cholinergic ligand, inhibited endogenous membrane protein phosphorylation by 75%. This effect of
carbachol was observed at concentrations as low as 10^{-6}M.

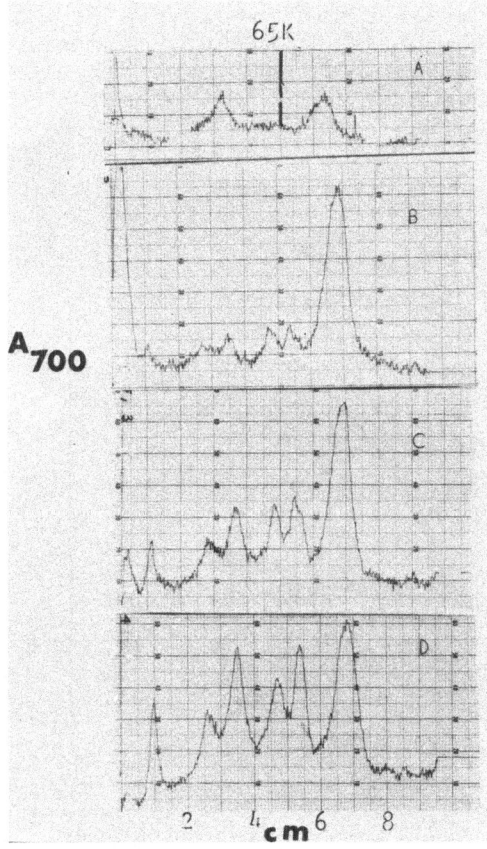

Fig. 3: Endogenous phosphorylation of membrane fractions
    A-D. Densitometric scan at 700 nm of autoradiograms
    of dried SDS polyacrylamide gels. Membranes were
    incubated for 0.5 minutes at 0° with $(\gamma-^{32}P)$ATP
    and Mg2+. The reaction was stopped by the addition
    of SDS and the phosphorylated polypeptides were
    separated by SDS polyacrylamide gel electrophoresis
    (GORDON, DAVIS & DIAMOND, 1977b).

Carbachol inhibition of protein phosphorylation had an
absolute requirement for K+, with an optimal concentration of 100 mM (GORDON, DAVIS & DIAMOND, 1977b).

In the experiments described above, we found phosphorylation of several major polypeptides which appeared
to have the same molecular weight as subunits of the AChR.
In addition, phosphorylation of these polypeptides appeared
to be regulated by K+ and cholinergic ligands.   These
results suggested that the membrane-bound AChR itself might
be a substrate for the endogenous protein kinase.   In order
to answer this question definitively, we used antibodies
prepared against the purified AChR to demonstrate that
several of the phosphorylated polypeptides are components
of the AChR (GORDON et al., 1977a).   The technique we

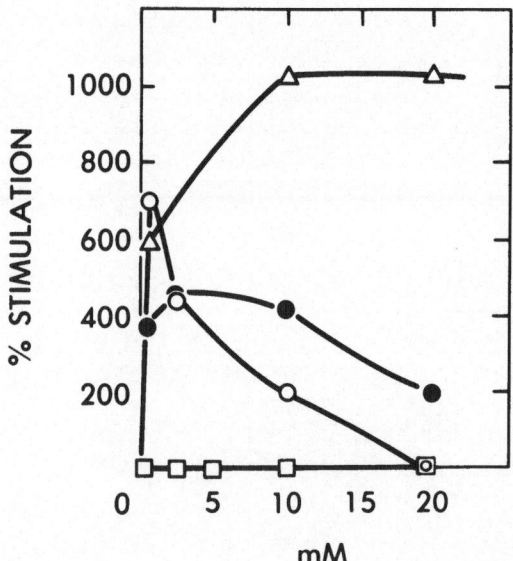

Fig. 4:   Effect of divalent cations on phosphorylation
          of the 65,000 M.W. polypeptide in AChR-enriched
          membranes. Ordinate is absorbance at 700 nm of
          autoradiograms of membranes phosphorylated in the
          presence of a given divalent cation relative to the
          absorbance of membranes phosphorylated in the absence
          of that cation: Δ, Mg2+; O, Zn2+; ● Mn2+; □ , Ca.

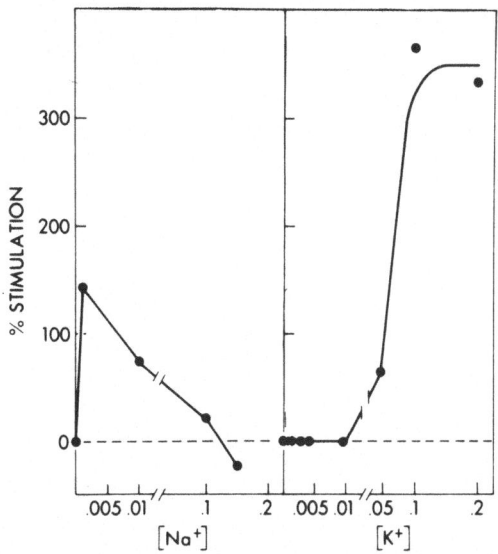

Fig. 5: Phosphorylation of the 65,000 M.W. polypeptide as a function of Na+ and K+ concentration.

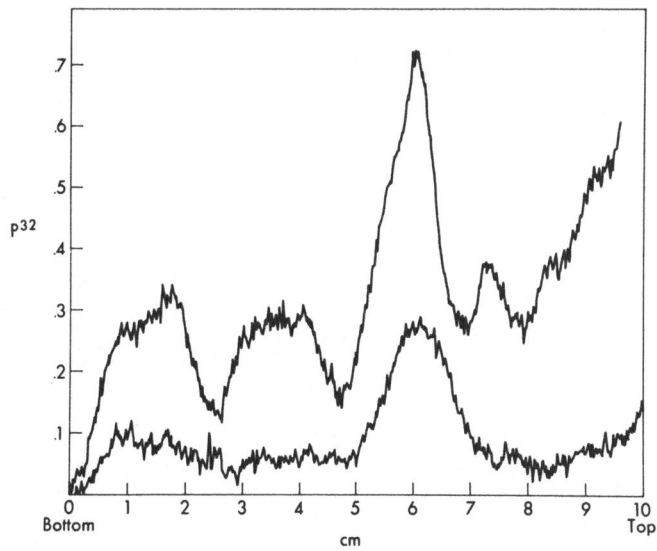

Fig. 6: Effect of 10^{-4}M carbachol on endogenous phosphorylation of AChR-enriched membranes in the presence of 100 M K+. Scan of autoradiogram of dried SDS gel. Ordinate is absorbance at 700 nm. Upper curve, no additions· lower curve, 10^{-4}M carbachol.

employed was two-dimensional immunoelectrophoresis as described by CONVERSE & PAPERMASTER (1975). Acetylcholine receptor-enriched membranes were incubated with $(\gamma-^{32}P)$ ATP and Mg2+ and the reaction terminated by the addition of SDS. The solubilized membranes were then electrophoresed in duplicate in SDS polyacrylamide gels (Fig. 7a). The gels were cut into longitudinal strips along the sample wells; one gel was stained for protein, another was subjected to immunoelectrophoresis at right angles into two layers of agarose. The first layer contained 1.5% Lubrol PX to remove excess SDS and the second contained goat anti-AChR antiserum.

Polypeptides derived from the AChR which are present in optimal concentrations form immunoprecipitates with the anti-AChR antiserum and are recognized as "rockets". Fig. 7b shows a major Coomassie blue-stained rocket corresponding to the 65,000 M.W. polypeptide of the AChR. Control gels run against pre-immune serum show no rockets. Autoradiography of the dried gel (Fig. 7c), demonstrates that the same polypeptide precipitated by anti-AChR antiserum contains ^{32}P. This experiment provided unambiguous evidence that the 65,000 M.W. component of the AChR was phosphorylated in situ by an endogenous membrane protein kinase present in AChR-enriched membranes. This is an exciting finding since it is the first time a component of a membrane receptor protein has been identified as a substrate for an endogenous membrane protein kinase. Unlike work with other membrane preparations, the phosphorylated substrate in our studies is a subunit of a post-synaptic membrane protein whose function is known and which has been biochemically characterized. This is an important first step in correlating phosphorylation of membrane proteins with synaptic function.

We have found that phosphorylation in AChR-enriched membranes is stimulated by K+ and inhibited by cholinergic ligands. However, we still do not know the site for regulation of AChR phosphorylation. It seems likely that K+ and carbachol might interact with the membrane-bound receptor altering its availability as a substrate for phosphorylation. Alternatively, it is also possible that the endogenous membrane protein kinase itself is regulated by K+ and cholinergic ligands. This latter possibility could be tested by measuring membrane protein kinase activity with exogenous protein substrates.

PROPERTIES OF THE EXOGENOUS PHOSPHORYLATION REACTION

Histone and casein have both been used as substrates
to study membrane-bound protein kinase activity. Histone
appears to be phosphorylated optimally by cAMP-dependent
protein kinases (RUBIN & ROSEN, 1975). Fig. 8 shows that
both histone and casein can be phosphorylated by the membrane-bound protein kinase present in AChR-enriched membranes. The amount of $(^{32}P)-PO_4$ incorporated into either
protein was proportional to the amount of membrane protein.
We have previously shown that endogenous phosphorylation
in these membranes is cAMP-independent (GORDON, DAVIS &
DIAMOND, 1977b). Therefore, as expected, casein was a
better substrate than histone for the enzyme. cAMP did not
stimulate phosphorylation of casein (not shown) or histone
(Fig. 9). Fig. 9 also shows that phosphorylation of casein
was linear to 30 minutes. These results indicated that
casein could be used effectively as an exogenous substrate
to study the properties of the membrane protein kinase in
AChR-enriched membranes.

To determine whether K+ and cholinergic ligands alter
membrane protein kinase activity, we studied the effect of
these agents on phosphorylation of casein by AChR-enriched
membranes. Fig. 10 shows the effects of Na+ and K+ on
exogenous phosphorylation. Low concentrations of K+ or
Na+ produced a small stimulation of casein phosphorylation

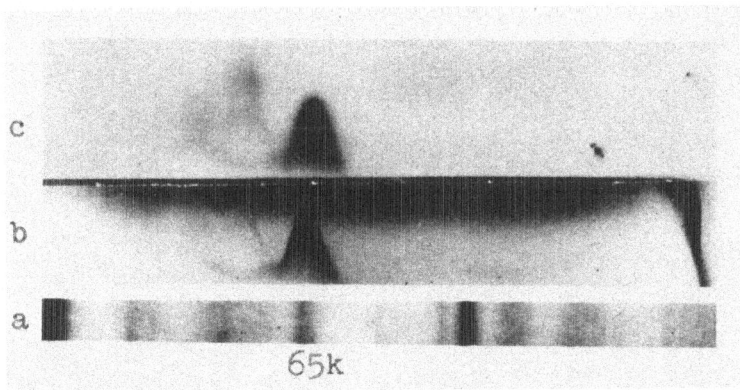

Fig. 7: Two-dimensional immunoelectrophoresis of AChR-
enriched membranes phosphorylated in situ (GORDON
et al., 1977a). (a) Coomassie blue stained SDS gel;
(b) Coomassie blue stained Agarose gel after immunoelectrophoresis of the SDS strip into goat anti-
AChR-containing Agarose; (c) Autoradiography of (b).

but high concentrations of either cation inhibited this
reaction.   Carbachol at 10^{-4}M did not have an effect on
phosphorylation of casein.   These effects of Na+, K+ and
carbachol on casein phosphorylation were very different
from their effects on endogenous membrane protein phosphorylation in receptor-enriched membranes (see Figs. 5
and 6).   Assuming there is only one membrane protein kinase
in this membrane preparation, these results suggest that
cholinergic ligands and K+ do not regulate receptor phosphorylation by regulating membrane protein kinase activity.

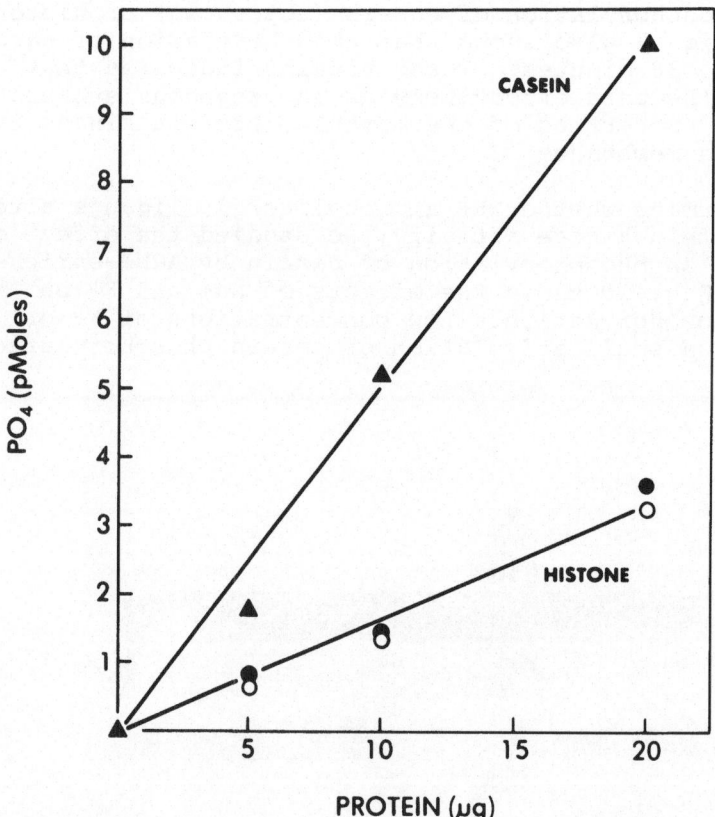

Fig. 8:   Phosphorylation of exogenous substrates by
      membrane protein kinase present in AChR-enriched
      membranes.   Reported values are corrected for endo-
      genous membrane protein kinase activity in the
      absence of exogenous substrate.   ▲ , casein;
      0, histone;   ● histone and 10 μM 3', 5' -cAMP.

Instead, it seems likely that these agents interact directly
with the AChR to alter its availability as a substrate for
phosphorylation. Cholinergic ligands and K+ have been
shown to produce conformational changes in the AChR
(CHANGEUX et al., 1975· GIBSON, JUNI & O'BRIEN, 1977).
This suggests that the conformational state of the
membrane-bound AChR could determine its level of phosphorylation. In an analogous manner, phosphorylation of
membrane bound rhodopsin is also dependent on its conformation (KUHN, COOK & DRYER, 1973; WELLER, VIRMAUX &
MANDEL, 1975). Therefore, this could be a general mechanism for the regulation of receptor function, i.e. the
conformational state of the receptor protein regulates

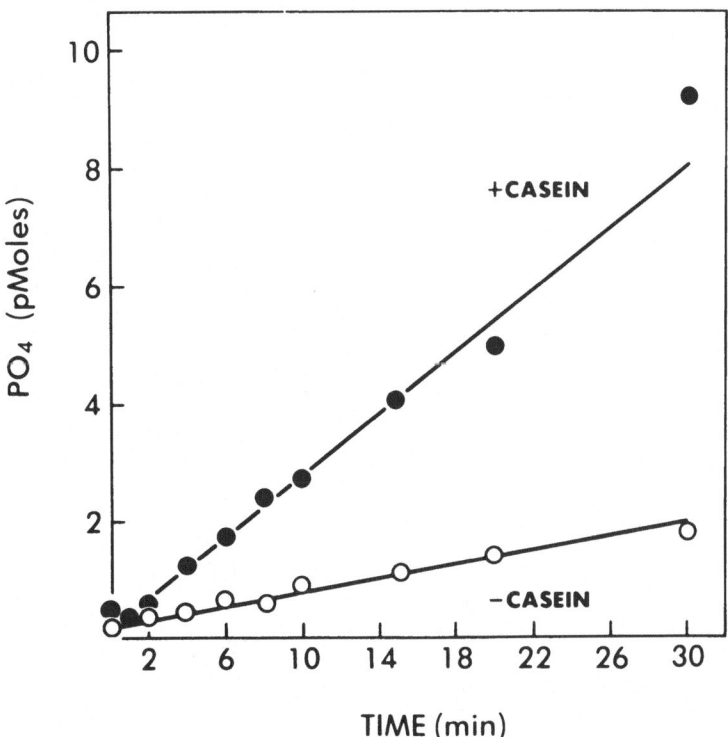

TIME (min)

Fig. 9: Time course of phosphorylation of casein by
        protein kinase in AChR-enriched membranes. Membrane
        protein concentration is 5μgm. 0, endogenous phos-
        phorylation in the absence of casein; ● exogenous
        phosphorylation.

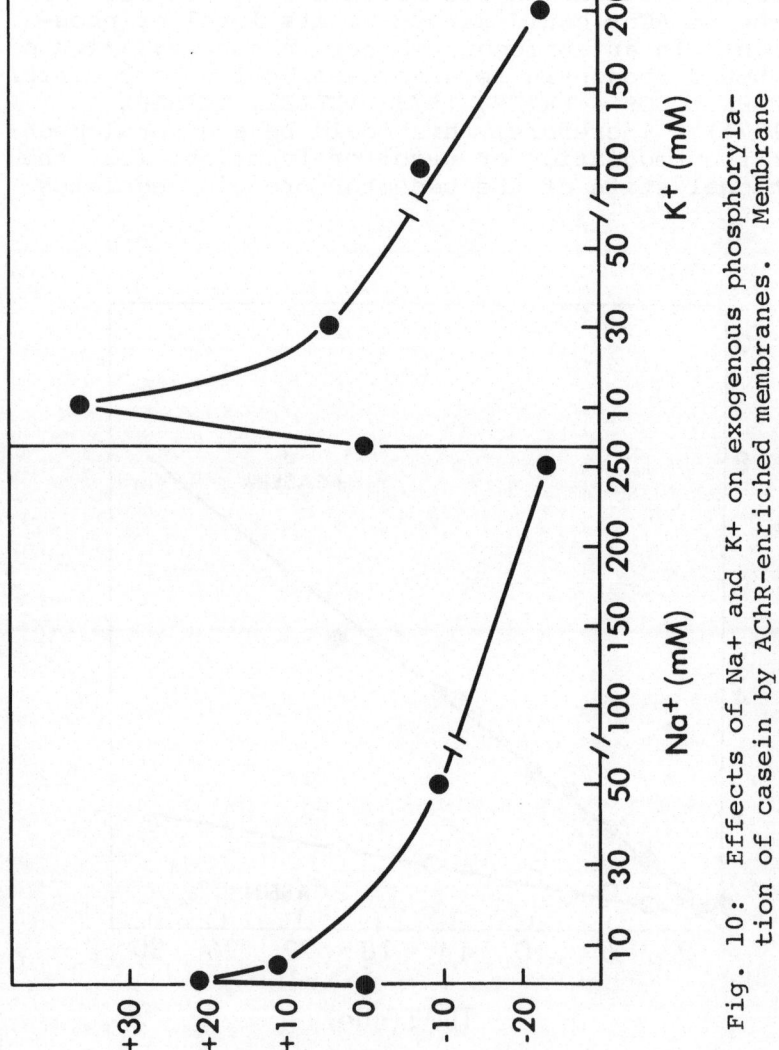

Fig. 10: Effects of Na+ and K+ on exogenous phosphorylation of casein by AChR-enriched membranes. Membrane
protein concentration is 5 µgm.

its level of phosphorylation which in turn determines the
functional state of the protein.

PROPERTIES OF THE DEPHOSPHORYLATION REACTION

If phosphorylation of the AChR is an important regulatory event at the synapse, then dephosphorylation of
the AChR must also occur in situ. Therefore, the AChR-
enriched membranes must have an endogenous membrane phosphoprotein phosphatase activity which can dephosphorylate
the AChR. We first studied phosphoprotein phosphatase
activity using (^{32}P)-PO_4 labeled casein as exogenous substrate. Fig. 11 shows the time course for release of ^{32}P
from phosphorylated casein after incubation with AChR-
enriched membranes. Boiled membrane controls or phosphorylated casein without membranes did not show release
of radioactivity over this same time interval. Release
of ^{32}P from casein was proportional to the concentration
of membrane protein (Fig. 12). ^{32}P released from casein
after incubation with membranes could be due to proteolytic breakdown of casein into labeled amino acids. To
investigate this possibility, we found that (^{32}P)-PO_4
casein migrated as one band on SDS acrylamide gel electrophoresis both by Coomassie blue staining and autoradiography. These patterns were unchanged after incubation
of the casein with membranes. In addition, all of the
^{32}P released from casein was extractable into isobutanol
in the presence of ammonium molybdate and sulfuric acid.
This confirms that the label was inorganic phosphate.
The effect of activators and inhibitors on protein phosphatase activity in AChR-enriched membranes is shown
in Table I. The data indicate that this membrane protein
phosphatase is similar to other cAMP-independent protein
phosphatases.

If phosphatase activity is important in regulating
the level of AChR phosphorylation, then the enzyme should
dephosphorylate the AChR in situ. We used two different
assay systems to study this question. In the first, we
measured time-dependent release of (^{32}P)-PO_4 from AChR-
enriched membranes that had been phosphorylated in the
presence of $(\gamma-^{32}P)$ ATP and Mg2+. Table IIA shows that
^{32}P labeled inorganic phosphate was released from the phosphorylated membranes after incubation at 37°. This reaction was also inhibited by fluoride ion. Thus, the enzyme
does dephosphorylate endogenous membrane protein.

In the second assay, we measured changes in the level of phosphorylation of the membrane-bound AChR directly after incubation at 37° for 10 min. SDS gel electrophoresis of the phosphorylated membranes shows that the 65,000 M.W. subunit of the AChR has been dephosphorylated

TABLE I

Casein Phosphatase Activity

Additions	mM	Relative Activity (%)
None	--	100
NaF	10	50
NaF	100	16
NaCl	100	77
DTT	4	340 *
Mg++	10	77
Mn++	10	58
cAMP	10^{-3} – 10^{-6}	100
cGMP	10^{-3} – 10^{-6}	100
ATP	1	11
GTP	1	15
P	10	46

AChR-enriched membranes containing 0.1 mg protein were incubated in duplicate with 50 μg of (^{32}P)-PO$_4$ casein at 37° for 20 min in a total volume of 100 μl containing 4 mM dithiothreitol (DTT), 0.1% Triton and 15 mM Tris-HCl, pH 6.8. The reaction was stopped with 3 ml of an ice cold solution containing Norit A (40 mg/ml), 0.1 M HCl, 0.2 mg/ml bovine serum albumin, 1 mM NaP$_i$, and 1 mM NaPP . After holding on ice for 10 min the suspension was filtered on Millipore filters (0.45 μm) and the residue washed twice with 3 ml of 0.01 N HCl-1 mM NaP$_i$. Releasable ^{32}P not absorbed to the charcoal was determined by measurement of Cerenkov radiation in the filtrate.
* Relative to a control without DTT. All other samples contained 4 mM DTT.

since there is less covalently bound (^{32}P)-PO_4 (Table IIB). As before, endogenous dephosphorylation of the AChR was also inhibited by 0.1 M NaF.

In summary, we have shown that the AChR-enriched membranes contain both endogenous protein kinase and phosphatase activities to regulate the level of phosphorylation of the AChR in situ. Such a reversible covalent biochemical modification of a receptor protein has important functional implications. For example, changes in the level of receptor phosphorylation might regulate receptor sensitivity to acetylcholine.

TABLE II

Dephosphorylation of the Membrane Bound AChR

Condition	A pmoles (^{32}P) PO_4 released	B (^{32}P)-PO_4 in the 65,000 M.W. subunit of the AChR A_{700}
zero time control	$0.57 \pm .03$	4.5
NaCl (0.1 M)	$0.74 \pm .02$	2.8
NaF (0.1 M)	$0.50 \pm .03$	4.2

Membranes were initially phosphorylated as described. The phosphorylated membranes were then incubated for 10 min at 37° in either 200 mM NaCl or NaF. (A) Reaction was stopped by addition of trichloroacetic acid. (^{32}P) PO_4 released was measured by extraction into isobutanol and liquid scintillation counting of the organic layer. (B) Reaction was stopped by addition of SDS and mercaptoethanol to final concentrations of 3.7% and 4% respectively. The samples were then subjected to SDS gel electrophoresis and autoradiography as described in the text.

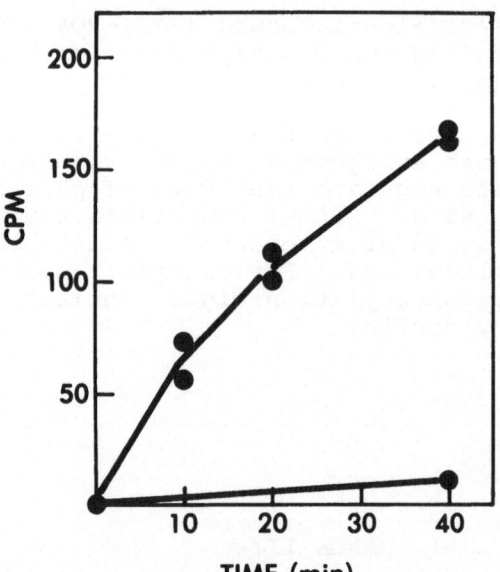

Fig. 11:   Time course of phosphoprotein phosphatase
           activity in AChR-enriched membranes using (^{32}P)
           PO$_4$-casein as substrate.   Reaction carried out as
           described in Table I.   Lower curve is a control
           in the absence of membranes.

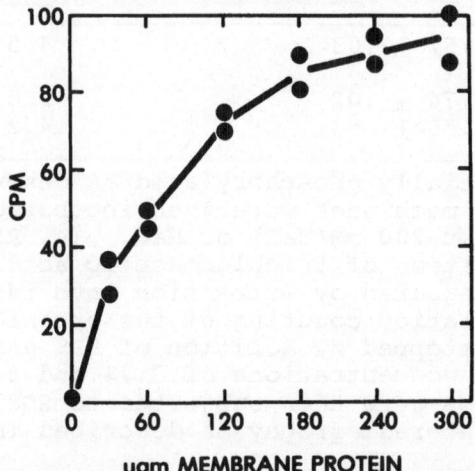

Fig. 12:   Phosphoprotein phosphatase activity as a func-
           tion of membrane protein concentration.   Blank values
           in the absence of membranes have been subtracted.

IDENTIFICATION OF ATP BINDING PROTEINS IN THE MEMBRANE

In order to correlate the level of AChR phosphorylation with the regulation of receptor function, it is
necessary to maintain the receptor at a given level of
phosphorylation when studying such functional properties
as affinity for agonists or Na+ efflux. However, since
phosphatase activity is always present in these membranes
it is difficult to maintain the receptor in a phosphorylated state to compare to the functional properties of the
dephosphorylated receptor. We therefore developed an
alternative approach to overcome this methodologic limi-
tation. We wanted to identify the kinase and phosphatase
in the membrane in order to purify each of these enzymes,
prepare antibodies against them and use the antibodies to
inhibit the corresponding enzyme activity. Such an approach
would allow us to maintain the receptor in a given phosphorylated state. Since our data showed that both the
kinase and phosphatase had ATP-binding sites, we used
arylazido-β-alanyl ATP as a photoaffinity ligand to identify these polypeptides. These experiments were done in
collaboration with Drs. Ferdinand Hucho and Richard
Guillory. Arylazido-β-alanyl ATP was synthesized as described by JENG & GUILLROY (1975) using either unlabeled
ATP or (α-^{32}P) ATP. The AChR enriched membranes were
incubated with the photoaffinity label and then irradiated
so that the label remained covalently bound to the polypeptide containing the ATP binding sites. The polypeptides were then separated by SDS polyacrylamide gel
electrophoresis, and the dried gels autoradiographed
to determine which polypeptides reacted with the ATP
analog. Fig. 13 shows that the AChRenriched membranes
used in this study contained about 8 major polypeptides
including the four subunits of the AChR (α, β, γ, δ).
After photoirradiation, the ATP photoaffinity analog,
arylazido-β-alanyl (α-^{32}P) ATP, reacted with only three
polypeptides and none of these were components of the
AChR. In the absence of irradiation, there was no incorporated radioactivity (data not shown). Thus, comparison
of the Coomassie blue stained gel (Fig. 13b) with the
autoradiogram (Fig. 13a) of the gel shows that the affinity
ligand labeled bands migrate with molecular weights of
45,000, 55,000 and 100,000.

The specificity of binding of the photoaffinity label
for ATP binding sites is shown in Fig. 14. Unlabeled ATP
at increasing concentrations correspondingly inhibited
the reaction of the labeled photoaffinity compound with
the membranes. On the other hand, arylazido-β-alanine,
i.e. the photolabel without the ATP moiety, did not cause

a significant decrease in reactivity of the radioactive
affinity label (data not shown). These results strongly
suggest that the analog is acting as a photoaffinity probe
of adenine nucleotide binding sites in AChR-enriched
membranes.

The 100,000 dalton band probably is the Na-K-ATPase
which has the same molecular weight (JEAN, ALBERS & KOVAL,
1975). This band is not a component of the AChR but is
present in variable amounts in receptor-enriched membrane
preparations (see Fig. 2) and does have an ATP binding
site. The 45,000 and 55,000 labeled bands probably are
related to the protein kinase and protein phosphatase
activities present in the receptor-enriched membranes.

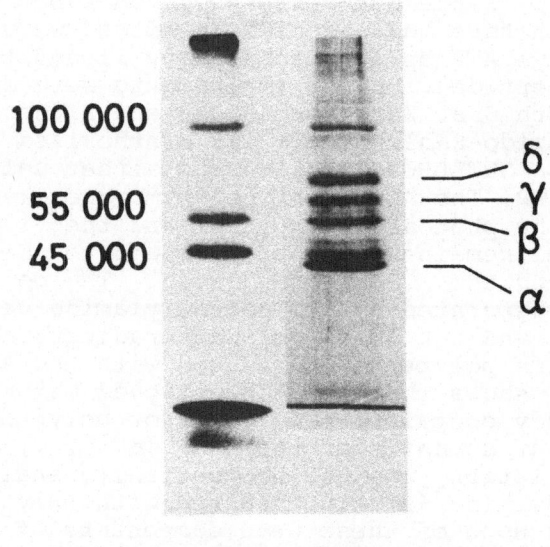

Fig. 13: Photoaffinity labeling of AChR-enriched mem-
    branes with arylazido-β-alanyl ATP. Protein (1 mg/ml)
    was suspended in 15 mM Tris-HCl, pH 7.5 containing
    10^{-4}M ouabain and 0.225 mM arylazido-β-alanyl (α-P^{32})
    ATP. Irradiation was performed 4 times for 15 seconds
    each at room temperature. The samples were chilled
    in ice following each 15 sec irradiation. SDS electro-
    phoresis was carried out according to GORDON, DAVIS
    & DIAMOND (1977b) in 7.5% acrylamide gels: auto-
    radiograph (a) and Coomasie blue-stained SDS poly-
    acrylamide gel (b).

ATP is a substrate for the protein kinase reaction and
therefore this enzyme must have an ATP binding site.
As we have shown in Table I, the phosphatase is inhibited
by ATP and therefore probably also has a specific binding
site for the nucleotide. If the photoaffinity ATP analog
reacts with protein kinase in the membrane it might be
expected to inhibit protein kinase activity. Fig. 15
shows that the photoaffinity label has a striking inhibitory effect on AChR phosphorylation in receptor-enriched
membranes. The membrane-bound AChR was phosphorylated
in situ with (γ-^{32}P) ATP as described in the figure
legend. Increasing concentrations of unlabeled photoaffinity ligand progressively inhibited receptor phosphorylation. Since we know that the ATP analog does not bind
to the AChR (Fig. 13) then the ATP photoaffinity label
must be reacting with the membrane kinase to inhibit
phosphorylation of the membrane-bound AChR.

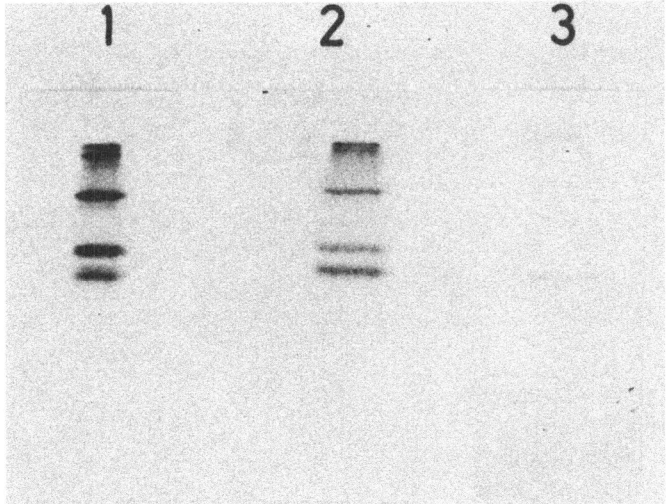

Fig. 14: Inhibition of photolabeling with ATP. Autoradio-
      grams of SDS gels.   (1) Control as in Fig. 1.   (2)
      and (3): Incubation mixture also contained 50 mM
      (2) and 100 mM ATP (3).

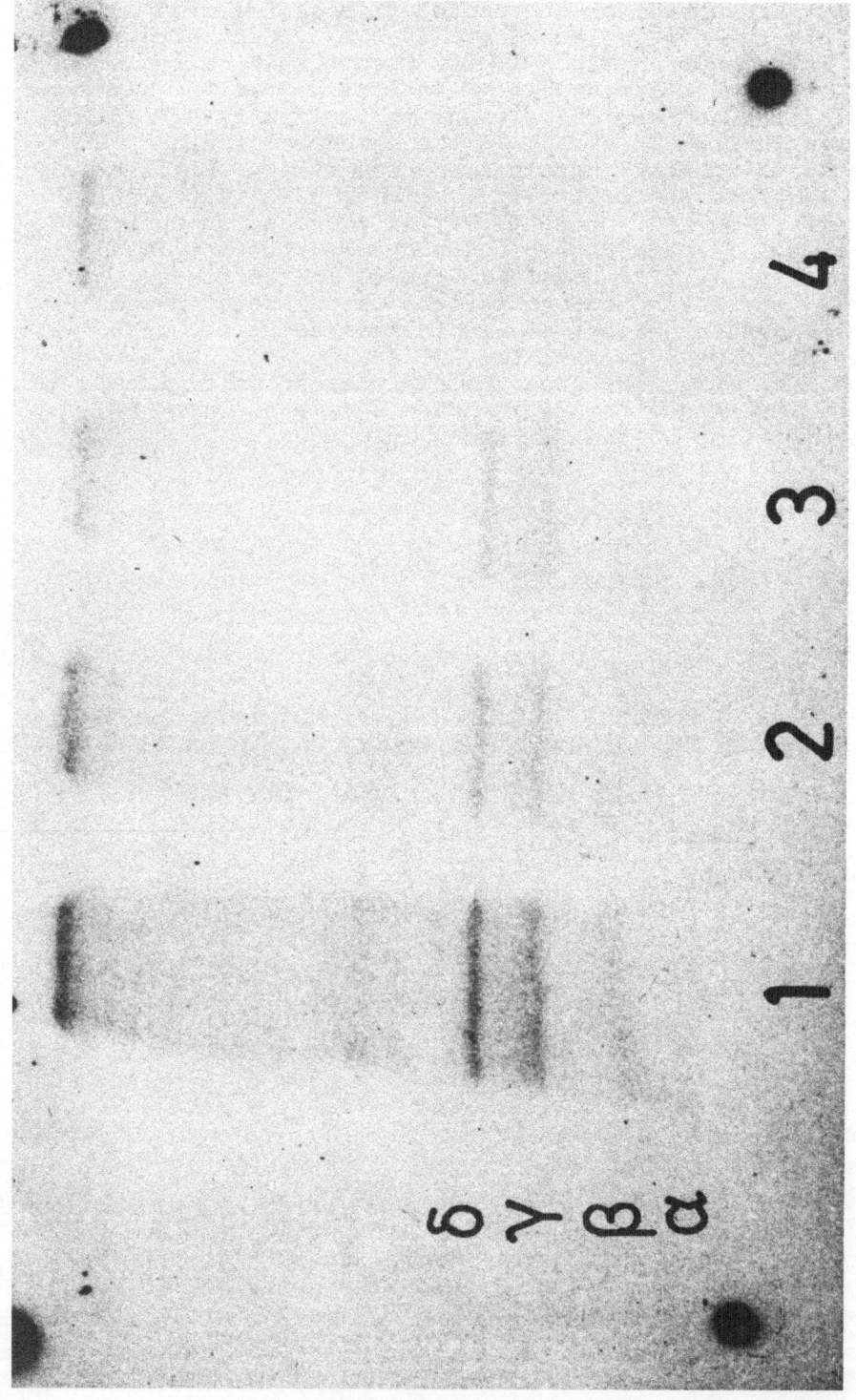

We have determined that ATP inhibits phosphatase activity in the same membrane preparation. However, because of methodologic limitations we have not been able to study the interaction of the ATP photoaffinity label with this endogenous membrane phosphoprotein phosphatase. Therefore, we cannot assign a specific function to the 45,000 and 55,000 M.W. bands labeled by the ATP photo-affinity label. We can conclude however that at least one of these bands appears to be the kinase and that neither is a component of the purified AChR. We are now in the process of purifying the 45,000 and 55,000 M.W. polypeptides with the use of the radioactive photoaffinity label so that we can generate specific antibodies against these polypeptides. We anticipate that the antibodies will allow us to determine the function of each of these polypeptides. Once we establish that these antibodies inhibit enzyme activity we will be able to maintain the level of phosphorylation of the AChR in situ. This will allow us to correlate receptor phosphorylation with recep-tor function in the membrane.

Fig. 15: Inhibition of endogenous protein kinase activity in AChR-enriched membranes by arylazido-β-alanyl ATP. Photoaffinity labeling of membranes carried out as in Fig.13 using unlabeled arylazido-β-alanyl ATP. Endogenous kinase activity was then determined (GORDON, DAVIS & DIAMOND, 1977b). Membranes con-taining 50 μg photoaffinity-labeled protein were incubated in 0.1 ml for 0.5 min at 0°C with 5 μM $(\gamma-P^{32})$ATP (1-2 μCi per tube), 0.25 mM EGTA, 10 mM $MgAc_2$, 0.0625 M Tris-HCl, pH 6.8 and 100 mM KCl. The reaction was stopped by the addition of 15 μl of 20% SDS and electrophoresis was carried out ac-cording to GORDON, DAVIS & DIAMOND (1977b). (1) Autoradiogram of a 7.5% SDS acrylamide gel showing phosphorylation of the γ (MW 62,000) and δ (MW 68,000) bands of the AChR by an endogenous protein kinase. (2) Phosphorylation in the presence of 0.03 mM arylazido-β-alanyl ATP in the dark. (3) Same con-ditions as before, but with 4 X 15 sec irradiation. (4) Phosphorylation after photoaffinity labeling with 0.3 mM arylazido-β-alanyl ATP.

CONCLUSIONS

These studies open a new avenue for investigating
biochemical regulatory mechanisms at the synapse. In
most studies of synaptic membrane phosphorylation the identity of the phosphorylated proteins is unknown. In our
studies, however, the AChR substrate for the phosphorylation-dephosphorylation reaction is an identified protein
and the function of this receptor has been well studied
in many systems. A major advantage to using T. californica
receptor-enriched membranes to study phosphorylation of the
acetylcholine receptor in situ is that the level of phosphorylation can be correlated with receptor function.
The AChR in these membranes has been well-characterized
with respect to its affinity for cholinergic agonists and
antagonists (MICHAELSON et al., 1974). Changes in binding
affinity (desensitization) also occur as a result of
preincubation with agonist (WEBER, DAVID-PFEUTY & CHANGEUX,
1975). In addition, this same preparation exhibits agonist-
induced Na+ flux and "densensitization" of this response
has also been observed in vitro (SUGIYAMA, POPOT & CHAN-
GEUX, 1976). Therefore, it will be possible to determine
the effects of phosphorylation and dephosphorylation of
the AChR on receptor affinity for agonists as well as
receptor-mediated ion fluxes. Such an approach offers a
unique advantage to study the functional significance of
receptor phosphorylation and dephosphorylation at the
synapse.

Covalent reversible modification of a post-synaptic
receptor protein by an endogenous membrane regulatory
mechanism must have important pathophysiologic implications. In clinical medicine we often encounter disturbances
of neurologic function which are not necessarily associated
with structural damage or neuropathologic findings. This
suggests that we are dealing with functional and potentially
reversible alterations of neural activity. One of the
best contemporary examples is tardive dyskinesia. This
disorder is produced by neuroleptic drugs which block
dopamine receptors and appears to be associated with
changes in the number and sensitivity of the dopamine
receptor. The signal which regulates such changes is
unknown but probably involved biochemical regulatory
mechanisms affecting the dopamine receptor. Unfortunately
it is not yet possible to study covalent biochemical modification of most neurotransmitter receptors. We are
studying a model system in which an associated membrane
kinase and phosphatase regulate the level of phosphorylation of the AChR. We are hopeful these studies will provide
new insights into the regulation of receptor function in

the nervous system and its alterations in neurologic disease.

ACKNOWLEDGMENTS

The labeling experiments with arylazido-β-alanine ATP were done in collaboration with Dr. Ferdinand Hucho, Universitat Konstanz and Dr. Richard Guillory, University of Hawaii. We thank Dale Milfay and Geoffrey Davis for excellent technical assistance. This work was supported by grants from the NIH, NSF, Muscular Dystrophy Associations of America and the Los Angeles and California Chapters of the Myasthenia Gravis Foundation.

REFERENCES

CHANGEUX, J.-P., BENEDETTI, L., BOURGEOIS, J.-P., BRISSON, A., CARTAUD, J., DEVAUX, P., GRUNHAGEN, H., MOREAU, M., POPOT, J.-L. & WEBER, M. (1975) Some structural properties of the cholinergic receptor protein in its membrane environment relevant to its function as a pharmacological receptor. Cold Spring Harbor Symp. Quant. Biol. 40, 211-230.

CONVERSE, C.A. & PAPERMASTER, D. (1975) Membrane protein analysis by two dimensional immunoelectrophoresis. Science 189, 469-472.

DUGUID, J. & RAFTERY, M.A. (1973) Fractionation and partial characterization of membrane particles from Torpedo californica electroplax. Biochemistry 12, 3593-3597.

GIBSON, R.E., JUNI, S. & O'BRIEN, R.D. (1977) Monovalent ion effects on acetylcholine receptor from Torpedo californica. Arch. Biochem. Biophys. 179, 183-188.

GORDON, A.S., DAVIS, C.G., MILFAY, D. & DIAMOND, I. (1977a) Phosphorylation of acetylcholine receptor by endogenous membrane protein kinase in receptor-enriched membranes of Torpedo californica. Nature 267, 539-540.

GORDON, A.S., DAVIS, C.G. & DIAMOND, I. (1977b) Phosphorylation of membrane proteins at a cholinergic synapse. Proc. Natl. Acad. Sci. USA 74, 263-267.

JEAN, D.H., ALBERS, R.W. & KOVAL, G.J. (1975) Sodium-potassium-activated adenosine triphosphatase of Electrophorus electric organ. J. Biol. Chem. 250, 1035-1040.

JENG, S.T. & GUILLORY, R.T. (1975) The use of arylazido ATP analogs as photoaffinity labels for myosin ATPase. J. Supramolec. Structure 3, 448-468.

KARLIN, A., MCNAMEE, M.G., WEILL, C.L. & VALDERRAMA, R.
    (1975) Facets of the structures of acetylcholine
    receptors from Electrophorus and Torpedo. Cold
    Spring Harbor Symp. Quant. Biol. 40, 203-210.
KUHN, H., COOK, J.H. & DRYER, J. (1973) Phosphorylation
    of rhodopsin in bovine photoreceptor membranes. A
    dark reaction after illumination. Biochemistry 12,
    2495-2502.
MICHAELSON, D., VANDLEN, R., BODE, J., MOODY, T., SCHMIDT,
    J. & RAFTERY, M.A. (1974) Some molecular properties
    of an isolated acetylcholine receptor ion-transloca-
    tion protein. Arch. Biochem. Biophys. 165, 796-804.
PATRICK, J. & LINDSTROM, J. (1973) Autoimmune response
    to acetylcholine receptor. Science 180, 871-872.
POPOT, J.-L., SUGIYAMA, H. & CHANGEUX, J.-P. (1976) Studies
    on the electrogenic action of acetylcholine with
    Torpedo marmorata electric organ. J. Mol. Biol.
    106, 469-483.
RUBIN, C.S. & ROSEN, O.M. (1975) Protein phosphorylation.
    Ann. Rev. Biochem. 44, 831-887.
SUGIYAMA, H., POPOT, J.-L. & CHANGEUX, J.-P. (1976)
    Studies on the electrogenic action of acetylcholine
    with Torpedo marmorata electric organ. J. Mol. Biol.
    106, 485-496.
WEBER, M., DAVID-PFEUTY, T. & CHANGEUX, J.-P. (1975)
    Regulation of binding properties of the nicotinic
    receptor proteins by cholinergic ligands in membrane
    fragments from Torpedo marmorata. Proc. Natl. Acad.
    Sci. USA 72, 3443-3447.
WELLER, M., VIRMAUX, N. & MANDEL, P. (1975) Light-
    stimulated phosphorylation of rhodopsin in the
    retina: The presence of a protein kinase that is
    specific for photobleached rhodopsin. Proc. Natl.
    Acad. Sci. USA 72, 381-385.

THE BEHAVIORALLY ACTIVE NEUROPEPTIDE ACTH AS NEUROHORMONE
AND NEUROMODULATOR: THE ROLE OF CYCLIC NUCLEOTIDES AND
MEMBRANE PHOSPHOPROTEINS

W. H. Gispen, H. Zwiers, V. M. Wiegant, P.
Schotman and J. E. Wilson*

Division of Molecular Neurobiology, Rudolf Magnus
Institute for Pharmacology, Laboratory of Physio-
logical Chemistry, Medical Faculty, Institute of
Molecular Biology, University of Utrecht,
Padualaan 8, Utrecht, The Netherlands and
*Department of Biochemistry and Nutrition,
Medical School, UNC, Chapel Hill, N.C., USA.

INTRODUCTION

Investigations on the role of the pituitary-adrenal
system in the adaptation of the organism to environmental
stimuli, led de Wied to postulate that the pituitary would
manufacture peptides which modulate behavior by a direct
action on the brain (neuropeptides; DE WIED, 1969). Indeed, fragments of hormones of both the posterior pituitary (vasopressin, oxytocin) and the anterior pituitary
(ACTH, MSH, β-LPH) were found to possess strong behavioral
activity which was clearly dissociated from effects on the
endocrinon. Much effort was given to characterize the
effects of the peptides on animal behavior and the significance of the research on behaviorally active peptides
became eminent when independent reports in the literature
pointed to a physiological role of known and hitherto unknown peptides in normal and abnormal human behavior
(DE WIED and GISPEN, 1977; DE WIED, 1978). To date research on neuropeptides is a multidisciplinary struggle to
unravel their location in the brain and their mechanism of
action in terms of electrophysiological and neurochemical
events and to assess their usefulness in the treatment of
brain and behavioral diseases.

Acquisition of new behavioral patterns is generally believed to occur through changes in interneuronal connectivity (plasticity) so that (functionally) new pathways are formed. It is difficult to conceive of such permanent connectivity changes occurring in the absence of underlying chemical changes. Phosphorylation could alter the conformation of proteins thus affecting essential membrane properties at the synaptic cleft and hence interneuronal connectivity. Evidence is accumulating that phosphorylation of synaptic proteins may play a key role in nervous system functions (HEALD, 1962; GREENGARD, 1976) and behavior (GLASSMAN, et al., 1973; ROUTTENBERG, et al., 1975; EHRLICH, et al., 1977; PERUMAL, et al., 1977; HOLMES, et al., 1977).

The present paper focuses on the neurochemical mechanism of action of the behaviorally active peptide ACTH with special reference to its effects on brain cyclic nucleotides and synaptosomal plasma membrane phosphorylation.

Origin and function of ACTH in the CNS

Recent evidence indicates that the peptide hormones ACTH and LPH originate from a large molecule (prohormone) with a molecular weight in the range of 31,000 (31K, MAINS et al., 1977; ROBERTS and HERBERT, 1977). Enzymatic cleavage may release $ACTH_{1-39}$ and it is hypothesized that smaller fragments (α-MSH = $Ac-ACTH_{1-13}-NH_2$, $ACTH_{4-10}$, $ACTH_{4-7}$) may be produced by further enzymatic degradation of the parental molecule. The classical target for pituitary ACTH is the adrenal cortex. Demonstration of its behavioral activity has suggested that the brain also serves as a target for ACTH and its active fragments (DE WIED, 1964; GISPEN, et al., 1977).

Although circulating ACTH reaches brain structures involved in the regulation of adaptive behavior (VERHOEF, et al., 1977a,b), new evidence points to a possible direct transport of ACTH from pituitary to septum (MEZEY, et al., 1978). That the brain itself could be a source of ACTH was first suggested on the basis of data on levels of ACTH in human cerebrospinal fluid (ALLEN, et al., 1974). This suggestion was further supported by the demonstration of immuno-reactive ACTH in several brain structures (KRIEGER, et al., 1977). At present, immunocytochemical studies strongly suggest the presence of a 31K prohormone system in hypothalamic neurons and possibly of ACTH as a free peptide (AKIL, et al., 1978). Thus the regulatory action

of ACTH on behavior could be envisaged to originate from
two alternative peptide sources, i.e. pituitary cells and/
or peptidergic neurons.

As pointed out before, ACTH and its fragments are
supposed to play an essential role in homeostasis. Its
regulatory role in central nervous functions and behavior
was discovered in rats whose pituitary-adrenal system was
disrupted or functionally suppressed (DE WIED, et al.,
1972). Over the years numerous studies using a variety
of behavioral paradigms and subjects have added to the
knowledge about the functional significance of the regulatory role of ACTH in the brain. In an attempt to unify
the known behavioral and neurophysiological effects in one
hypothesis, Bohus and de Wied (1978) favor a mechanism
involving motivation. They postulate that ACTH and its
fragments increase the state of arousal in midbrain limbic
structures, which may determine the motivational influence
of environmental stimuli and thereby the probability of
the generation of stimulus-specific behavioral responses
(BOHUS and DE WIED, 1978).

If ACTH, MSH or fragments from LPH are injected directly into the cerebrospinal fluid of mammals, a stretching and yawning syndrome occurs (FERRARI, et al., 1963).
In rodents this syndrome is preceded by the display of
excessive grooming. The so-called grooming response is
characterized by the enhanced display of vibrating of the
front paws, washing, grooming, scratching, paw licking,
tail licking, etc. In Fig. 1 A,B,C three elements of the
grooming response are shown: washing, grooming and scratching. Fig. 1D and E show examples of the stretch and
yawn usually seen following a long episode of excessive
grooming. Fig. 1D depicts a stretch which is usually seen
following a long episode of excessive grooming. The behavior is recorded, using a 15 sec sampling technique,
establishing whether or not one of these behavioral elements is displayed by the animal. Previously, it has been
shown that the data obtained by this procedure are in good
agreement with data collected using duration as an Index
of activity (GISPEN, et al., 1975). From the behavioral
elements constituting the excessive behavioral response,
the grooming element itself is by far the predominant
seen (70% of total). Therefore, it was decided to refer
to excessive grooming in keeping with previous reports.

The behavioral significance of the grooming response
is not entirely established. Grooming in rodents is sometimes interpreted as representative of "displacement activities" but other investigators think of grooming as a col-

A: Washing

B: Grooming

C: Scratching

D: Stretching

E: Yawning

Fig. 1:   The behavioral response after interventricular
          injection of $ACTH_{1-24}$ in rats.

lateral act, i.e. behavior associated with, but not part
of, a goal-directed activity (JOLLES, et al., 1979).
While feedback from the periphery is usually important for
the development and maintenance of integrated movement re-
pertoire like grooming, there is evidence for a strong internal control of grooming (FENTRESS, 1973).   The peptide-
induced grooming is not observed when the peptides are administered peripherally.   In view of the route of administration, the short onset latency of the effect and its
independence of endocrine activity, the induction of excessive grooming behavior seems the result of a direct
effect of ACTH on the CNS (GISPEN, et al., 1975).

The induction of excessive grooming by intraventricular administration of $ACTH_{1-24}$ was studied in rats with lesions in midbrain-limbic structures. Such areas have been reported to be implicated in mediating ACTH-induced effects on avoidance behavior, sexual excitement or stretching and yawning, (see COLBERN, et al., 1977). Electrolytic lesions in the septal complex, the anterior hypothalamic/preoptic area, the mammillary bodies, the amygdala, the posterior thalamus and dorsal or ventral hippocampus did not interfere with ACTH-induced excessive grooming. Lesioning of the hippocampal complex by aspiration led to an inhibition of excessive grooming depending on the degree of hippocampal damage. Amygdala and hippocampal lesions enhanced the display of stretching and yawning activity after treatment with the peptide. These data indicate differences in the neural substrates mediating the effect of ACTH on extinction of conditioned avoidance behavior, excessive grooming, sexual excitement and stretching and yawning (COLBERN, et al., 1977).

Neuropharmacological studies also pointed to differences between the effects of ACTH on adaptive processes as discussed above and those on the induction of excessive grooming. It was demonstrated that pre-treatment of rats with the opiate antagonist naltrexone did not abolish the effect of ACTH or β-LPH fragments on avoidance extinction (DE WIED, et al., 1978), but completely suppressed the induction of excessive grooming (GISPEN and WIEGANT, 1976). Further characterization of the neural substrate involved in the induction of excessive grooming pointed to the essential role of dopaminergic fibers originating in the substantia nigra and projecting in the neostriatum and nucleus accumbens (WIEGANT, et al., 1977a; COOLS, et al., 1978). The grooming behavior could be elicited from the substantia nigra by local injections of $ACTH_{1-24}$. In addition local application of dopamine antagonists in the neostriatum and n. accumbens interferred with the induction of excessive grooming. Despite the difference in neural substrate, it was hypothesized that in the case of excessive grooming also peptide-induced arousal is a causative stimulus. Thus, the grooming response would reflect a de-arousal mechanism triggered by the arousal brought about by ACTH (JOLLES, et al., 1979).

Neurochemical mechanism of action of ACTH

There is sufficient evidence available to indicate that ACTH or its fragments affect cerebral RNA and protein synthesis, cyclic nucleotide metabolism, protein phosphorylation and neurotransmitter turnover (acetylcholine,

dopamine and noradrenaline; DUNN and GISPEN, 1977). The
scarcity of knowledge about the first steps in the pep-
tide-nerve cell interaction (recognition, activation of
receptors, etc.) greatly hampers our understanding of the
neurochemical mechanism of ACTH. For instance, knowing
the localization of ACTH receptors on nerve cells might
help to decide whether ACTH should be considered to be a
neuromodulator or a putative neurotransmitter (BARCHAS,
et al., 1978). The complexity of the known behavioral,
electrophysiological and neurochemical responses to ACTH
implies that most likely more than one type of mechanism
is involved (differences in receptor, neural substrate,
neurochemical response, etc.). In an attempt to bring
most of the known neurochemical events in a causal relationship, we took the view that ACTH would not only act
as neuromodulator but also would exert trophic influences
on its target cells in brain as it does in other tissues
(DUNN and GISPEN, 1977; GISPEN, et al., 1977b). By doing
so, the peptide would enhance protein metabolism and alter
intercellular communication (interneural connectivity),
but would not be responsible for direct transynaptic transfer of a nerve signal. The proposed mechanisms of ACTH
action on neurons is represented in Fig. 2. ACTH interacts with specific receptors on the outside of the cell
membrane, thus activating adenylate cyclase (or perhaps
guanylate cyclase). Cyclic nucleotide dependent protein
kinases then may activate mRNA in the nucleus, activate
protein synthesis by polyribosomes in the perikaryon,
modify the properties of post-synaptic receptors, stimulate neurotransmitter synthesis or modulate neurotransmitter release, or any combination of these. Not all
cells respond to ACTH and even responsive cells will only
show one or a few of these responses (DUNN and GISPEN,
1977).

ACTH and brain nucleotides

Up to now, few reports have been published on the
relation of central effects of peptides and the cyclic
nucleotide systems of the brain. Occasionally, ACTH has
been included in studies on the effects of putative transmitters and hormones on brain adenylate cyclase in cell-
free membrane preparations. Burkhard and Gey (1968) and
Von Hungen and Roberts (1973) were unable to detect an
effect of ACTH on adenylate cyclase in such preparations.
Forn and Krishna (1971) did not observe an effect of ACTH
on cAMP accumulation in rat brain cerebral cortex slices.
On the other hand, indirect indications that ACTH-like
peptides might affect brain cyclic nucleotide levels

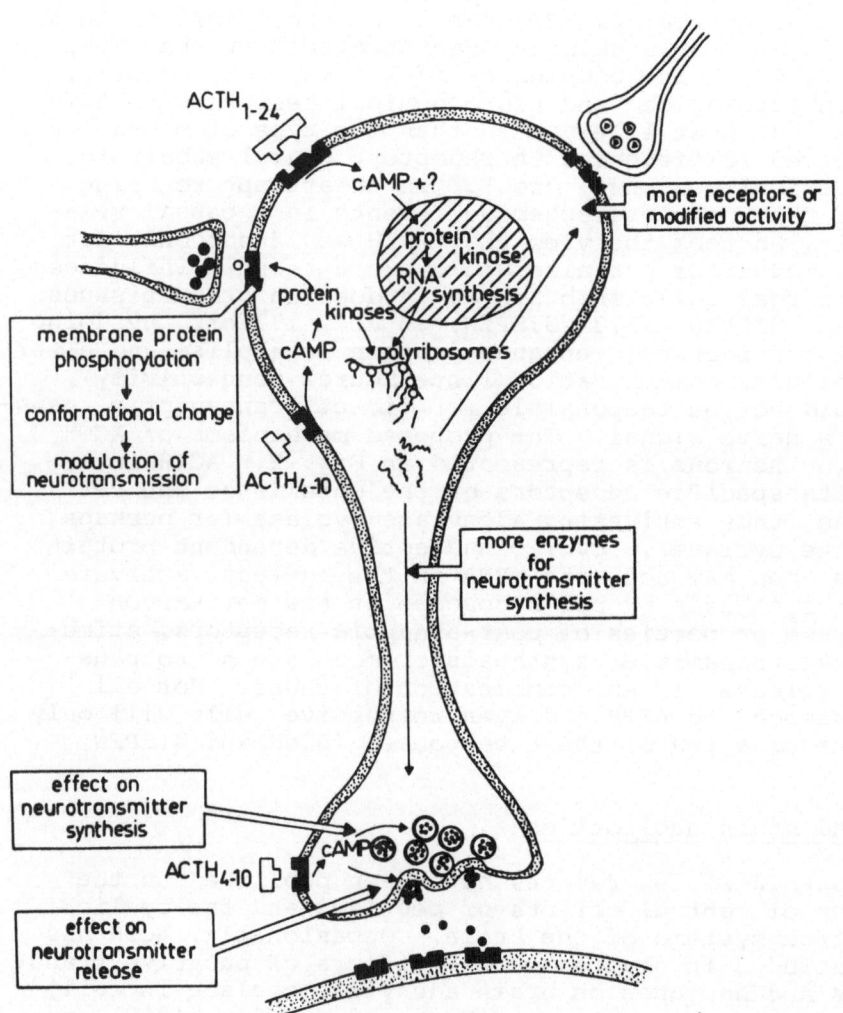

Fig. 2: Neurochemical mechanism of action of ACTH,
        (see DUNN and GISPEN, 1977).

in vivo, were presented by Rudman and co-workers (RUDMAN and ISAACS, 1975; RUDMAN, 1976). They showed that intra-cisternal injection of ACTH or β-MSH in rabbits increased the cAMP but not the cGMP concentration in cerebrospinal fluid. In a preliminary study it was reported that chronic treatment of rats with α-MSH (= Ac-Ser^1|ACTH$_{1-13}$-NH$_2$) in-creased the level of cAMP in the occipital cortex (CHRIS-TENSEN, et al., 1976). Similar treatment left the level of cGMP unaltered in all brain regions studied (SPIRTES, et al., 1978).

Recently, we investigated the influences of N-terminal fragments of ACTH on the accumulation of cAMP in rat brain using three different approaches viz. broken cell prepara-tions (adenylate cyclase activity), slices from posterior thalamus and neostriatum and in vivo levels (WIEGANT and GISPEN, 1975; WIEGANT, et al., 1979). ACTH$_{1-24}$ had a bi-phasic effect on the activity of adenylate cyclase in broken cell preparations of rat brain subcortical tissue; concentrations below 25 μM stimulated, whereas concen-trations of 0.1 mM and higher inhibited adenylate cyclase activity (Fig. 3). The magnitude of the stimulation was dependent on the concentrations of ATP and Mg^{++} in the incubation medium. Under optimal conditions in synapto-somal plasma membrane fractions, maximal stimulation of adenylate cyclase occurred at 0.1 - 1.0 μM ACTH$_{1-24}$. In membrane preparations derived from peripheral target cells (adrenal cortex, fat cells) maximal stimulation of the enzyme also required ACTH-concentrations in the mi-cromolar range (BIRNBAUMER, et al., 1969; GLOSSMAN and GIPS, 1975; LANG, et al., 1976; SCHLEGEL and SCHWYZER, 1977).

In our hands the magnitude of the stimulation was extremely variable. This prevented pharmacological char-acterization of this stimulatory effect of ACTH$_{1-24}$. The inhibition of brain adenylate cyclase in vitro by concen-trations of ACTH$_{1-24}$ higher than 5 μM could not be explain-ed as a secondary effect resulting from an action of the peptide on the residual activity of phosphodiesterase, or from peptide-induced changes in pH or ATP concentration. Therefore, it was concluded that this inhibition resulted from a direct action of ACTH on the rate of formation of cAMP catalyzed by adenylate cyclase. Structure activity studies revealed that in a concentration of 10^{-4} M ACTH$_{1-16}$ -NH$_2$ and ACTH$_{4-7}$ also inhibit the activity of adenylate cyclase, whereas ACTH$_{11-24}$, ACTH$_{1-10}$, ACTH$_{4-10}$, |D-Phe7| - ACTH$_{1-10}$ and |D-Phe7| ACTH$_{4-10}$ are inactive in this res-pect.

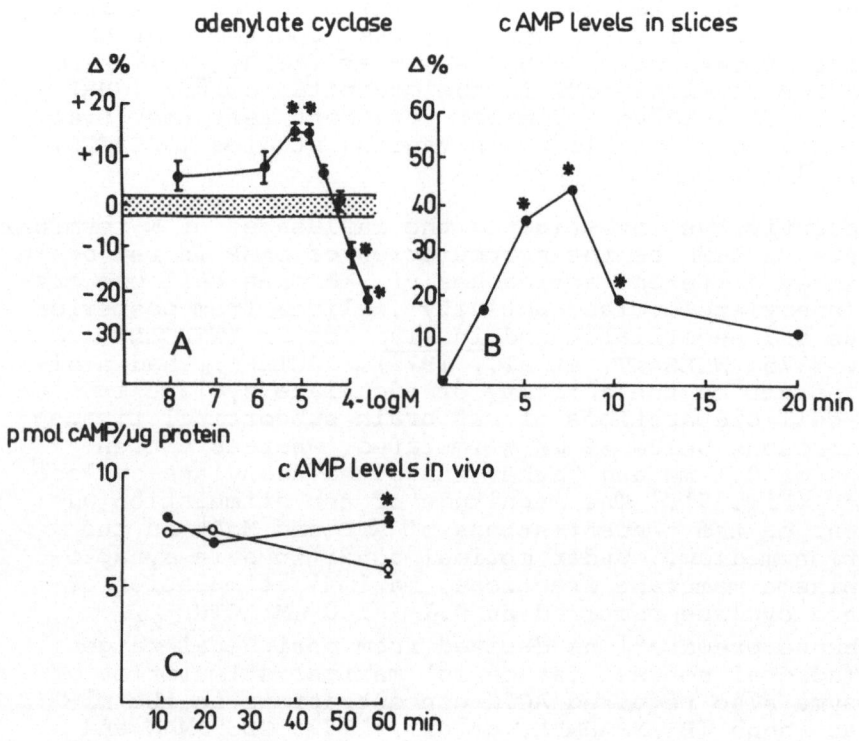

Fig. 3: Effects of ACTH on brain cyclic nucleotides.

A.   Dose response curve for the effect of $ACTH_{1-24}$ on the
      activity of adenylate cyclase in a homogenate of rat
      brain subcortical tissue (n=5). Basal activity in
      this preparation was 189 $\pm$ 5 pmol cAMP/mg protein/
      min.

B.   Time course of the effect of 10^{-4} M $ACTH_{1-24}$ on the
      accumulation of cAMP in slices of rat neostriatum
      (n=5). Basal level = 2 $\pm$ 0.1 pmol cAMP/mg protein.

C.   The effect of intraventricular injection of $ACTH_{1-16}$
      (1 μg) on the level of cAMP in rat septum as deter-
      mined after microwave fixation of the brain.

*   p <0.05 (Student t-test, two-tailed). For further
      experimental details see WIEGANT, et al., 1979.

$ACTH_{1-24}$ dose-dependently enhanced the accumulation of cAMP in slices from rat brain neostriatum and posterior thalamus, but did not influence the concentration of cGMP in striatal slices. The effect of ACTH on cAMP occurred rapidly, and was of short duration (Fig. 3). Isobutyl-methylxanthine (IBMX), an inhibitor of phosphodiesterase, potentiated the effect of $ACTH_{1-24}$ suggesting that the increase in cAMP concentration in the striatal slices was not the result of inhibition of the hydrolysis of cAMP but rather of an $ACTH_{1-24}$-induced activation of adenylate cyclase activity, perhaps by an interaction of an ACTH receptor with adenylate cyclase.

Intraventricular administration of a low dose of $ACTH_{1-16}-NH_2$ (1 μg) resulted in a significant 27% increase of septal cAMP concentration, 60 min. after the injection (Fig. 3). No effect of the peptide could be detected in the other brain regions studied, including the neostria-tum. Others have also reported relatively small changes in levels of cAMP after electrical stimulation of nerve cell bodies (KORFF, Personal Commun.). Therefore, the in-crease in septal cAMP was taken to reflect a significant alteration in synaptic activity in this region. In view of the importance of the septal complex to the expression of the behavioral activity of ACTH (VAN WIMERSMA, GREIDA-NUS, et al., 1975; VERHOEF, et al., 1977a,b), a direct effect of ACTH-like peptides on septal cells resulting in an increase in cAMP seems possible. From the present re-sults, however, we cannot exclude indirect effects such as neurotransmitter mediated effects, etc.

In summary, N terminal fragments of ACTH modulate the activity of brain adenylate cyclase in broken cell preparations, stimulate the accumulation of cAMP in slices of brain tissue containing intact cells through activation of adenylate cyclase, and increase septal cAMP in vivo. These findings evidence a role of ACTH-like neurotrophic peptides as modulators of brain adenylate cyclase although, clearly, cAMP cannot be taken as a possible second mes-senger for all the effects of ACTH-like peptides on the central nervous system (see below).

Phosphorylation of synaptic plasma membrane proteins

As pointed out in the introduction, phosphorylation of synaptic membrane proteins may affect the transmission of information between neurons. As early as 1962, HEALD speculated that the change in protein phosphorylation that occurred in response to electrical stimulation of res-piring brain slices altered the conformation of neuronal

membrane proteins, and that such changes might be involved
in the regulation of ion movements through cell membranes
(HEALD, 1957, 1962). Trevor and Rodnight (1965) demonstrated that the protein-bound phosphoryl-serine groups
that responded to electrical stimulation were indeed in
cerebral membrane fractions. In a more defined approach,
BROWNING, et al. (1977, 1978) reported that synaptic
potentiation of the Schaeffer collaterals in rat hippocampal slices led to an altered Ca^{++}-dependent phosphorylation of a protein band, with a molecular weight of about
40 K, associated with synaptic plasma membrane fraction.

It was recently suggested that phosphorylation of
specific protein substrates is involved in the release
of neurotransmitters. Such a release is calcium-dependent and experiments were reported describing a calcium-
dependent phosphorylation in relation to the exocytosis
of neurotransmitters (KATZ and MILEDI, 1967; DOUGLAS,
1973; DE LORENZO, 1976; REDBURN, et al., 1976; DE LORENZO
and FREEDMAN, 1977; KRUEGER, et al., 1977; HERSHKOWITZ,
1978; SCHULMAN and GREENGARD, 1978; SIEGHART, et al.,
1978).

With respect to cAMP-sensitive brain membrane protein phosphorylation it was hypothesized that the enhanced phosphorylation of the substrate protein accounts
for the membrane hyperpolarization observed after the
exposure of the nerve cell to dibutyryl cAMP (GREENGARD,
1976; NATHANSON and GREENGARD, 1977). The involvement
of protein phosphorylation in brain function is further
evidenced by reports on correlative changes in membrane
phosphoproteins with a variety of behavioral experiences
(GLASSMAN, et al., 1973; PERUMAL, et al., 1975, 1977;
EHRLICH, et al., 1977; GISPEN, et al., 1977c; HOLMES, et
al., 1977).

The first demonstration of a correlate between behavior and protein phosphorylation concerned the acquisition of a conditional avoidance response in mice and the
enhanced incorporation in vivo of radioactive phosphate
into total protein of synaptosome-enriched fractions of
their brains (GLASSMAN, et al., 1973). The increase
could only be detected in the particulate fraction of the
synaptosomes. Initial neurochemical characterization
confirmed that the radioactivity was covalently bound to
amino acids in membrane proteins (PERUMAL, et al., 1975,
1977). Further behavioral studies revealed that the increased phosphorylation of these proteins is specific to
the conditioning experience, since mice that were merely

exposed to the conditioned and unconditioned stimuli or
performed the avoidance after they had been previously
trained, did not show the response. However, mice that
extinguished the learned behavior did show such increased
phosphorylation (GISPEN, et al., 1977c).

     Using a different experimental paradigm Routtenberg
and Ehrlich (ROUTTENBERG, et al., 1975; EHRLICH, et al.,
1977) have shown that the rate of in vitro phosphorylation of rat and mouse brain membrane proteins is altered
by behavioral experience. Interestingly, the proteins
whose phosphorylation was maximally altered by the preceding training appeared to be different than those whose
phosphorylation was maximally stimulated by cAMP in vitro.
Recently, this in vivo/in vitro approach was also used
to demonstrate that in rats the anxiety accompanying an
electroshock procedure can induce an increase in the in
vitro activity of a membrane-bound protein phosphorylation system, probably involved in synaptic function
(HOLMES, et al., 1977). Thus both in vivo/in vitro studies point to a behaviorally induced alteration of brain
membrane protein phosphorylation. Direct comparison of
the studies mentioned above remains, however, difficult
as it was recently reported that the proteins that preferentially accumulate radioactive phosphate in vivo are
not necessarily the ones that preferentially accept phosphate in vitro (BERMAN, et al., 1977; WILSON, 1977).
Further studies by these authors indicate that proteins
associated with the postsynaptic density, although difficult to phosphorylate under in vitro conditions, readily
incorporate phosphate in vivo and thus may represent a
major proportion of the labelled proteins of the synaptic
membranes (BERMAN, et al., 1978).

     We have been studying the endogenous phosphorylation
of proteins from rat brain synaptosomal plasma membranes
in vitro with or without the behaviorally active peptide
$ACTH_{1-24}$ in the incubation medium. Many factors may influence this in vitro process (RODNIGHT, et al., 1975).
Important determinants are: the activity of ATPase(s)
controlling the availability of the phosphate donor (ATP),
the activity of the protein phosphatase(s) and of course
that of the protein kinase(s). As was shown for total
membrane protein (RODNIGHT, et al., 1975), the phosphorylation of a given SPM protein band depends very much on
the ATP/SPM ratio used in the incubation system (WIEGANT,
et al., 1978b). Therefore, as a routine, a low and a
high ATP/SPM ratio were used simultaneously, to study the
effects of ACTH on endogenous SPM phosphorylation.

At first we reported that $ACTH_{1-24}$ when added <u>in vitro</u>
reduced the incorporation of phosphate into a number of
SPM protein bands which, based on their electrophoretic
mobility in SDS-PAGE, have molecular weights smaller
than those previously reported to be affected by the <u>in
vitro</u> addition of cAMP (ZWIERS, et al., 1976). In a
number of protein bands a biphasic effect of $ACTH_{1-24}$ was
observed; in concentrations of 10^{-4}- 10^{-5} M $ACTH_{1-24}$, a
reduced incorporation was apparent, in concentrations of
10^{-6}- 10^{-7}M hardly any effect could be detected, whereas
at 10^{-8} M a consistent but small decrease was observed.
This latter effect could only be detected when the SPM
was freshly prepared (ZWIERS, et al., 1976). It was
shown that the effect of ACTH was caused by an inhibition
of SPM protein kinase(s) and not by a stimulation of SPM
protein phosphatase (ZWIERS, et al., 1976, 1978). Using
N-terminal fragments of ACTH, the structure activity requirements for the inhibition were studied. A rather
complex interaction of the ACTH fragments with endogenous
SPM phosphorylation was observed (ZWIERS, et al., 1978b).
The effects were not only dependent on the primary structure of the peptide studied, but also on the ATP/SPM ratio
and varied for the various protein bands. In Fig. 4,
the structure activity relationships for the inhibition
of the endogenous phosphorylation of a phosphoprotein
band B-50 (MW 48 K) and those for the induction of excessive grooming in the rat are shown. It is clear that
the requirements for both these CNS effects of ACTH are
very similar. One of the important findings is that in
neither instance does the combination of the sequences
$ACTH_{1-10}$ and $ACTH_{11-24}$ act as the total sequence $ACTH_{1-24}$
(GISPEN, et al., 1975; ZWIERS, et al., 1978b).

Furthermore, preliminary evidence indicates that the
endogenous <u>in vitro</u> phosphorylation of specific SPM proteins prepared from rats treated intraventricularly with
$ACTH_{1-24}$ differs from <u>in vitro</u> phosphorylation of SPM proteins from saline-treated animals (ZWIERS, et al., 1977).
There appeared to be a U-shaped dose-response curve with
significantly enhanced incorporation of phosphate into
SPM protein bands after injection of 30 or 300 ng $ACTH_{1-24}$.
This effect of <u>in vivo</u> ACTH treatment on <u>in vitro</u> endogenous phosphorylation was not only dose- but also time-
dependent, giving the largest effects 30 min. after the
peptide injection (ZWIERS, et al., 1977).

In Fig. 5 the effect of magnesium and calcium on the
endogenous phosphorylation of protein band B-50 is shown.
Using a low ATP/SPM ratio, the B-50 protein kinase is
stimulated by both Ca^{++} and Mg^{++} . At relatively high

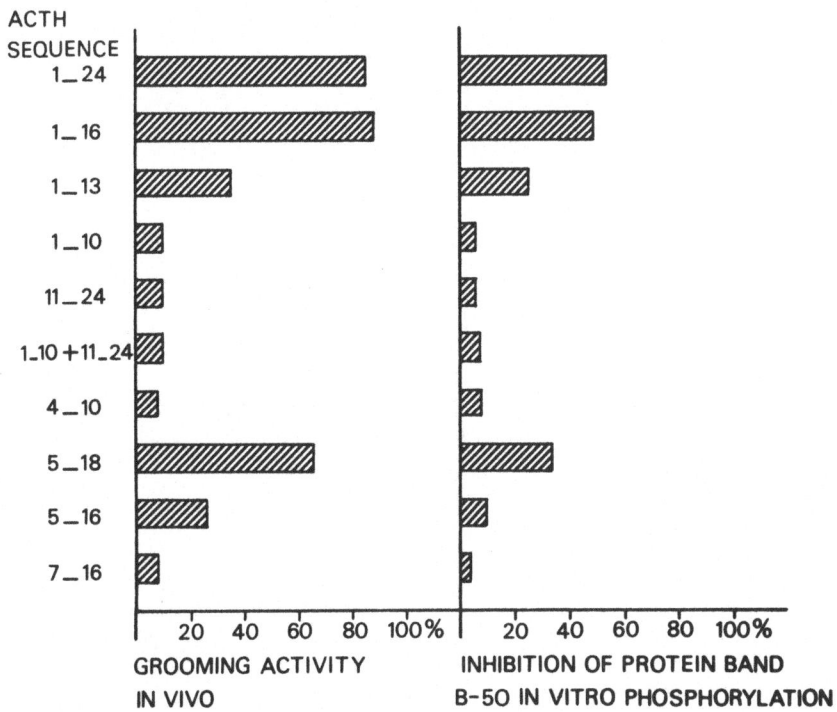

Fig. 4: Structure activity studies on the effect of ACTH
on excessive grooming in the rat and inhibition of
SPM protein phosphorylation in vitro. The grooming
activity was elicited by intraventricular administration of doses equimolar to 3 μg $ACTH_{1-24}$. The
inhibition of phosphorylation was studied in the
presence of 10^{-4} M peptide.

Ca^{++}a marked drop in enzyme activity was observed. The
optimal conditions for Mg^{++} and Ca^{++} are 10^{-2}M and 10^{-3} M
respectively. Under optimal conditions also the largest
effect of ACTH was observed. Interestingly at 5 x 10^{-2}M
Mg^{++} ACTH was not effective anymore. It may be that ACTH
competes with Mg·ATP for a site at the catalytic unit
of the B-50 protein kinase. Further study with regard
to this aspect is in progress. Although apparently the
divalent cation sensitivity of B-50 protein kinase differs
from that reported by HERSHKOWITZ (1978) for his protein C,

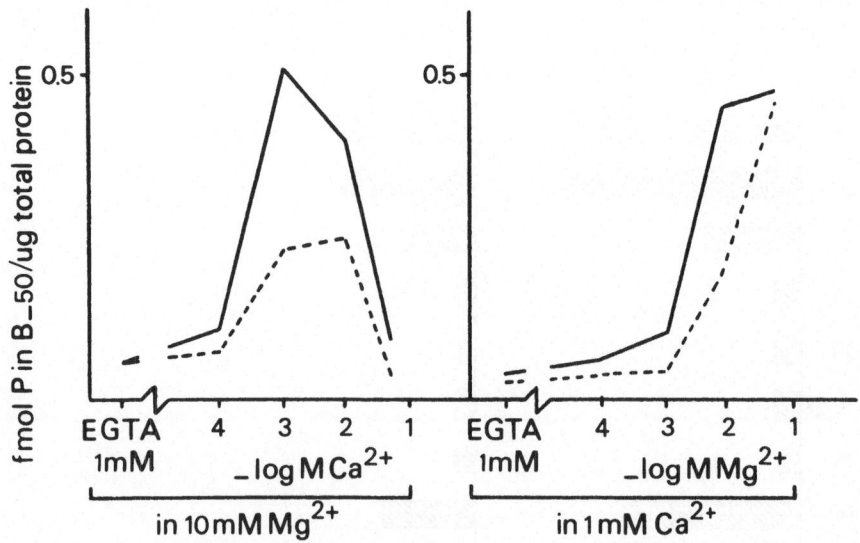

Fig. 5: Effect of Ca^{++} and Mg^{++} on the endogenous phosphorylation of SPM protein band B-50. All incuba-
tions were carried out in 1 mM EGTA, 7.5 μM ATP,
3 μCi α-^{32}P ATP and 50 μg SPM protein. The dotted
line represents the phosphorylation in the presence
of 10^{-5}M ACTH$_{1-24}$. For experimental details see
ZWIERS, et al. (1976, 1978).

it is of interest that the phosphorylation of a Ca^{++}-de-
pendent membrane protein band is susceptible to the behaviorally active peptide ACTH$_{1-24}$. It may be that this
direct effect on SPM phosphorylation underlies the modulatory role ACTH has on neurotransmission and behavior
(see below).

Recently, we demonstrated that treatment of the rat
brain SPM with 0.5% Triton X-100 in 75 mM KCl solubilized
15% of the total membrane B-50 protein kinase activity and
more importantly, preserved the sensitivity of the enzyme
to ACTH$_{1-24}$ (ZWIERS, et al., 1978a). The solubilized membrane material also phosphorylated exogenous B-50 proteins
which were previously extracted from similar rat brain
membranes. Column chromatography of the solubilized material over DEAE-cellulose pointed to the presence of
multiple protein kinase activities in rat brain membranes
(MIYAMOTO, et al., 1978), one being the ACTH-sensitive

B-50 protein kinase (ZWIERS, et al., 1979). Further
study is in progress to characterize and purify this
ACTH-sensitive membrane-bound protein kinase. So far,
we have demonstrated the existence of a cAMP-independent,
ACTH-sensitive phosphorylation of SPM proteins in vitro.
Structure-activity studies and experiments on the effect
of in vivo treatment of rats with ACTH revealed that most
likely this effect on SPM phosphorylation is related to
the behavioral activity of the ACTH peptide.

Working hypothesis: neurohormone and neuromodulator

     There is no doubt that ACTH or related peptides in-
fluence or modulate the functioning of the central nervous system. The diversity of the presently known neurophysiological and neurochemical effects makes it unlikely
that one single mechanism of action underlies the CNS
effects of ACTH. In Fig. 6, three possible modes of
action are visualized:

A. In keeping with the observation of peptidergic (ACTH)
neuronal pathways and ACTH-sensitive cells in iontophoretic studies, one could speculate that ACTH or its fragments
should be considered as putative neurotransmitters. The
effects of the peptides on cyclic nucleotide matabolism
are then analogous to those reported for many other neurotransmitters. However, at present, very little if any
evidence is available to further support this notion.
Therefore, we will focus on the two other possible mechanisms of action, i.e. ACTH as neurohormone or as neuromodulator.

B. Viewing the brain as a target for circulating ACTH
(blood, CSF), one may expect to find a brain-ACTH interaction analogous to that documented for peripheral target
tissues (SAYERS, et al., 1974). Thus, in addition to responses of the target cells as steroidogenesis or lyposly-
sis, one will find trophic effects on the metabolism of
macromolecules (cell growth and cell division). It is not
clear whether this latter response involving, for instance,
effects on target cell protein metabolism is mediated by
the same peptide-cell interaction responsible for steroidogenesis. It may be that so-called "internalisation" of
hormone-receptor complexes (see KOLATA, 1978) is a key
element in the mechanism by which peptide hormones affect
their target cell-protein metabolism. This would provide
the possibility that the peptide directly could affect its
target cell protein synthesis machinery. Thus, viewing
the brain as target of the neurohormone ACTH, it may be

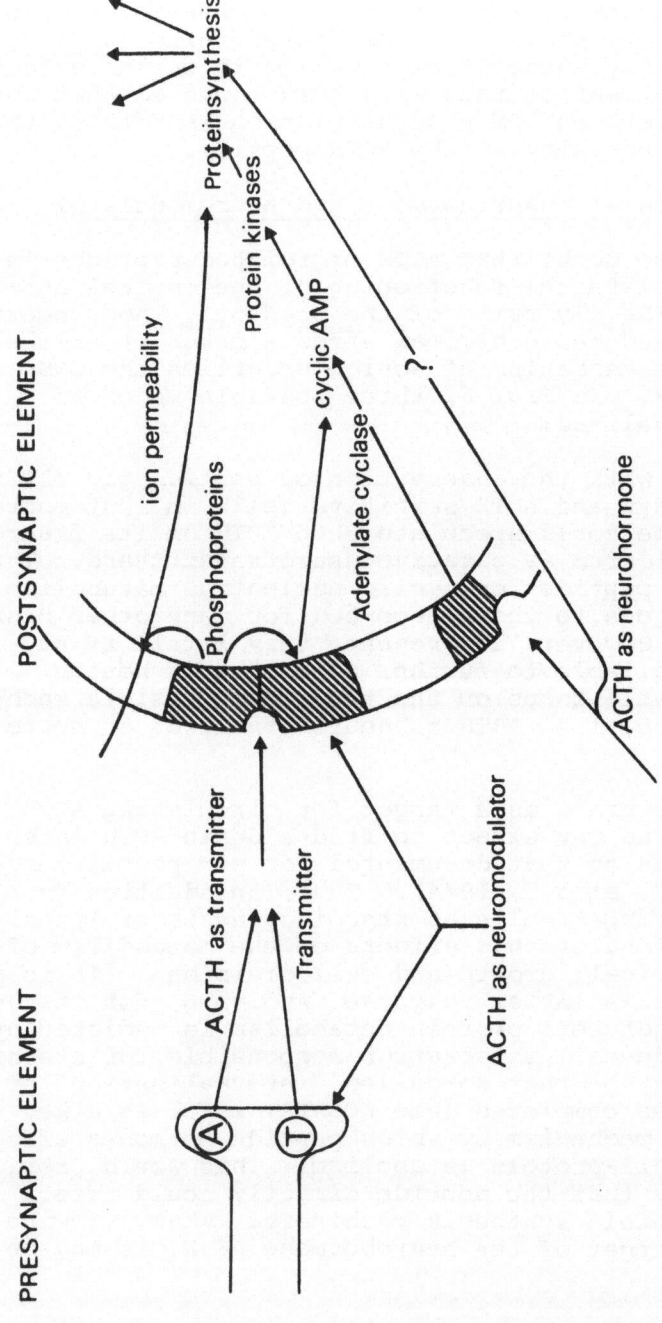

Fig. 6: Three possible modes of action of ACTH: Neurotransmitter, neurohormone and neuromodulator.

that the observed effects on brain RNA and protein synthesis represent the neurotrophic influence exerted by ACTH.
Whether or not changes in levels of cyclic nucleotides
function as regulatory signals in this mechanism, is not
clear.

Of course, one should keep in mind that all hitherto
described effects of ACTH on brain protein synthesis in
vivo and in vitro could have been the result of indirect
stimulation through, for instance, the release of neurotransmitters affecting postsynaptic neuronal metabolism.
The effects on protein synthesis would then be envisaged
to be brought about by ACTH acting as a neuromodulator.

C.   Neuromodulators are entities which alter the flow of
information between neurons.   The mechanism of action
should be rapid in view of the physico-chemical events
underlying neurotransmission.   It may be that they affect
processes which are related directly to transmitter availability (release, re-uptake, turn-over, etc.).   Modulation
of neurotransmission may also be brought about by altering
the electrical properties of synaptic membranes.   One of
the mechanisms by which the latter could be achieved, is
the alteration of ion permeability.

We propose that the direct effect of ACTH on a membrane-bound protein kinase may underlie a modulatory role
in neurotransmission.   Since the membranes used in our
studies, mainly will have been of presynaptic origin, it
may well be that this effect on membrane phosphorylation
represents a presynaptic event.   As the B-50 protein
kinase is stimulated by magnesium and calcium, it may be
that the function of the B-50 substrate protein resembles
that described for other Ca-dependent phosphoproteins
(see above), i.e., Ca-dependent exocytosis of neurotransmitters.   If at the other hand the B-50 protein kinase-
substrate protein comples affected by ACTH primarily
would be of postsynaptic nature, the observed inhibition
of phosphorylation may alter ion permeability and hence
sensitivity to presynaptic input as discussed for the role
of cAMP in neurotransmission (NATHANSON AND GREENGARD,
1977).   Currently we are preparing antibodies to a rather
pure protein-B-50 preparation in order to be able to localize B-50 in nerve cells and tissue.

It is clear that more work is necessary before one can start to unravel the mechanism by which ACTH affects the functioning of the central nervous system. The absence of a proven specific binding site (receptor) at the neuron outer membrane greatly hampers our present understanding. Thus it is vital to identify the first step in the ACTH-nerve cell interaction. The important role neuropeptides play in animal and human behavior prompts us to further study their exact mechanism of action.

ACKNOWLEDGEMENT

Part of the research discussed in this chapter was supported by The Netherlands Organization for the Advancement of Pure Research, The Hague.

REFERENCES

ALLEN, J. P., KENDALL, J. W., MCGILVRA, R. & VANCURA, C. (1974). Immunoreactive ACTH in cerebrospinal fluid. J. Clin. Endocrinol. 38, 586-593.

AKIL, H., WATSON, S. J., LEVY, R. M. and BARCHAS, J. D. (1978). β-Endorphin and other 31 K fragments: pituitary and brain systems. In: Characteristics and Functions of Opioids. (J. van Ree and L. Terenius, eds.) Elsevier/ North Holland Biomed. Press, Amsterdam. pp. 123-134.

BARCHAS, J. D., AKIL, H., ELLIOTT, G. R., HOLMAN, R. B. & WATSON, S. J. (1978). Behavioral Neurochemistry: Neuroregulators and behavioral states. Science. 200, 964-973.

BERMAN, R. F., HULLIHAN, J. P. & WILSON, J. E. (1977). Comparison of in vivo and in vitro phosphorylation of synaptic membranes. Soc. Neuroscience Abstr., 3, 984.

BERMAN, R. F., KINNIER, W. J., HULLIHAN, J. P. & WILSON, J. E. (1978). In vivo phosphorylation of postsynaptic density proteins. Soc. Neuroscience Abstr. 4, 987.

BIRNBAUMER, L., POHL, S. L. & RODBELL, M. (1969). Adenylcyclase in fat cells. 1. Properties and the effects of adrenocorticotropin and fluoride. J. Biol. Chem. 244, 3468-3470.

BOHUS, B. & DE WIED, D. (1978). Pituitary-adrenal system hormones and adaptive behavior. In: General, Comparative and Clinical Endocrinology of the Adrenal Cortex. (J. CHESTER JONES & I. W. HENDERSON, eds.) Vol. 3, Acad. Press London, in press.

BROWNING, M., DUNWIDDIE, T., GISPEN, W. H. & LYNCH, G.
    (1977). Alterations in a specific repetitive stimu-
    lation of the hippocampus. Soc. Neuroscience Abstr.
    3, 1341.
BROWNING, M., DUNWIDDIE, T., BENNETT, W., GISPEN, W. H. &
    LYNCH, G. (1978). Synaptic phosphoproteins: Specific
    changes after repetitive stimulation of the hippo-
    campal slice. Science, 203, 60-62.
BURKHARD, W. P. & GEY, F. (1968). Adenyl cyclase in rat
    brain. Helv. Physiol. Pharmacol. Acta. 26, 197-198.
CHRISTENSEN, C. W., HARSTON, C. T., KASTIN, A. J., KOSTR-
    ZEWA, R.M. & SPIRTES, M. A., Preliminary investiga-
    tion on αMSH and MIF-I effects on cyclic AMP levels
    in rat brain. Pharmac. Biochem. Behav. 5, Suppl. 1,
    117-120.
COLBERN, D., ISAACSON, R., BOHUS, B. & GISPEN, W. H. (1977).
    Limbic-midbrain lesions and ACTH-induced excessive
    grooming. Life Sci. 21, 393-402.
COOLS, A. R., WIEGANT, V. M. & GISPEN, W. H. (1978). Dis-
    tinct dopaminergic systems in ACTH-induced grooming.
    Eur. J. Pharmacol. 50, 265-268.
DELORENZO, R. J. (1976). Calcium-dependent phosphorylation
    of specific synaptosomal fraction proteins: possible
    role of phosphoproteins in mediating neurotransmitter
    release. Biochem. Biophys. Res. Comm. 71, 590-597.
DELORENZO, R. J. & FREEDMAN, S. D. (1977). Calcium-depen-
    dent phosphorylation of synaptic vesicle proteins and
    its possible role in mediating neurotransmitter re-
    lease and vesicle function. Biochem. Biophys. Res.
    Comm. 77, 1036-1043.
DE WIED, D. (1964). Influence of anterior pituitary on
    avoidance learning and escape behavior. Amer. J.
    Physiol. 207, 255-259.
DE WIED, D. (1969). Effects of peptide hormones on be-
    havior. In: Frontiers in Neuroendocrinology
    (GANONG, W. F. & MARTINI, L. eds.). Oxford Univer-
    sity Press, New York. 1969, 97-140.
DE WIED, D., VAN DELFT, A. M. L., GISPEN, W. H., WEIJNEN,
    J. A. W. M. & VAN WIMERSMA GREIDANUS, Tj. B. (1972).
    The role of pituitary-adrenal system hormones on
    active-avoidance conditioning. In: Hormones and
    Behavior (S. LEVINE, ed.). Academic Press, Inc.
    New York: 135-171.
DE WIED, D. & GISPEN, W. H. (1977). Behavioral effects of
    peptides. In: Peptides in Neurobiology (GAINER,
    H., ed.). Plenum Press, New York, 397-448.
DE WIED, D., BOHUS, B., VAN REE, J. M. & URBAN, I. (1978).
    Behavioral and electrophysiological effects of pep-
    tides related to lipotropin (β-LPH). J. Pharm. Exp.
    Ther. 204, 570-580.

DE WIED, D. (1978). Psychopapthology as a neuropeptide
    dysfunction. In: Characteristics and Function of
    Opioids (VAN REE, J. & TERENIUS, L. eds.) Elsevier/
    North Holland Biomedical Press, 123-134.
DOUGLAS, W. W. (1973). How do neurones secrete peptides?
    Exocytosis and its consequences, including 'synaptic
    vesicle' formation, in the hypothalamo-neurohypophy-
    seal system. In: Drug Effects on Neuroendocrine
    Regulation, Progress in Brain Research, Vol. 39,
    (E. ZIMMERMANN, W. H. GISPEN, B. H. MARKS, & D. DE
    WIED, eds.) Elsevier, Amsterdam, pp. 21-39.
DUNN, A. J. & GISPEN, W. H. (1977). How ACTH acts on the
    brain. Biobehav. Reviews. 1, 15-23.
EHRLICH, Y. H., RABJOHNS, R. H. & ROUTTENBERG, A. (1977).
    Experiential input alters the phosphorylation of
    specific proteins in brain membranes. Pharmacol.
    Biochem. Behav. 6, 169-175.
FENTRESS, J. C. (1973). Development of grooming in mice
    with amputated forelimbs. Science, 179, 704-705.
FERRARI, W., GESSA, G. L. & VARGIU, L. (1963). Behavioral
    effects induced by intracisternally injected ACTH and
    MSH. Ann. New York Acad. Sci. 104, 330-345.
FORN, J. & KRISHNA, G. (1971). Effect of norepinephrine,
    histamine and other drugs on cyclic 3'5'-AMP forma-
    tion in brain slices of various animal species.
    Pharmacology. 5, 193-204.
GISPEN, W. H., WIEGANT, V. M., GREVEN, H. M. & DE WIED, D.
    (1975). The induction of excessive grooming in the
    rat by intraventricular application of peptides de-
    rived from ACTH: Structure-activity studies. Life
    Sci. 17, 645-652.
GISPEN, W. H. & WIEGANT, V. M. (1976). Opiate antagonists
    suppress $ACTH_{1-24}$ induced excessive grooming in the
    rat. Neurosci. Lett. 2, 159-164.
GISPEN, W. H., VAN REE, J. M. & DE WIED, D. (1977a).
    Lipotropin and the central nervous system. Int. Rev.
    Neurobiol. 20, 209-250.
GISPEN, W. H., REITH, M. E. A., SCHOTMAN, P., WIEGANT,
    V. M., ZWIERS, H. & DE WIED, D. (1977b). CNS and
    ACTH-like peptides: Neurochemical responses and
    interaction with opiates. In:Neuropeptide Influences
    on the Brain and Behavior. (L. H. MILLER, C. A.
    SANDMAN & A. J. KASTIN, eds.). Raven Press, New
    York, pp. 61-80.
GISPEN, W. H., PERUMAL, R., WILSON, J. E. & GLASSMAN, E.
    (1977c). Phosphorylation of proteins of synapto-
    some-enriched fractions of brain during short-term
    training experience: The effects of various be-
    havioral treatments. Behav. Biol. 21, 358-363.

GLASSMAN, E., GISPEN, W. H., PERUMAL, R., MACHLUS, B. & WILSON, J. E. (1973). The effect of short experiences on the incorporation of radioactive phosphate into synaptosomal and non-histone acid-extractable nuclear proteins from rat and mouse brain. In: Proceedings of 5th Int. Congress Pharmacol. San Francisco, 1972. Vol. 4, pp. 14-17.

GLOSSMAN, H. & GIPS, H. (1975). Bovine adrenal cortex adenylate cyclase: properties of the particulate enzyme and effects of guanyl nucleotides. Naunyn-Schmiedeberg's Arch. Pharmacol. 289, 77-97.

GREENGARD, P. (1976). Possible role for cyclic nucleotide and phosphorylated membrane proteins in postsynaptic actions of neurotransmitters. Nature. 260, 101-108.

HEALD, P. J. (1957). The incorporation of phosphate into cerebral phosphoprotein promoted by electrical impulses. Biochem. J. 66, 659-663.

HEALD, P. J. (1962). Phosphoprotein metabolism and ion transport in nervous tissue: a suggested connexion. Nature (Lond.) 193, 451-454.

HERSHKOWITZ, M. (1978). Influence of calcium on phosphorylation of a synaptosomal protein. Biochim. Biophys. Acta. 542, 274-283.

HOLMES, H., RODNIGHT, R. & KAPOOR, R. (1977). Effect of electroshock and drugs administered in vivo on protein kinase activity in rat brain. Pharmacol. Biochem. Behav. 6, 415-420.

JOLLES, J., ROMPA-BARENDREGT & GISPEN, W. H. (1979). Novelty and grooming in the rat. Behav. Biol. in-press.

KATZ, B. & MILEDI, R. (1967). A study of synaptic transmission in the absence of nerve impulses. J. Physiol. 192, 407-436.

KOLATA, G. B. (1978). Polypeptide hormones: What are they doing in cells? Science, 201, 895-897.

KRIEGER, D. T., LIOTTA, A., SUDA, T., PALKOWITS, M. & BROWNSTEIN, M. J. (1977). Presence of immunoassayable β-lipotropin in bovine brain and spinal cord; Lack of concordance with ACTH concentrations. Biochem. Biophys. Res. Comm. 76, 930-936.

KRUEGER, B. K., FORN, J. & GREENGARD, P. (1977). Depolarization-induced phosphorylation of specific proteins, mediated by calcium ion influx, in rat brain synaptosomes. J. Biol. Chem. 252, 2764-2773.

LANG, U., FAUCHERE, J. L., PELICAN, G. M., KARLAGANIS, G. & SCHWYZER, R. (1976). Hormone-receptor interactions. Adrenocorticotrophin-(7-24)- octadeca peptide stimulates adipocyte membrane adenylate cyclase without causing lipolysis in fat cells. FEBS Letters 66, 246-249.

MAINS, R., EIPPER, B. A. & LING, N. (1977).   Common pre-
     cursor to corticotropins-endorphins. Proc. Natl.
     Acad. Sci., USA 74, 3014-3018.
MEZEY, E., PALKOVITS, M., DE KLOET, E. R., VERHOEF, J. &
     DE WIED, D. (1978).   Evidence for pituitary-brain
     transport of a behaviorally potent ACTH analog.
     Life Sci. 22, 831-838.
MIYAMOTO, E., MIYAZAKI, K., HIROSE, R. & KASHIBA, A. (1978).
     Multiple forms of protein kinases in myelin and
     microsomal fractions of bovine brain. J. Neurochem.
     31, 269-275.
NATHANSON, J. A. & GREENGARD, P. (1977).   "Second mes-
     sengers" in the brain. Sci. American. 237 (2)
     108-119.
PERUMAL, R., GISPEN, W. H., WILSON, J. E., GLASSMAN, E.
     (1975). Phosphorylation of proteins from the brains
     of mice subjected to short-term behavioral experiences.
     In: "Hormones Homeostasis and the Brain". (GISPEN,
     W. H., VAN WIMERSMA GREIDANUS, TJ. B., BOHUS, B. &
     DE WIED, D. eds.) Progress in Brain Res. Vol. 42,
     201-207.
PERUMAL, R., GISPEN, W. H., GLASSMAN, E. & WILSON, J. E.
     (1977). Phosphorylation of proteins of synaptosome-
     enriched fractions of brain during short term train-
     ing experiences: the effects of various behavioral
     treatments. Behavioral Biol. 21, 341-357.
REDBURN, D. A., SHELTON, D. & COTMAN, C. W. (1976).
     Calcium-dependent release of exogenously loaded
     γ-amino- ($U-^{14}C$) butyrate from synaptosomes: time
     course of stimulation by potassium, veratridine, and
     the calcium ionphore A 23187. J. Neurochem. 26,
     297-303.
REITH, M. E. A., SCHOTMAN, P. & GISPEN, W. H. (1974).
     Hypophysectomy, $ACTH_{1-10}$ and in vitro protein synthe-
     sis in rat brain stem slices. Brain Res. 81, 571-575.
ROBERTS, J. L. & HERBERT, E. (1977). Characterization of
     a common precursor to corticotropin and β-lipotropin:
     Identification of β-lipotropin peptides and their
     arrangement relative to corticotropin in the pre-
     cursor synthesized in a cell-free system. Proc.
     Natl. Acad. Sci. USA. 74, 5300-5304.
RODNIGHT, R., REDDINGTON, M. & GORDON, M. (1975). Methods
     for studying protein phosphorylation in cerebral
     tissues. In: Research Methods in Neurochemistry
     (N. MARKS & RODNIGHT, R. eds.). Plenum Press, New
     York. 3, 324-367.
ROUTTENBERG, A., EHRLICH, Y. H. & RABJOHNS, R. H. (1975).
     Effect of a training experience on phosphorylation
     of a specific protein in neocortical and subcortical
     membrane preparations. Fed. Proc. 34, 17.

RUDMAN, D. (1976). Injection of melatonin into cisterna magna increases concentrations of 3',5'-cyclic guanosine monophosphate in cerebrospinal fluid. Neuroendocrinology. 20, 235-242.

RUDMAN, D. & ISAACS, J. W. (1975). Effects of intrathecal injection of melanotropic-lipolytic peptides on the concentration of 3', 5'-cyclic adenosine monophosphate in cerebrospinal fluid. Endocrinology. 97, 1476-1480.

SAYERS, G., BEALL, R. J. & SEELIGS (1974). Modes of action of ACTH. In: Biochemistry of Hormones (RICHENBERG, H. V., ed.) Butterworths Univ. Park Press, London, 25-60.

SCHLEGEL, W. & SCHWYZER, R. (1977). Purification of bovine adrenal-cortex plasma membrane vesicles containing a higher corticotrophin-sensitive adenylate cyclase system and angiotensin-II-binding sites. Eur. J. Biochem. 72, 415-424.

SCHOTMAN, P. & GISPEN, W. H. (1978). Neuropeptides and brain protein synthesis. Neuroscience Lett. Suppl. 1, S 228.

SCHULMAN, H. & GREENGARD, P. (1978). Stimulation of brain membrane protein phosphorylation by calcium and an endogenous heat-stable protein. Nature, Vol. 271. p. 478-479.

SIEGHART, W., THEOHARIDES, T. C., ALPER, S. L., DOUGLAS, W. W. & GREENGARD, P. (1978). Calcium-dependent protein phosphorylation during section by exocytosis in the mast cell. Nature. 279, 329-331.

SPIRTES, M. A., CHRISTENSEN, C. W., HARSTON, C. T. & KASTIN, A. J. (1978). α-MSH and MIF-I effects on cGMP levels in various rat brain regions. Brain Research. 144, 189-193.

TREVOR, A. J. & RODNIGHT, R. (1965). The subcellular localization of cerebral phosphoproteins sensitive to electrical stimulation. Biochem. J. 95, 889-896.

VAN WIMERSMA GREIDANUS, Tj. B., BOHUS, B. & DE WIED, D. (1975). CNS sites of action of ACTH, MSH and vasopressin in relation to avoidance behavior. In: Anatomical Neuroendocrinology (STUMPF, W. E., GRAND, L. D., eds.). S. Karger, A.G. Basel, 284-289.

VERHOEF, J., PALKOVITS, M. & WITTER, A. (1977a). Distribution of a behaviorally highly potent $ACTH_{4-9}$ analog in rat brain after intraventricular administration. Brain Research. 126, 89-104.

VERHOEF, J., WITTER, A. & DE WIED, D. (1977). Specific uptake of a behaviorally potent (^3H)-$ACTH_{4-9}$ analog in the septal area after intraventricular injection in rats. Brain Research. 131, 117-128.

VON HUNGEN, K. & ROBERTS, S. (1973). Adenylate-cyclase receptors for adrenergic neurotransmitters in rat cerebral cortex. Eur. J. Biochem. 36, 391-401.

WIEGANT, V. M. & GISPEN, W. H. (1975). Behaviorally active ACTH analogs and brain cyclic AMP. Exp. Brain Res. 23, Suppl. 219.

WIEGANT, V. M., ZWIERS, H., SCHOTMAN, P. & GISPEN, W. H. (1977). Endogenous phosphorylation of rat brain synaptosomal plasma membranes in vitro: some methodological aspects. Neurochem. Res. 3, 443-453.

WIEGANT, V. M., COOLS, A. R., GISPEN, W. H. (1977). ACTH-induced excessive grooming involves brain dopamine. Eur. J. Pharmacol. 41, 343-345.

WIEGANT, V. M., DUNN, A. J., SCHOTMAN, P. & GISPEN, W. H. (1979). ACTH-like peptides: possible regulators of rat brain cyclic AMP: Brain Research, in-press.

WILSON, J. E. (1979). Protein phosphorylation involvement in brain function. In: Biochemistry of Brain Vol. II. (KUMAR, S. ed.) Pergamon Press. In-press.

ZWIERS, H., VELDHUIS, D., SCHOTMAN, P. & GISPEN, W. H. (1976). ACTH, cyclic nucleotides and brain protein phosphorylation in vitro. Neurochem. Res. 1, 669-677.

ZWIERS, H., WIEGANT, V. M., SCHOTMAN, P. & GISPEN, W. H. (1977). Intraventricular administered ACTH and changes in rat brain protein phosphorylation: a preliminary report. In: Mechanisms, Regulation and Special Function of Protein Synthesis in the Brain (ROBERTS, S., LAJTHA, A. & GISPEN, W. H. eds.) Elsevier/North Holland Biomed. Press, Amsterdam, 267-272.

ZWIERS, H., WIEGANT, V. M., OESTREICHER, A. B., SCHOTMAN, P. & GISPEN, W. H.(1978a). Peptides and rat brain membrane phosphoproteins. Proc. Eur. Soc. Neurochem. Vol. 1 (NEUHOFF, V. ed.). Verlag Chemie, Weinheim, New York, 463.

ZWIERS, H., WIEGANT, V. M., SCHOTMAN, P. & GISPEN, W. H. (1978b). ACTH-induced inhibition of endogenous rat brain protein phosphorylation in vitro: structure-activity. Neurochem. Res. 3, 455-463.

ZWIERS, H., TONNAER, J., WIEGANT, V. M., SCHOTMAN, P. & GISPEN, W. H. (1979). ACTH-sensitive protein kinase from rat brain membranes. Submitted.

OPIOID PEPTIDES AS MODULATORS OF CYCLIC AMP LEVELS

Werner A. Klee

Laboratory of General and Comparative Biochemistry
National Institute of Mental Health
Bethesda, Maryland   20014

INTRODUCTION

One of the most clearly defined biochemical actions
of the opiates is the receptor mediated inhibition of
adenylate cyclase. Collier and Roy (1974) showed that
cAMP accumulation in rat brain homogenates is inhibited
by opiates, that the inhibition is blocked by the specific
inhibitor, naloxone, and that the potencies of a series
of opiate agonists as blockers of cAMP formation are similar
to their potencies as analgesic agents. Soon thereafter,
Sharma, et al, (1975a) found that the adenylate cyclase of
neuroblastoma X glioma cell (NG108-15) membranes is inhibited by opiates in a similar, receptor mediated, fashion.
The neuroblastoma X glioma hybrid cell NG108-15 homogenates
provide an easily accessible source of opiate receptors
(KLEE and NIRENBERG, 1974) which remain coupled to adenylate
cyclase more tenuously than those in brain homogenates
seem to be. The NG108-15 system is therefore more easily
reproducible (VAN INWAGEN, et al, 1975). In a parallel
series of studies, Traber, et al, (1975a), using the same
neuroblastoma X glioma hybrid cell line, have also shown
that opiates decrease cellular cAMP levels.

Not surprisingly, the endogenous opioid peptides,
the endorphins (and enkephalins) also inhibit adenylate
cyclase of NG108-15 homogenates in receptor mediated reactions (KLEE and NIRENBERG, 1976), lower cAMP levels in
intact NG108-15 cells (BRANDT, et al, 1977; WAHLSTROM,
et al, 1977) and also can sometimes be shown to inhibit

adenylate cyclase in brain homogenates (COLLIER & FRANCIS, 1978). The enkephalins are particularly potent inhibitors of NG108-15 adenylate cyclase, as may be seen in Table 1.

TABLE 1. Concentrations of endorphins and opiates required for half-maximal inhibition of adenylate cyclase in NG108-15 cell homogenates.

OPIATE	Ki (nM)
Met-Enkephalin	10
α-Endorphin	250
γ-Endorphin	300
β-Endorphin	150
Morphine	1500
Etorphine	10

Interestingly, the longer endorphins are appreciably less potent as inhibitors of the enzyme (as is also true of the mouse vas deferens (LORD, et al, 1976) although they remain more potent than morphine in this assay. In vivo and in other in vitro assays, β-endorphin is usually found to be much more potent than is enkephalin. It seems likely that one reason for these differences is that the NG108-15 adeny-late cyclase assay can readily be performed without compli-cations due to the degradation of enkephalins by pepti-dases which are found in great abundance in brain.

Exposure of NG108-15 cells to opiates and to opioid peptides for 12 or more hours results in an increase in adenylate cyclase activity which effectively compensates for the inhibition of activity observed as an immediate opiate action (SHARMA, et al, 1975b; TRABER, et al, 1975b; LAMPERT, et al, 1976). Both basal and hormone stimulated adenylate cyclase activities are increased as a result of chronic opioid treatment. Thus, changes in hormone coupling are not exclusively responsible for this phenomenon. These experiments led to a model for opiate tolerance and

dependence based upon the dual regulation of adenylate
cyclase activities, (SHARMA, et al, 1975b). Opiates produce an immediate drop in cellular cAMP levels, due to an
inhibition of adenylate cyclase activity, which may be a
mechanism for the acute effects of these substances. Upon
continued exposure to opiates, cellular cAMP levels return
to the normal state (SHARMA, et al, 1975b) as a result of
the slow increase in adenylate cyclase activity elicited by
the opiates. At this stage, the cells are tolerant to
opiates since cAMP levels are normal even in their presence.
The cells are also in a dependent state since rapid withdrawal of opiates, by addition of naloxone for example,
results in an immediate, large rise in cAMP due to the expression of the hitherto masked enzyme activity. Continued
culture of the cells in the absence of opiates for a few
hours leads to a loss of adenylate cyclase activity until
normal levels are reached. It is thus possible to account
for the major features of opiate action in animals by this
simple dual regulation model. Very similar proposals have
been put forward by Collier, et al (1975), and the general
features of this model conform to the homeostatic regulation first proposed by Himmelsbach (1943) and more recently
by Goldstein and Goldstein (1961) and Shuster (1961).
Since basal and hormone stimulated enzyme activity are both
increased in opiate conditioned cells, hormone receptor
coupling does not seem to be appreciably changed in the
process. Interestingly, however, Sharma, et al, (1977)
found that when the enzyme assays are performed in the
presence of F^- or the stable GTP analogue guanylylimidodi-
phosphate, agents which activate the enzyme but also un-
couple it from receptors, there were no differences between
enzyme prepared from control or from opiate conditioned
cells. These data imply that the increased enzyme activity
elicited by opiates is due to an activation of pre-existing
enzyme rather than to the production of new enzyme. Con-
sistant with this interpretation is the observation that
cycloheximide, a protein synthesis inhibitor, blocks only
a portion of the increase in enzyme activity (SHARMA, et al,
1977).

Dual regulation of adenylate cyclase activity is not
a unique property of opiates. Agents which act as agonists
at each of three inhibitory receptors in NG108-15 cells
induce a slow compensatory increase in adenylate cyclase
activity. Representative data, from separate experiments,
with α-adrenergic, muscarinic and opiate agonists are compiled in Table 2. Thus, dual regulation of the enzyme is a
general property of inhibitory receptors. A degree of
tolerance to and dependence upon neuro-effectors, including

TABLE 2. Long term effects of inhibitory receptor ligands
          on NG108-15 cells.

Culture Conditions	Adenylate Cyclase Activity	Hours of Growth	
Control	13.9		
1 µM Norepinephrine	23.8	16	Sabol & Niren-berg (1979)
Control	5		
100 µM Carbachol	16	48	Nathanson, et al, (1978)
Control	10.5		
1 µM Etorphine	18.5		
10 µM Met-Enkephalin	19.3	25	Lampert, et al, (1976)

the endorphins, may be the normal state of the organism
and may therefore play an important role in controlling
adenylate cyclase activity and cyclic AMP levels. Changed
cAMP levels will be reflected in alterations of protein
phosphorylation pattern within the cell. The nature and
ramifications of cAMP dependent protein kinase reactions
are amply discussed elsewhere in this volume. Suffice it
to say here that many aspects of neuronal function must be
controlled by such reactions.

     Measurements of adenylate cyclase activity of neuro-
blastoma X glioma hybrid cell homogenates provide a rapid,
sensitive and specific assay of opioid peptide activity.
The homogenates may be prepared from large quantities of
cells and stored at -80° for many months without change.
Since methionine-enekphalin inhibits the enzyme half-
maximally at 10 nM concentrations, the assay can reliably
quantitate sub-picomole quantities of opioid peptide. A
particularly important feature of the assay is the specific
blockade of opioid activity by (-)-naloxone (IIJIMA, et al,
1978), which allows the facile demonstration of opioid
specificity. A number of recent studies have employed the
adenylate cyclase assay as the primary tool in the measurement of opioid peptide activities. In a particularly elegant
study, Giagnoni, et al, (1977) have used this assay procedure to study opioid peptide production in a number of
cultured cell lines. Changes in pituitary endorphin levels

in response to modulation of vasopressin function have been
measured by Mata, et al (1977), by this technique. Recently,
Zioudrou, et al, (1979) have used the adenylate cyclase
assay to monitor the purification of a new class of opioid
peptides, the exorphins, which are derived from food proteins upon digestion with the stomach proteinase, pepsin
(KLEE, et al, 1978). Because of report linking wheat gluten,
and other cereal proteins, with schizophrenia (DOHAN, 1966;
DOHAN and GRASBERGER, 1973) these workers wondered whether
such proteins could serve as precursors of opioid activity.
Preliminary experiments showed that degradation products
of some food proteins, including notably wheat gluten and
α-casein, prepared under conditions which mimic digestion
in the stomach, do indeed show true, naloxone reversible,

TABLE 3. Concentrations of Exorphins (nanograms/ml) for
          Half-Maximal Inhibitions in Several Opiate Assays.

	Brain receptor binding	Adenylate cyclase	Vas deferens assay
Gluten Exorphin	2	30	2
Casein Exorphin	3500	320	8000
Met-Enkephalin	20	7	7[a]
Morphine Sulfate	1	570	190[a]

[a]Calculated from the data of Lord, et al, (1976).

opiate activities in a number of assays. After extensive
purification, a peptide isolated from gluten was found to
be approximately equipotent with enkephalin in three different assay systems (Table 3). The table also shows that
an equally highly purified exorphin, isolated from casein,
has a much lower intrinsic activity. Although the physiological and potential pathological significance of the
exorphins is still only a matter of speculation, these
experiments show that exogenous sources should be considered
for neurpeptides in general.

REFERENCES

BRANDT, M., BUCHEN, C., and HAMPRECHT, B. (1977) Endor-
    phins exert opiate-like action on neuroblastoma X
    glioma hybrid cells. FEBS Lett., 80, 251-254.
COLLIER, H. O. J., and ROY, A. C. (1974) Morphine-like
    drugs inhibit stimulation by E prostaglandins of
    cyclic AMP formation by rat brain homogenate. Nature,
    248, 24-26.
COLLIER, H. O. J. and FRANCIS, D. L. (1978) A Pharmacolo-
    gical Analysis of Opiate Tolerance/Dependence in:
    The Bases of Addiction, edited by J. Fishman, Dahlem
    Konferenzen, Berlin. p. 281-298.
COLLIER, H. O. J., FRANCIS, D. L., MCDONALD-GIBSON, W. J.,
    ROY, A. C., and SAEED, S. A. (1975) Prostaglandins,
    cyclic AMP. and the mechanism of opiate dependence.
    Life Sci., 17, 86-90.
DOHAN, F. C. (1966) Cereals and schizophrenia: data and
    hypothesis. Acta Psychiat. Scand., 42, 125-152.
DOHAN, F. C., and GRASBERGER, J. C., (1973) Relapsed
    schizophrenics: earlier discharge from the hospital
    after cereal-free, milk-free diet. Amer. J. Psychiat.,
    130, 685-688.
GIAGNONI, G., SABOL, S. L., and NIRENBERG, M. (1977) Syn-
    thesis of opiate peptides by a clonal pituitary tumor
    cell line. Proc. Natl. Acad. Sci. USA, 74, 2259-2263.
GOLDSTEIN, D. B., and GOLDSTEIN, A. (1961) Possible role
    of enzyme inhibition and repression in drug tolerance
    and addiction. Biochem. Pharmacol., 8, 48.
HIMMELSBACH, C. K. (1943) The morphine abstinence syndrome.
    Fed. Proc., 2, 201-203.
IIJIMA, I., MINAMIKAWA, J., JACOBSON, A. E., BROSSI, A.,
    RICE, K. C., and KLEE, W. A. (1978) Synthesis and
    biological properties of (+)-naloxone. J. Med. Chem.,
    21, 398-400.
KLEE, W. A., and NIRENBERG, M., (1974) A neuroblastoma X
    glioma hybrid cell with morphine receptors. Proc.
    Nat. Acad. Sci. USA, 71, 3474-3477.
KLEE, W. A., and NIRENBERG, M., (1976) Mode of action of
    endogenous opiate peptides. Nature, 263, 609-612.
KLEE, W. A., ZIOUDROU, C., and STREATY, R. A. (1978)
    Exorphins: peptides with opioid activity isolated
    from wheat gluten and their possible role in the
    etiology of schizophrenia. In: Endorphins in Mental
    Health Research, eds. E. Usdin, W. E. Bunney, and
    N. S. Kline, pp. 209-218. MacMillan, New York.
LAMPERT, A., NIRENBERG, M., and KLEE, W. A. (1976) Toler-
    ance and dependence evoked by an endogenous opiate
    peptide. Proc. Natl. Acad. Sci. USA, 73, 3165-3167.

LORD, J. A. H., WATERFIELD, A. A., HUGHES, J., and KOSTER-
    LITZ, H. W. (1976) Multiple opiate receptors.
    In: Opiates and Endogenous Opioid Peptides, edited
    by H. W. Kosterlitz, pp. 275-280. Elsevier, North
    Holland.
MATA, M. M., GAINER, H., and KLEE, W. A. (1977) Effect
    of dehydration on the endogenous opiate content of
    the rat neuro-intermediate lobe. Life Sci., 21,
    1159-1162.
NATHANSON, N. M., KLEIN, W. L., and NIRENBERG, M. (1978)
    Regulation of adenylate cyclase activity mediated by
    muscarinic acetylcholine receptors. Proc. Natl. Acad.
    Sci. USA, 75, 1788-1792.
SABOL, S. L., and NIRENBERG, M. (1979) Regulation of adeny-
    late cyclase of neuroblastoma X glioma hybrid cells by
    α-adrenergic receptors. II. Long-lived increase of
    adenylate cyclase activity mediated by α-receptors.
    J. Biol. Chem., in-press.
SHARMA, S. K., NIRENBERG, M., and KLEE, W. A. (1975a)
    Morphine receptors as regulators of adenylate cyclase
    activity. Proc. Nat. Acad. Sci. USA, 72, 590-594.
SHARMA, S. K., KLEE, W. A., and NIRENBERG, M. (1975b)
    Dual regulation of adenylate cyclase account for narco-
    tic dependence and tolerance. Proc. Natl. Acad. Sci.
    USA, 72, 3092-3096.
SHARMA, S. K., KLEE, W. A., and NIRENBERG, M. (1977)
    Opiate-dependent modulation of adenylate cyclase.
    Proc. Natl. Acad. Sci., USA, 74, 3365-3369.
SHUSTER, L. (1961) Repression and de-repression of enzyme
    synthesis as a possible explanation of some aspects
    of drug action. Nature, 189, 314-315.
TRABER, J., FISCHER, K., LATZIN, S., and HAMPRECHT, B.
    (1975a) Morphine antagonizes action of prostaglandin
    in neuroblastoma and neuroblastoma X glioma hybrid
    cells. Nature, 253, 120-122.
TRABER, J., GULLIS, R., and HAMPRECHT, B. (1975b) In-
    fluence of opiates on the levels of adenosine 3'-5'
    cyclic monophosphate in neuroblastoma X glioma hybrid
    cells. Life Sci., 16, 1863-1868.
VAN IUWEGEN, R. G., STRADA, S. J. and ROBISON, G. A. (1975)
    Effects of Prostaglandins and Morphine on Brain Adeny-
    late Cyclase. Life Sci. 16, 1875-1876.
WAHLSTROM, A., BRANDT, M., MORODER, L., WUNSCH, E., LINDE-
    BERG, G., RAGUARSSON, U., TERENIUS, L., and HAMPRECHT,
    B. (1977) Peptides related to β-lipotropin with
    opioid activity. FEBS Lett., 77, 28-32.
ZIOUDROU, C., STREATY, R. A., and KLEE, W. A. (1979) Opioid
    peptides derived from food proteins: the exorphins.
    J. Biol. Chem., in press.

OPIOID PEPTIDES AND PROTEIN PHOSPHORYLATION

Leonard G. Davis and Yigal H. Ehrlich

Missouri Institute of Psychiatry, University of
Missouri-Columbia, School of Medicine,
5400 Arsenal Street
St. Louis, Missouri   63139

INTRODUCTION

The identification and isolation of naturally occurring neuropeptides with morphine-like activity was first
reported by Hughes (1975) and co-workers (HUGHES, et al,
1975). These compounds were named methionine-enkephalin
and leucine-enkephalin and have the amino-acid sequences
of Tyr-Gly-Gly-Phe-Met and Tyr-Gly-Gly-Phe-Leu, respectively. Subsequently, a 31 amino-acid peptide with similar
pharmacological and biochemical properties was described
(COX, et al, 1975; BRADBURY, et al, 1976; GUILLEMAN, et al,
1976). Sequential amino-acid analysis of this larger peptide, named β-endorphin, demonstrated a similarity to the
C-terminal portion (#61-91) in the sequence of β-lipotropin
(LI and CHUNG, 1976: see Figure 1). Using the mouse vas
deferens, the guinea pig ileum, or the opiate receptor
assays, it was shown that these neuropeptides act in vitro
similar to alkaloid opiates like morphine (HUGHES, 1975;
HUGHES, et al, 1975; LI and CHUNG, 1976; SIMANTOV and
SNYDER, 1976; GRAF, et al, 1976). These endogenous opioid-
like neuropeptides have been shown to interact differen-
tially with the opiate receptors of brain (LORD, et al,
1976). Upon administration to intact animals, they produce tolerance (WEI and LOH, 1976; TSENG, et al, 1976),
analgesia (FELDBERG and SMYTH, 1976; MEGLIO, et al, 1977)
and dependence (WEI and LOH, 1976). These endogenous peptides have been collectively named endorphins since they
mimic the action of opiate agonists. The above findings
have greatly stimulated investigations into the role of

1
NH₂-Glu-Leu-Thr-Gly-Gln-Arg-Leu-Arg-Gln-Gly-Asp-Gly-Pro-Asn-Ala-Gly-Ala-Asn-

19
Asp-Gly-Glu-Gly-Pro-Asn-Ala-Leu-Glu-His-Ser-Leu-Leu-Ala-Asp-Leu-Val-Ala-

37
Ala-Glu-Lys-Lys-Asp-Glu-Gly-Pro-Tyr-Arg-Met-Glu-His-Phe-Arg-Try-Gly-Ser-

55
Pro-Pro-Lys-Asp-Lys-Arg-Tyr-Gly-Gly-Phe-Met-Thr-Ser-Glu-Lys-Ser-Gln-Thr-

73   91
Pro-Leu-Val-Thr-Leu-Phe-Lys-Asn-Ala-Ile-Lys-Asn-Ala-Tyr-Lys-Lys-Gly-Glu-OH

Fig. 1: Structure of human β-lipotropin according to Li and Chung (1976).
The sequence of human β-endorphin (⟶), methionine-enekphalin
(══), and ACTH₄-10 (----) are indicated.

endorphins in brain function and to understanding their
biochemical mechanisms of action.

The first insight into the biochemical mechanisms that
mediate the action of opiates was provided in studies using
NG108-15 neuroblastoma X glioma hybrid cells grown in culture (SHARMA, et al, 1975; TRABER, et al, 1975). With these
cells, which are highly enriched in stereospecific opiate
receptors (see A. BLUME, this volume), the alkaloid opiate
agonists inhibit basal and prostaglandin E_1-stimulated
formation of cyclic AMP by adenyl cyclase. Subsequent
studies (KLEE and NIRENBERG, 1976; BRANDT, et al, 1976;
GOLDSTEIN, et al, 1977) provided evidence that the enkephalins and endorphins also inhibit adenyl cyclase in the
neuroblastoma-glioma hybrid cells. Minneman and Iverson
(1976) demonstrated similar effects for the enkephalins on
adenyl cyclase activity using brain tissue. The possibility
that the mode of action of opiate agonists may also involve
protein phosphorylation systems was first indicated by
Clark and co-workers (1972). They reported that chronic
morphine treatment of rats resulted in a 30% decrease in
the protein kinase activity present in brain membranes. We
have demonstrated that in rats treated for three weeks with
incremental doses of morphine, and thus in a state of protracted narcotic dependence, there is a 50% decrease in
striatal, membrane-bound protein kinase activity towards its
endogenous substrates (EHRLICH, et al, 1977a; BONNET, et al,
1978). Furthermore, examination of the endogenously phosphorylated substrates by SDS polyacrylamide gel electrophoresis established that this affect was selective towards
the specific phosphoproteins we (EHRLICH and ROUTTENBERG,
1974) have designated F and H (apparent molecular weight
of 47K and 15-20K, respectively). The long-term narcotic
exposure resulted in a 60-70% decrease in the phosphorylation of these specific proteins (EHRLICH, et al, 1978).

Based on the above observations, we have suggested
that the mechanism whereby long-term morphine treatment
causes decreased phosphorylation of proteins may be mediated
in vivo by alterations in the interactions between endogenous opioid peptides and the enzymatic systems that phosphorylate these specific proteins (EHRLICH, et al, 1977a;
1978). The confirmation of the suggestion that neuropeptides alter endogenous phosphorylation, when put together
with the previous finding that the endorphins inhibit adenylate cyclase, would indicate a dual but interrelated mechanism of action for these neuropeptides. Indeed, it has
been previously demonstrated (ZWIERS, et al, 1976; ZWIERS,
et al, 1978) that adrenocorticotropin (ACTH) peptides affect
the phosphorylation of specific protein components in

synaptic membranes (see also W. GISPEN, et al, this volume).
Furthermore, specific ACTH fragments have been shown to
interact with opiate receptors (TERENIUS, 1976; JACQUET,
1978) and to stimulate adenylate cyclase in thalamic slices
(WIEGANT and GISPEN, 1975; A report (FORN and KRISHNA, 1971)
using other brain regions has also appeared in which no
affects of ACTH on adenyl cyclase activity were found.).
Thus, it was reasonable to assume that opioid-peptides may
alter protein phosphorylative activity in addition to
their ability to inhibit adenylate cyclase. We have, therefore, initiated studies to determine whether the mode of
action of opioid-peptides, such as met-enkephalin and leu-
enkephalin, involve interactions with neuronal membrane-
protein phosphorylation systems.

The methods used in these studies have been described
in detail elsewhere (EHRLICH, et al, 1977b). Briefly,
fractions enriched in synaptosomal membranes are prepared
from rat brain by differential centrifugation and hypotonic
treatment. These membranes are used in the assay of endogenous phosphorylation. Aliquots containing 60 μg protein
and the appropriate concentration of peptide are preincu-
bated in a shaking water bath (30°C) for 5 minutes. The
reaction is initiated by adding ATP. The reaction medium
contains 50 mM Na -acetate buffer (pH 6.5), 10 mM $MgCl_2$,
and 3 μM γ-^{32}P-ATP. At various times the reaction is terminated by adding an SDS solution to a final concentration
of 3%, containing 1 mM EDTA, 2% β-mercaptoethanol, 6% sucrose,
and 0.1% bromophenol blue. The solubilized membrane proteins are applied to linear gradients of acrylamide (7-16%)
in a slab gel electrophoresis unit, and run for 26 hours.
Separated proteins are identified by Coomaisee blue staining. Proteins containing radioactive phosphate, i.e. phosphoproteins, are identified by autoradiography of dried gels.

In our initial efforts to evaluate the influences of
opioid peptides on protein phosphorylation in vitro, we
observed a decrease in the amount of radiophosphate incorporated into synaptic membrane proteins, but there was a
great deal of sample to sample variation in the magnitude
of this inhibition. We hypothesized that this variability
could be caused by multiple degrees of RECEPTOR OCCUPANCY
in the membranes isolated from different animals. Indeed,
it has been reported that the level of endogenous ligands
bound to opiate receptors varies, based upon past experiences, in preparations from individual animals (PERT and
BOWIE, 1978). We have, therefore, dialyzed the membrane preparations overnight to dissociate and remove receptor-bound
endogenous ligands. The level of endogenous phosphorylative

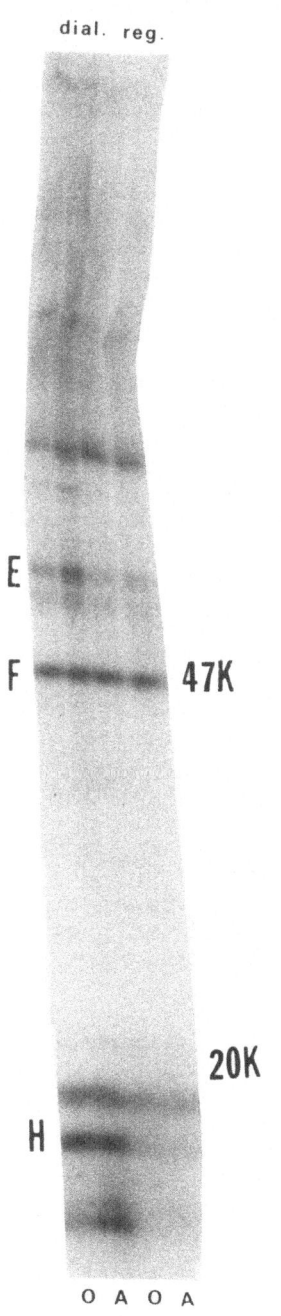

dial. reg.

E

F   47K

20K

H

O A O A

Fig. 2: Autoradiogram of ^{32}P-phosphate incorporation into SDS-solubilized and polyacrylamide gel electrophoresed synaptosomal proteins. The left pair of gel patterns (dial.) is representative of a dialyzed preparation; the right pair (reg.) is an aliquot of the same preparation prior to dialysis. A=5 μM cAMP added to the phosphorylation assay; O=equivalent volume of water added.

activity (i.e. the amount of phosphate incorporated into
proteins) increased after this dialysis indicating that
endogenous inhibitors of protein phosphorylation had been
removed by dialysis (Figure 2). The pattern of endogenously
phosphorylated proteins in dialyzed membranes when assayed
in the presence of a stochiometric amount of freeze dried,
reconstituted dialyzate was similar to that of nondialyzed
membranes. The addition of a tenfold excess of this reconstituted dialyzate to dialyzed membranes almost completely
abolished their endogenous phosphorylative activity (Figure
3). We have initiated studies to characterize the inhibitory component(s) in this dialyzate. Preliminary results
indicate that the active component is possibly a peptide(s)
with apparent molecular weight of about 2000 daltons.

We then tested whether these endogenous inhibitors of
protein phosphorylation were dissociated from opiate receptors. Freshly prepared, non-dialyzed membranes were incubated with morphine or naloxone (10^{-5}M) for two hours under
the binding conditions described by Pert and Bowie (1978).
The added opiate agonist (morphine) or antagonist (naloxone)
was expected to displace any endogenous ligands that might
be present by competitively binding to opiate receptors.
The membranes were separated from the incubation medium by
centrifugation (100,000g x 60 minutes) and then assayed
for endogenous phosphorylation activity. After this treatment, the phosphorylative activity increased when compared
to appropriate controls (incubation without opiates). This
relative increase in endogenous activity was similar to
that observed after dialysis of membranes. Apparently,
isolated membrane preparations exist in a state of opiate
receptor occupancy. Upon treatments such as dialysis or
displacement, the inhibitor(s) is removed. However, additional research is necessary to establish whether the endogenous inhibitor(s) of protein phosphorylation is an endogenous opioid-peptide(s). It should be emphasized that
the dialysis also minimized the variability between preparations.

Using either preparation, that is membranes after
dialysis or isolated after binding conditions with naloxone,
both met-enkephalin and leu-enkephalin inhibited the phosphorylation of the specific proteins designated F and H
(Figures 4 and 5). To date, inhibition of protein phosphorylation by met-enkephalin has been observed in all brain
regions that we have tested, except, possibly, the hippocampus.

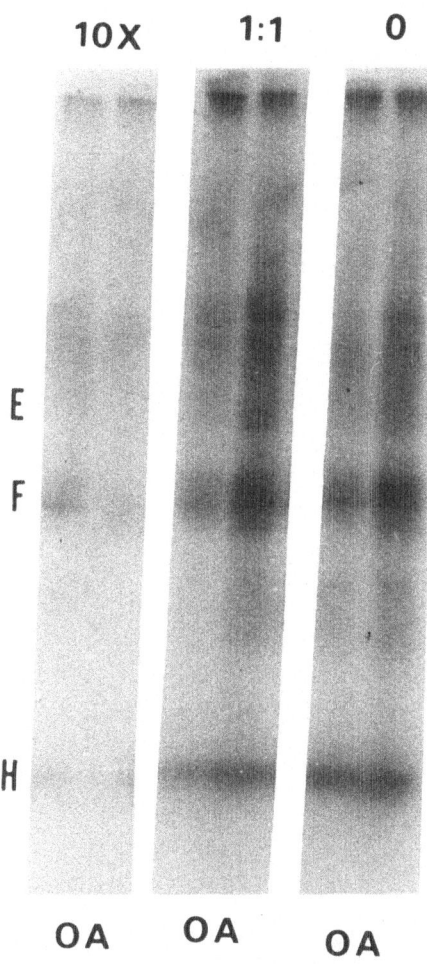

Fig. 3: Autoradiogram of ^{32}P-phosphate incorporation into
SDS-solubilized and polyacrylamide gel electrophoresed
dialyzed synaptosomal proteins. Additions to the
phosphorylation assay were: 10X=a tenfold concentrate
of reconstituted (by freeze drying) dialyzate; 1:1=
a stochiometric amount of reconstituted dialyzate;
O=equivalent volume of reconstituted, freeze dried
water; A and O as in Figure 2.

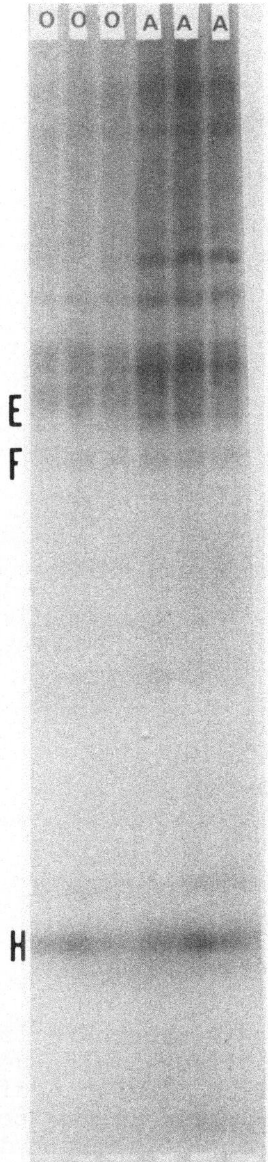

Fig. 4: Autoradiogram of dialyzed synaptosomal membranes
         as in Figure 3. Additions to the phosphorylation
         assay included: L=leucine-enkephalin (10^{-4}M);
         M=methionine-enkephalin (10^{-4}M); A and O as in
         Figure 2.

The inhibition of protein phosphorylation by enkephalins was found to be concentration dependent. Each of these opioid-peptides was most effective in inhibiting the phosphorylation of the phosphoproteins designated F and H at a concentration of 10^{-4}M, and ineffective at 10^{-10}M. A curvilinear gradient of inhibitory strength was observed between 10^{-5}M to 10^{-9}M. It would appear that the enkephalins affect protein kinases rather than phosphoprotein phosphatases: At short reaction times (5-20 seconds), when protein kinases activity predominates, the enkephalins are more potent inhibitors than at longer reaction times (5-10 minutes) when dephosphorylation activity predominates. However, inhibition at these longer reaction times was still observed.

Naloxone, a pure opiate antagonist, has been tested for its ability to block the enkephalin effect on protein phosphorylation. In the presence of met-enkephalin and naloxone at equimolar concentrations, met-enkephalin inhibition of phosphorylative activity was still in evidence but at a lower level than that caused by met-enkephalin alone. We have found, however, a few preparations in which naloxone was ineffective in blocking the met-enkephalin induced inhibition of phosphorylation. It may be worthwhile mentioning that at least two types of opiate receptors have been suggested to exist in brain (LORD, et al, 1976) and to co-interact (JACQUET, 1978). Interestingly, one of these opiate receptors is apparently selective for ACTH and not reversible by naloxone (JACQUET, 1978). The possibility thus exists that this type receptor could be coupled to a protein phosphorylation system rather than cyclases since it has been shown that ACTH (ZWIERS, et al, 1976; ZWIERS, et al, 1978) and the enkephalins do decrease the phosphorylation of the same proteins when these ligands are added to brain membranes. We are presently investigating the multiple opiate receptor types in an attempt to elucidate the mechanism of naloxone action and its role in enkephalin inhibition of protein phosphorylation.

In summary, the present results indicate that opioid-peptides may exert some of their effects on neuronal function by affecting the phosphorylation of specific proteins endogenous to synaptic membranes. The findings that chronic morphine treatment in vivo and met-enkephalin in vitro selectively affect the phosphorylation of the same synaptic-membrane proteins support our suggestion that "endorphin" regulated phosphorylation may play a role in mechanisms underlying narcotic dependence. Moreover, the present results and those of others (for example, see W. GISPEN, et al, this volume) suggest that, in general, neuropeptides

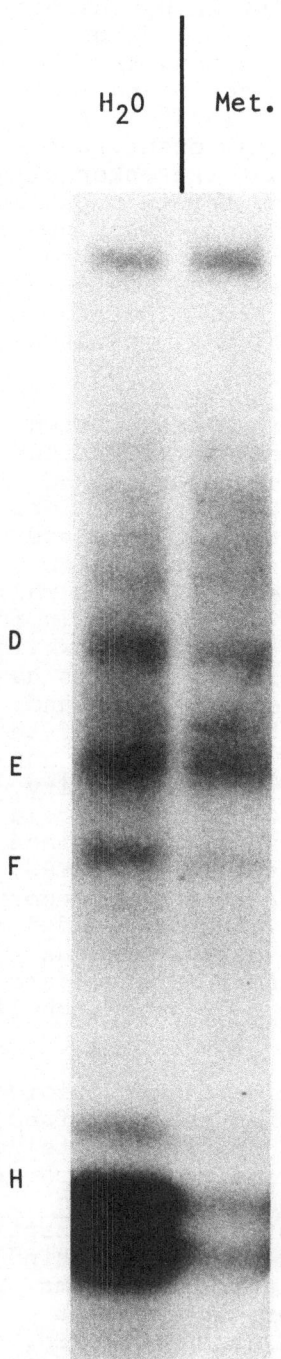

Fig. 5: Autoradio-
gram of ^{32}P-phosphate
incorporation into
SDS-solubilized and
polyacrylamide gel
electrophoresed sy-
naptosomal membrane
proteins isolated by
centrifugation after
two hours incubation
with naloxone (10^{-5}M).
Additions to the
phosphorylation assay
were methionine-
enkephalin (10^{-4}M)
or water (M and O,
respectively).

may exert their modulatory effects on neuronal activity
through mechanisms that involve the phosphorylation of
specific proteins. This effect may be in addition to their
direct effects on adenylate and quanylate cyclase activities.
The possibility of such a dual mechanism of action for the
opioid peptides may parallel the findings indicating that
neuroactive peptides play both modulatory and mediatory
roles in synaptic function.

ACKNOWLEDGEMENTS

The authors wish to thank Mrs. Linda Lee Schneider
for typing the original manuscript; the Missouri Institute
of Psychiatry, University of Missouri-Columbia, School of
Medicine for intramural funds; and Doctor Eric G. Brunngraber
for valuable discussions during these studies. We also wish
to thank ENDO LABS (Garden City, N.Y.) for the supply of
naloxone.

REFERENCES

BONNET, K. A., BRANCHEY, L. B., FRIEDHOFF, A. J., EHRLICH,
    Y. H. (1978) Life Sci. 22, 2003-2008.
BLUME, A., BOONE, G. and LICHTSHTEIN, D. Modulators,
    Mediators and Specifiers in Brain Function. (this
    volume).
BRADBURY, A. F., SMYTH, D. G., SNELL, G. R., BIRDSALL,
    N.J.M. and HULME, E. C. (1976) Nature 260, 793-795.
BRANDT, M. GULLIS, R. J., FISCHER, K., BUCHEN, C.,
    HAMPRECHT, B., MORODER, L. and WUNSCH, E. (1976)
    Nature 262, 311-313.
CLARK, A. G., TONIC, R., ORNELLAS, M. R., and WELLER, M.
    (1972) Bioch. Pharm. 21, 1989-1990.
COX, B. M., OPHEIM, K. E., TECHEMACHER, H., and GOLDSTEIN,
    A. (1975) Life Sci. 16, 1777-1782.
EHRLICH, Y. H., BONNET, K. A., DAVIS, L. G. and BRUNNGRABER,
    E. G. (1977a) In: Mechanisms, Regulation and Special
    Functions of Protein Synthesis in the Brain. eds:
    Roberts, S., Lajtha, A., Gispen, W. H. Elsevier,
    Amsterdam, 273-278.
EHRLICH, Y. H., BONNET, K. A., DAVIS, L. G., and BRUNNGRABER,
    E. G. (1978) Life Sci. 23, 137-146.
EHRLICH, Y. H., DAVIS, L. G., GILFOIL, T. and BRUNNGRABER,
    E. G. (1977b) Neurochem. Res. 2, 533-548.
EHRLICH, Y. H. and ROUTTENBERG, A. (1974) FEBS Lett. 45,
    237-243.
FELDBERG, W. S. and SMYTH, D. G. (1976) J. Physiol.
    (London) 260, 30-31 P.
FORN, J. and KRISHNA, G. (1971) Pharmacology 5, 193-204.

GISPEN, W. H., ZWIERS, H., WIEGANT, V. M., SCHOTMAN, P.
    and WILSON, J. E. Modulators, Mediators and Specifiers
    in Brain Function. (this volume).
GOLDSTEIN, A., COX, B. M., KLEE, W. A. and NIRENBERG, M.
    (1977) Nature 265, 362-363.
GRAF, L., SZEKELY, J. I., RONAI, A. Z., DUNAI-KOVACS, Z.
    and BAJUSZ, S. (1976) Nature 263, 240-242.
GUILLEMAN, R., LING, N. and BURGUS, R. (1976) C. R. Acad.
    Sci. 282, 783-785.
HUGHES, J. (1975) Brain Research 88, 295-306.
HUGHES, J., SMITH, T. W., KOSTERLITZ, H. W., FOTHERGILL,
    L. A., MORGAN, B. A. and MORRIS, H. R. (1975) Nature
    258, 577-579.
JACQUET, Y. F. (1978) Science 201, 1032-1034.
KLEE, W. A. and NIRENBERG, M. (1976) Nature 263, 609-611.
LI, C. H. and CHUNG, D. (1976) Proc. Nat. Acad. Sci. 73,
    1145-1148.
MEGLIO, M., HOSOBUCHI, Y., LOH, H. H., ADAMS, J. E., and
    LI, C. H. (1977) Proc. Nat. Acad. Sci. 74, 774-776.
MINNEMAN, K. P. and IVERSON, L. L. (1976) Nature 262,
    313-314.
PERT, C. B. and BOWIE, D. L. (1978) In: Endorphins in
    Mental Health Research. eds: Usdin, E., Bunney, Jr.,
    W. E., and Kline, N. S. MacMillian Press, London,
    in press.
SHARMA, S. K., NIRENBERG, M. and KLEE, W. A. (1975) Proc.
    Nat. Acad. Sci. 72, 590-594.
SIMANTOV, R. and SNYDER, S. H. (1976) Life Sci. 18, 781-
    788.
TERENIUS, L. (1976) Europ. J. Pharmacol. 38, 211-213.
TRABER, J., FISCHER, K., LATZIN, S. and HAMPRECHT, B.
    (1975) Nature, 253, 120-122.
TSENG, L. G., LOH, H. H., and LI, C. H. (1976) Proc. Natl.
    Acad. Sci. 73, 4187-4189.
WEI, E. and LOH, H. H. (1976) Science 193, 1262-1263.
WIEGANT, V. M. and GISPEN, W. H. (1975) Exp. Brain Res.
    23 Suppl. p. 219.
ZWIERS, H., VELDHUIS, H. D., SCHOTMAN, P., GISPEN, W. H.
    (1976) Neurochem. Res. 1, 669-677.
ZWIERS, H., WIEGANT, V. M., SCHOTMAN, P., GISPEN, W. H.
    (1978), Neurochem. Res. 3, 455-463.

S E C T I O N   I I I

Clinical   Implications

INTRODUCTION

This section is devoted to studies of receptor dysfunction and changes of neuropeptide systems in certain
neuropsychiatric disorders. A conceptual framework is
provided by Bonnet. He reviews studies which suggest
that certain mental disorders result from long-lasting
changes in receptor sensitivity, and points out how
neuropeptides, cyclic nucleotides, and phosphoproteins
may be involved in these alterations. Moreover, evidence
is presented that purposeful manipulation of these systems
may provide a novel means for the treatment of such disorders.
This general review is followed by examples of recent studies
in the area. Terenius and Wahlstrom present an overview
on the role of endorphins in clinical pain. Margules
relates the obesity of middle age to changes in ACTH and
β-endorphin activity. Finally, two different approaches,
utilized in studies on the role of endorphins in psychiatry,
are presented. Volavka and his co-workers provide a
critical evaluation of the strategy utilized in the administration of opiate antagonists. Emrich and colleagues
present new data on the measurement of β-endorphin-like
immunoreactivity in CSF and plasma.

ADAPTIVE ALTERATIONS IN RECEPTOR MEDIATED PROCESSES AND

THEIR IMPLICATIONS FOR SOME MENTAL DISORDERS

Kenneth A. Bonnet

Millhauser Laboratories
Department of Psychiatry
New York University School of Medicine
New York, New York   10016

INTRODUCTION

The locus and mechanisms of long-lasting adaptation in neural systems have been the subject of intensive research efforts for many years. Prominent among early efforts was the demonstration of an increased sensitivity to acetylcholine in denervated skeletal muscle (CANNON and ROSENBLUTH, 1949). More recently, the technology for the direct study of high affinity receptor binding specific to individual neurotransmitters has evolved and with it has grown our knowledge of the complexity of receptor-ligand interactions and subsequent coupling to effector sites in the target cell. It has become clear that cell surface receptors for neurotransmitters can undergo long-lasting changes in sensitivity to ligand that are reciprocal to the frequency of ligand encounter (FRIEDHOFF, 1977). It is further apparent that in some instances the elements coupling cell surface receptors to an effector complex within the cell are capable of independent adaptation. Finally, it is evident that the purposeful manipulation of these receptor systems, by pharmacological strategies, may provide novel means by which to produce long-lasting remediation of some clinical disorders possibly resulting from receptor dysfunction.

Adaptation in Hormone Receptor Systems in Peripheral Cells.

Cell surface receptor systems in a number of peripheral tissues have proven instructive with regard to the principal of receptor adaptation in response to the frequency of

ligand encounter. These include the nicotinic acetylcholine
(ACh) receptor of skeletal muscle, the insulin receptor,
the β-adrenergic receptors of the erythrocyte and of the
pineal gland, and the receptor for thyrotrophin releasing
hormone (TRH).

The sensitivity of the neuromuscular junction to ACh
is increased when the cholinergic afferents to skeletal
muscle are interrupted. Cannon called this phenomenon
denervation supersensitivity (CANNON and ROSENBLUTH, 1949).
Denervation supersensitivity at the neuromuscular junction
is now well known to be mediated by an increase in the
number of nicotinic receptor sites. Less well known is the
demonstration of Katz and Thesleff (1957) that ACh receptor
desensitization occurs when the levels of ACh in the neuromuscular junction are increased for prolonged periods.
Supersensitivity provides a means by which the target cell
receptor system can adapt to overcome a persistent insufficiency of ligand encounter. Subsensitivity in the same
receptor system indicates that receptor cells are capable
of the reverse adaptation to a persistently high level of
ligand encounter.

The number of insulin receptors is decreased in mono-
nuclear leukocytes of hyperinsulinemic, hyperglycemic
diabetic patients (GOLDSTEIN, et al, 1975). Recently, at
least six diseases characterized by insulin resistance have
been demonstrated to have a significantly reduced number
of insulin receptor sites in several types of target cells
(e.g. liver, thymic lymphocytes, muscle and fat cells).
The remaining insulin receptor sites on these cells demonstrate normal affinity, temperature dependence and biological
specificity. A similar receptor subsensitivity accompanied
by hyperinsulinemia is demonstrable in genetically obese
animals, and in animals with acquired obesity (SOLL, et al,
1974; SOLL, et al, 1975). When insulin-resistant patients,
or obese rodents, were fasted for 24 hours the circulating
levels of insulin fell and there was a compensatory increase in the number of insulin receptors. This bidirectional regulation of insulin-receptor number by insulin was
most clearly demonstrated in vitro. Cultured lymphoblastoid
cells grown in the presence of insulin exhibit a reduction
of insulin receptors in direct proportion to the concentration of added insulin (GAVIN and ROTH, 1974). The insulin-
induced loss of receptors seems to be caused by a receptor
degradation process that requires protein synthesis. Removal of insulin from these preparations leads to as much as
a 3-fold recovery in receptor number.

The thyrotrophin-releasing-hormone (TRH) receptor system has been shown to exhibit a similar regulation of target cell receptor number. When TRH is added to a cell culture the number of TRH binding sites decreases. Like insulin-mediated regulation of insulin receptors, TRH-induced sub-sensitivity of the TRH receptor system requires protein synthesis and is presumably under genetic control (HINKLE and TASHJIAN, 1975).

Finally, the β-adrenergic receptor system of erythrocytes, the pinealocytes and several other cell types have been shown to adapt to both acute and chronic ligand encounter or blockade. The earliest systematic demonstration of β-adrenergic bidirectional receptor sensitivity modification was reported by Deguchi and Axelrod (1973). Reserpine-induced depletion of norepinephrine in the rat pineal produced supersensitivity to isoproterenol stimulation of adenylate cyclase and the subsequent induction of N-acetyltransferase. Repeated isoproterenol treatment reversed that increased sensitivity. In the previously untreated animal, repeated encounter with isoproterenol decreased the subsequent response to isoproterenol. Since that report, Axelrod and others have shown in a variety of other tissues that β-adrenergic receptor system, and its coupling to adenylate cyclase, is highly sensitive to regulation by the rate of ligand encounter (LEFKOWITZ, et al, 1977; VETULANI and SULSER, 1976; DEGUCHI and AXELROD, 1975). Chronic stimulation by agonists produces a subsensitivity whereas depletion of norepinephrine or chronic receptor blockade leads to supersensitivity.

Sites at Which Receptor-Coupled Systems Undergo Adaptive Changes.

Adaptive changes in sensitivity to agonists in receptor-mediated effects are best characterized in the β-adrenergic receptors of turkey erythrocytes and pineal gland, and in the dopamine receptors of the rat neostriatum (LEFKOWITZ and MUKHERJEE, 1977; AXELROD and ZATZ, 1977; FRIEDHOFF, BONNET and ROSENGARTEN, 1977). As in the case of the insulin receptor, changes in sensitivity of catecholamine receptors to their ligands is seen predominantly as changes in the number of binding sites rather than in the affinity of the receptor for its ligand (CREESE, BURT and SNYDER, 1977). Catecholamine receptor sites occupied by agonists facilitate the binding of membrane calcium ions to a calcium-binding regulator protein (CRP) to stimulate adenylate cyclase (GNEGY, UZUNOV and COSTA, 1976). This stimulation of adenylate cyclase requires GTP as an allosteric effector, in many systems (RODBELL, et al, 1971). Elevated 3',5'AMP levels

stimulate the activity of membrane-associated protein kinases
that phosphorylate specific proteins in the membrane complex
(EHRLICH, et al, 1977).   Such phosphorylation promotes the
release of the Ca-CRP complex into the cytoplasm to activate
the soluble, high Km phosphodiesterase (CHEUNG, et al, 1976;
BROSTROM, et al, 1977).   This phosphodiesterase degrades
3',5'AMP reducing its concentration to prestimulation levels.
The catecholamine receptor then releases the agonist and
remains refractory to further agonist binding until reacti-
vated by guanine nucleotides (LEFKOWITZ, 1978).

        Chronic blockade of dopamine receptors, coupled to
adenylate cyclase, with antagonists such as haloperidol or
chlorpromazine leads to an increase in the number of binding
sites in the neostriatum as measured by either ^{3}H-neurolep-
tic binding or ^{3}H-dopamine binding (ROSENGARTEN, et al, 1978;
EZRIN-WATERS and SEEMAN, 1977).   Dopamine sensitive adenylate
cyclase also becomes hypersensitive to dopamine stimulation.
However, the magnitude of induced hypersensitivity is not
equivalent to the increase in binding sites (IWATSUBO and
CLOUET, 1975; FRIEDHOFF, BONNET and ROSENGARTEN, 1977;
MISHRA, et al, 1978).   Moreover, the CRP that is thought to
couple the dopamine receptor to the adenylate cyclase has
been reported to increase in concentration in striata of
animals chronically treated with haloperidol (FREDHOLM,
1977; GNEGY, UZUNOV, and COSTA, 1977).   It is evident, then,
that the development of supersensitivity phenomena is not a
simple reflection of alterations in receptor number alone,
but more likely reflects the adaptation of several regulatory
sites in the receptor-effector coupling complex.

        The concept of multiple regulatory sites contributing
to overall adaptation has been most clearly demonstrated by
the work of Zata (1977).   The pineal gland responds to
β-adrenergic stimulation of adenylate cyclase by activation
of a protein kinase, synthesis of m-RNA, protein synthesis
and an eventual increase in levels of the enzyme N-acetyltransferase.   In dark-adapted animals, there is increased
norepinephrine release into the pineal gland, and a characteristic increase in the N-acetyltransferase synthesis of
melatonin.   In light-adapted animals, norepinephrine release is decreased and the induction of N-acetyltransferase
is prevented.   Therefore, prolonged exposure of the animal
to light produces supersensitive glands, whereas prolonged
exposure to dark produces subsensitive glands (ROMERO and
AXELROD, 1975).   A similar supersensitivity may also be
produced by denervation of the pineal or by reserpine-
induced depletion of norepinephrine stores, and subsensitivity can also be produced by dark adaptation or by chronic

treatment with β-adrenergic agonists. The supersensitive glands show increased numbers of β-adrenergic binding sites and parallel increases in the sensitivity to isoproterenol-stimulated adenylate cyclase activity (KEBABIAN, ZATZ, ROMERO and AXELROD, 1975). Cholera toxin stimulates adenylate cyclase in a manner that bypasses the β-adrenergic receptor, and also elicits a greater response in adenylate cyclase activity in the supersensitive gland than in the subsensitive gland (ZATZ, 1977). Both basal and 3',5'AMP-stimulated protein kinase activity was greater in super-sensitive glands, and exogenous 3',5'AMP induced greater amounts of N-acetyltransferase synthesis in supersensitive glands. All of these phenomena appear to reflect changes in the level of maximal response rather than in the affinity for the stimulating agent.

The supersensitive and subsensitive states of a parti-cular receptor-coupled system may not simply reflect mirror images of a single adaptive response occurring at a parti-cular locus. Chronic isoproterenol treatment desensitizes frog erythrocyte membrane β-adrenergic binding sites and adenylate cyclase stimulation (LEFKOWITZ, et al, 1978). This subsensitivity is reversible, but this reversal re-quires no protein synthesis as does the development of supersensitivity. It is possible that the desensitized system may represent a conformationally altered state of the normally high affinity receptor at some sites, rather than an irrevocable loss of receptor sites from the mem-brane matrix. Some of these altered sites appear to tightly retain agonist molecules in the subsensitive state, but only a portion of the loss in binding sites is accounted for by this phenomenon.

The guanine nucleotides are a further example of the different processes by which super- or subsensitive states may be achieved. Guanine nucleotides play an essential role in receptor binding affinity, and in receptor coupling to adenylate cyclase in a number of receptor systems. Guanosine-triphosphate (GTP) selectively decreases the af-finity of adenylate cyclase-coupled receptor sites for agonists, but not for antagonists, of dopamine, β-norepine-phrine, glucagon, prostaglandin E_1 and opiate receptors (ZAHNISER and MOLINOFF, 1978; WILLIAMS and LEFKOWITZ, 1977; RODBELL, et al, 1971; LEFKOWITZ, et al, 1977; BLUME, 1978). The primary mode of action of guanine nucleotides in this effect is to accelerate the dissociation rate of agonists from the receptor. Zahniser and Molinoff (1978) suggested that this is an effect that may be restricted to receptors coupled to adenylate cyclase. GTP is also active in recoupling dopaminergic or β-adrenergic receptors to adenylate cyclase that have been uncoupled by treatment with Russell's viper venom (KEBABIAN, 1978; PECKER and HANOUNE, 1977). Thus, GTP regulation may act in the development of a sub-sensitive receptor state, but no opposite role for this regulation in the supersensitive state is known.

Changes in Receptor Sensitivity by Chronic Action at Secon-dary Neurotransmitter Systems.

The ramifications of receptor sensitivity modification can include changes in one receptor system resulting from sustained and marked changes in the activity of a second and interacting modulatory system. A clinical analogy might be the indirect treatment of hypertension by the use of the α-adrenergic agonist, clonidine, by modulation of the re-lease of norepinephrine. In such a system sudden withdrawal of clonidine results in a hypertensive crisis that can be life-threatening. The same result can be elicited with treatment by direct β-adrenergic blockade with propanolol; a sudden withdrawal from this blackade also results in a hypertensive crisis.

The opiate receptor system is known to interact with catecholamine neurotransmitter systems. Opioid agonist binding to the opiate receptor attenuates stimulation-induced release of norepinephrine or dopamine. Moreover, opioid agonists inhibit dopamine stimulation of adenylate cyclase (MOTOMATSU, et al, 1977). Similarly, we have recently re-ported that opioid agonists specifically displace calcium ions from thalamic membrane preparations and prevent sub-sequent β-adrenergic stimulation of adenylate cyclase (BONNET, GUSIK and SUNSHINE, 1978). Chronically, opiate treatment leads to greater than normal levels of calcium binding, and to increased sensitivity of adenylate cyclase to β-adrenergic and dopaminergic stimulation (CLOUET and IWATSUBO, 1975; PURI, et al, 1976; MERALI, et al, 1975; YAMAMOTO, et al, 1978; LLORENS, et al, 1978). However, protein kinase phosphorylation of specific membrane pro-teins is selectively decreased (CLARK, et al, 1973; BONNET, et al, 1978; EHRLICH, et al, 1978). Supersensitivity of receptor systems that interact with opiate receptors may not be restricted to the catecholamine systems (SCHULZ, et al, 1977), and may include supersensitivity to prosta-glandin E_1 and to adenosine stimulation of adenylate cyclase (SHARMA, et al, 1977). A similar state of supersensitivity has been proposed to underly withdrawal phenomena in opiate dependent individuals (COLLIER, 1966; BONNET, et al, 1977).

A common pathway for the effects of opiate withdrawal may
be supersensitivity of adenylate cyclase-coupled systems,
since opiate displacement seems to produce rapid and dramatic
elevation in 3',5'AMP levels in some brain regions, and is
potentiated by inhibitors of phosphodiesterase (BONNET and
GUSIK, 1978; COLLIER and FRANCIS, 1975).

Types of Receptors Demonstrated to Show Adaptive Alterations.

Adaptive alterations in receptors and in receptor-
coupled processes are known to occur in a number of receptor
systems in the central nervous system and in peripheral
tissue, as well. We have discussed earlier the alterations
known to occur in the TRH and insulin receptors in several
cell types, including nicotinic cholinergic receptors in
skeletal muscle, and β-adrenergic receptors in neuronal and
non-neuronal cells. The dopamine receptor system in the
neostriatum, and the above receptor systems, show both
super- and subsensitivity, or "up" and "down" regulation,
as a reciprocal function of marked and sustained changes in
ligand encounter.

The prostaglandin E_1 receptor also appears capable of
bidirectional regulation. Lefkowitz has demonstrated that
frog erythrocytes pre-exposed to PGE_1 show specific desensitization to PGE_1-stimulation of adenylate cyclase
(LEFKOWITZ, MULLIKIN, WOOD, GORE and MUKHERJEE, 1977).
Conversely, it appears that blockade of PGE_1 receptor stimulation in neuroblastoma-glioma hybrid cells results in hypersensitivity to subsequent PGE_1-stimulation of adenylate
cyclase (SHARMA, et al, 1977). Yet, not all receptor systems
can undergo bidirectional regulation. As super- and subsensitivity may not represent mirror images of a single
adaptive process, so three receptor types have been shown
to evidence subsensitivity, but not supersensitivity development. The muscarinic cholinergic receptor develops subsensitivity to repeated or sustained high levels of agonist
(MAAYANI, et al, 1977). Similarly, the benzodiazapine receptor in brain shows subsensitivity development after
several days of fluazepam treatment that is reflected in
both a decreased affinity for ligand and a reduced number of
binding sites (CHIU and ROSENBERG, 1978). In addition, the
α-adrenergic receptor stimulation by epinephrine in parotid
gland cells causes a specific loss of α-adrenergic binding
sites (STRIMWATTER, DAVIS and LEFKOWITZ, 1977). Yet, it is
not certain whether these receptor systems are incapable of
"up" regulation or if the proper technology is simply not
available to detect this phenomenon.

Particular emphasis has been placed on adaptations of the dopamine receptor system because of its putative central role in the action of antipsychotic drugs and in the development of tardive dyskinesia (CARLSSON, 1970; CARLSSON and LINDQUIST, 1963; KLAWANS, et al, 1970). However, it is now quite evident that there is more than one dopamine receptor type, and that our understanding of super- and subsensitivity development is largely restricted to the adenylate cyclase-coupled receptor of the neostriatum. The prototype of this type of dopamine receptor has been suggested to be the bovine parathyroid gland, and in the neostriatum this type of receptor appears to be localized predominantly to the post-synaptic membrane (SCHWARZ, et al, 1978; KEBABIAN and CALNE, 1979). In contrast, the high affinity neuroleptic binding receptors in the neostriatum may be largely presynaptically localized and are thought to be synonymous with the dopamine autoreceptor that is not coupled to adenylate cyclase. A number of other brain regions contain a dopamine receptor that is not coupled to adenylate cyclase, the prototype of which is the anterior pituitary regulator of prolactin secretion. All of these dopamine receptor types can be seen in the neostriatum (KEBABIAN and CALNE, 1979, CSERNANSKY, FRIEDHOFF and BONNET, in preparation). However, most radiolabelled ligands used for the study of dopamine receptors are not selective for one receptor type. Therefore, we are approaching the differentiation of receptor types that may or may not show adaptive regulation by available ligands, and we are only beginning to make clear the distinctive parameters of long-lasting adaptive responses in these different receptor systems (MULLER and SEEMAN, 1978).

Presently, there is no convincing evidence that opiate receptors can undergo adaptive changes that might explain the rapid and dramatic tolerance development attending chronic opiate treatment (BONNET, HILLER and SIMON, 1976). Further, we know of no direct evidence to date for super- or subsensitivity development in serotonin, GABA or glycine receptors.

Clinical Ramifications of Receptor Sensitivity Modification.

The clinical ramifications of receptor sensitivity modification can be illustrated in some select pathological states. Pharmacological strategy may sometimes seem contrary to acute symptom treatment. Insulin-resistant diabetes can be effectively treated by fasting, and by abstaining from exogenous insulin use. The resulting decrease in available insulin induces target cells to increase insulin receptor number and the individual can be maintained with a conservative diet but without exogenous insulin (BAR, et al, 1978).

Tardive dyskinesia results, in a number of patients, from long-term treatment with antipsychotic drugs. The dyskinesia has been thought to result from dopaminergic supersensitivity that develops in dopamine receptors after years of chronic blockade (CARLSSON, 1970; KLAWANS, et al, 1970). We have shown in animals that chronic antipsychotic drug administration results in compensatory increases in number of dopamine receptors and in dopamine stimulated adenylate cyclase activity that can be actively reversed with short term administration of L-DOPA (FRIEDHOFF, BONNET and ROSENGARTEN, 1977; WATERS and SEEMAN, 1978). Dr. Arnold Friedhoff and Dr. Murray Alpert of our group have treated tardive dyskinesia patients for five to eight weeks witn incrementing dosages of L-DOPA. L-DOPA produced an initial worsening of symptoms, followed by tolerance to the effects of L-DOPA. On cessation of L-DOPA treatment, several remitted patients have remained symptom-free for up to four years (ALPERT and FRIEDHOFF, 1978).

Gilles de la Tourette syndrome is characterized by outbursts of caprolalia and by uncontrolled tics that have been attributed to hyperdopaminergic activity in some brain regions, since haloperidol has been singularly effective in the management of the disorder. Recently, one patient showed only a temporary remission of symptoms, and the dose of haloperidol was increased repeatedly. On the hypothesis that this patient was developing uncommonly rapid dopaminergic supersensitivity to compensate for the haloperidol blockade, Dr. Friedhoff treated this patient with L-DOPA for several weeks without haloperidol. After the L-DOPA was withdrawn, the patient was clinically improved and currently remains free of medication, with the exception of a maintenance regimen of L-DOPA treatment for two weeks every six months to stabilize his symptoms.

In studies of schizophrenia, patients receiving no neuroleptic medication were treated with L-DOPA with increasing doses for up to eight weeks. If a hyperdopaminergic state predisposes to schizophrenia then desensitization of the dopaminergic receptor systems by increasing dopamine availability should remediate the symptoms. The initial activation by L-DOPA of locomotor activity, voice levels and expressions of anger were not accompanied by an apparent worsening of thought disorder or hallucinations. At the end of several weeks of incremental L-DOPA administration, patients were not worse than placebo or neuroleptic treated patients. On cessation of L-DOPA a number of patients were clinically improved (ALPERT, FRIEDHOFF, MARCOS & DIAMOND, 1978).

Narcotic dependence may result from opiate receptor-induced chronic modulation of catecholamine function that is postulated to cause catecholamine receptor supersensitivity. Therefore, on release from opiates the exaggerated receptor-mediated responses to catecholamine release may account for a number of the symptoms of narcotic withdrawal (COLLIER, 1966; BONNET, et al, 1977, 1978). It is of interest that haloperidol is effective in preventing some withdrawal symptoms such as aggression and sensory hypersensitivity. Propanolol has also been reported to be partially effective in the prevention of drug-seeking behavior in abstinent heroin-dependent individuals (GROSZ and BLACK, 1973). However, these catecholamine receptor antagonists may be too specific to provide satisfactory management of the broad range of symptoms attending narcotic withdrawal and abstinence. In this line, the α-adrenergic agonist clonidine which modulates central nervous system presynaptic release of catecholamines, has been reported to be effective as a detoxifying agent in methadone withdrawal, and to be preferred over methadone for post-withdrawal maintenance (GOLD, et al, 1978).

CONCLUSIONS

The phenomena of receptor sensitivity modifications indicate that there are mechanisms by which some classes of cell surface receptors and receptor mediated processes are regulated by the steady state supply of ligand. Pathological states involving putative abberrant function of neurotransmitter may be treated by novel pharmacological intervention strategies that can manipulate receptor sensitivity. This approach could form the basis for new and more discerning diagnostic strategies. However, the ramifications of receptor modification are different for prenatal encounter than for postnatal encounter on neurotransmitter systems in the central nervous system, and may also provide new insights into risks of perinatal or prenatal treatment of the mother (ROSENGARTEN, et al, 1979). The principles of receptor sensitivity modification promise to provide far greater specificity in our knowledge of pathological and normal adaptive states.

ACKNOWLEDGEMENTS

Grateful appreciation for discussions with A. J. Friedhoff, M. Alpert, J. Csernansky and J. Miller. Supported by Grants DA01113 and MH08618.

REFERENCES

ALPERT, M., FRIEDHOFF, A., MARCOS, L., & DIAMOND, F.
    (1978) Amer. J. Psychiat., 135, 1329-1332.
AXELROD, J. & ZATZ, M. (1977) In: Biochemical Actions
    of Hormones. Vol. 4, pp. 249-268. Acad. Press, N.Y.
BAR, R. S., HARRISON, L., MUGGEO, M., GORDEN, P., KAHN,
    C. R. & ROTH, J. (1979) In: Advances in Internal
    Medicine (in-press).
BONNET, K. A., HILLER, J. M. & SIMON, E. J. (1976) In:
    Opiates and Endogenous Opioid Peptides. (Kosterlits,
    H. W., ed) pp. 335-344, North Holland Publishing,
    Amsterdam.
BONNET, K. A., BRANCHEY, L., FRIEDHOFF, A. J. & EHRLICH,
    Y. H. (1977) Committee on Problems of Drug Dependence,
    Cambridge, Mass., July.
BONNET, K. A., BRANCHEY, L., FRIEDHOFF, A. J. & EHRLICH,
    Y. H. (1978) Life Sciences, 22, 2003-2008.
BONNET, K. A., GUSIK, S., & SUNSHINE, A. G. (1978) In:
    Characteristics and Functions of the Opioids.
    (Van Ree, J. & Terenius, L., eds.). pp. 453-464.
BLUME, A. J. (1978) Proc. Nat. Acad. Sci. USA, 75,
    1713-1717.
BROSTROM, C. O. & WOLFF, D. J. (1976) Arch. Bioch. Biophys.
    172, 301-311.
CANNON, W. B. & ROSENBLUTH, A. (1949) The Supersensitivity
    of Denervated Structures: A Law of Denervation.
    MacMillan, New York.
CARLSSON, A. (1970) In: L-DOPA and Parkinsonism (Barbeau,
    A. and McDowell, F., eds.) pp. 205ff. F.A. Davis Co.,
    Philadelphia.
CARLSSON, A. & LINDQUIST, M. (1963) Acta Pharmacol. Toxicol.
    20, 140-143.
CHEUNG, W. Y., LYNCH, T., & TALLANT, E. (1976) Bioch.
    Biophys. Res. Commun., 68, 616-621.
CHIU, T. & ROSENBERG, H. P. (1978) Life Sci., 23, 1153-
    1158.
CLARK, A. G., JOVIC, R., ORNELLAS, M. R. & WELLER, M. (1972)
    Bioch. Pharmacol., 21, 1989-1999.
CLOUET, D. H., & IWATSUBO, M. (1975) In: The Opiate
    Narcotics (Goldstein, A., ed.) pp. 189-194. Pergammon
    Press, New York.
COLLIER, H. O. J. (1966) In: Advances in Drug Research.
    3, 171-188.
COLLIER, H. O. J. & FRANCIS, D. L. (1975) Nature, 255,
    159-161.
CREESE, I., BURT, D. & SNYDER, S. H. (1977) Science, 197,
    596-598.
DEGUCHI, T. & AXELROD, J. (1973) Proc. Nat. Acad. Sci.
    USA., 70, 2411-2414.

EHRLICH, Y. H., DAVIS, L. G., GILFOIL, T. & BRUNNGRABER, E. G. (1977) Neurochem. Res., 3, 533-548.
EHRLICH, Y. H., BONNET, K. A., DAVIS, L. G. & BRUNNGRABER, E. G. (1978) Life Sciences, 23, 137-146.
EZIN-WATERS, C. & SEEMAN, P. (1977) Life Sci., 22, 1027-1032.
FREDHOLM, B. B. (1977) Medical Biology, 55, 61-65.
FRIEDHOFF, A. J. (1977) In: Psychopathology and Brain Dysfunction (Shagass, C., Gershon, S. and Friedhoff, A. J., eds.). pp. 139-148. Raven Press, New York.
FRIEDHOFF, A. J., BONNET, K. A. & ROSENGARTEN, H. (1977) Res. Commun. Chem. Pathol. Pharmacol., 16, 411-423.
FRIEDHOFF, A. J., and ALPERT, M. (1979) Proceedings of Fourth Annual Catecholamine Symposium, Pacific Grove, Calif. (in-press).
GAVIN, J. R., ROTH, J. & NEVILLE, D. M. (1974) Proc. Nat. Acad. Sci. USA., 71, 84.
GNEGY, M., UZUNOV, P., & COSTA, E. (1976) Proc. Nat. Acad. Sci. USA., 73, 3887-3890.
GNEGY, M., UZUNOV, P., & COSTA, E. (1977) J. Pharm. Exp. Ther., 202, 558-564.
GOLD, M. S., REDMOND, D. E. & KLEBER, H. D. (1978) Lancet September 16. 599-601.
GOLDSTEIN, S., BLECHER, B., BINDER, R., PERINNO, P. V. & RACANT, L. (1975) Endocrinol. Res. Commun. 2, 367-376.
GROSZ, H. J. (1973) Lancet, 2, 612.
HINKLE, P. M. & TASHJIAN, H., JR. (1975) Biochemistry, 14, 3845-3851.
IWATSUBO, K. & CLOUET, D. H. (1975) Bioch. Pharmacol., 24, 1499-1503.
KATZ, B. & THESLEFF, S. (1957) J. Physiol., 138, 63-80.
KEBABIAN, J. W., ZATZ, M., ROMERO, J. A. & AXELROD, J. (1975) Proc. Nat. Acad. Sci. USA, 72, 3735-3739.
KEBABIAN, J. W. (1978) Life Sciences, 23, 479-484.
KEBABIAN, J. W. & CALNE, D. B. (1979) Nature, 277, 93-96.
KLAWANS, H. L., SLAKI, M. M. & SHENKER, D. (1970) Acta. Neurol. Schand., 46, 409-411.
LEFKOWITZ, R. J. & MUKHERJEE, C. (1976) Proc. Nat. Acad. Sci. USA., 73, 1494-1498.
LEFKOWITZ, R. J., MULLIKIN, D., WOOD, C., GORE, T. & MUKHERJEE, C. (1977) J. Biol. Chem., 252, 5295-5303.
LEFKOWITZ, R. J., MULLIKIN, D., & WILLIAMS, L. T. (1978) Molec. Pharmacol., 14, 376-380.
LEFKOWITZ, R. J. (1978) In: Neuronal Information Transfer (Karlin, A., Tennyson, V. M. & Vogel, H. J., eds.) pp. 59-72. Academic Press, New York.
LLORENS, C., WATERS, M., BAUDRY, M., & SCHWARTZ, J. (1978) Nature, 274, 603-605.

MERALI, Z. & SINGHAL, R. (1975) Life Sci., 16, 1889-1894.
MISHRA, R. K., WONG, Y.-W., VARMUZE, S. L. & TUFF, L.
    (1978) Life Sciences, 23, 443-446.
MOTOMATSU, T., LIS, M., SEIDAH, N. & CHRETIEN, M. (1977)
    Bioch. Biophys. Res. Commun., 77, 442-447.
MULLER, P. & SEEMAN, P. (1978) Psychopharmacol. 60, 1-11.
PECKER, F. & HANOUNE, J. (1977) FEBS Lett., 83, 93-98.
PURI, S. K., COCHIN, J. & VOLICER, L. (1975) Life Sci.,
    16, 759-768.
RODBELL, M., KRANS, HM. J., PHOL, J. & BIRNBAUMER, L.
    (1971) J. Biol. Chem., 246, 1872-1876.
ROMERO, J. A., & AXELROD, J. (1975) Proc. Nat. Acad. Sci.
    USA. 72, 1661-1665.
ROSENGARTEN, H. & FRIEDHOFF, A. J. (1979) Science, (in-
    press).
SCHWARZ, R., FUXE, K., AGNATI, L. & GUSTAFSSON, J.-A.
    (1978) Life Sci., 23, 465-470.
SHARMA, S. K., KLEE, W. A. & NIRENBERG, M. (1977) Proc.
    Nat. Acad. Sci. USA. 74, 3365-3369.
SOLL, A. H., KAHN, C. R. & NEVILLE, D. M. (1975) J. Biol.
    Chem. 250, 4702-4707.
SOLL, A. H., GOLDFINE, I. D. & ROTH, J. (1974) J. Biol.
    Chem. 249, 4127-4131.
STRIMWATTER, W. J., DAVIS, J. N. & LEFKOWITZ, R. J. (1977)
    J. Biol. Chem., 252, 5472-5477.
VETULANI, J., STAWARZ, R., SULSER, F. (1976) J. Neurochem.
    27, 661-666.
WILLIAMS, L. T. & LEFKOWITZ, R. J. (1977) J. Biol. Chem.
    252, 7202-7213.
YAMAMOTO, H., HARRIS, R., LOH, H. H. & WAY, E. L. (1978)
    J. Pharm. Exp. Ther., 205, 255-264.
ZAHNISER, N. R. & MOLINOFF, P. B. (1978) Nature, 275,
    453-455.
ZATZ, M. (1977) Life Sciences, 21, 1267-1276.

ENDORPHINS AND CLINICAL PAIN, AN OVERVIEW

L. TERENIUS and A. WAHLSTRÖM

Department of Pharmacology
University of Uppsala
Uppsala, Sweden

INTRODUCTION

The existence of selective opiate receptors and the
finding that stimulation produced analgesia could be reversed by the selective opiate antagonist, naloxone, led
to the prediction and detection of endorphins. It was
subsequently found that the endorphins belonged to the
growing family of brain peptides. These peptides are
present in specific neurons of the CNS, suggesting that
they are neuromessengers[1]. The fact that the pharmacology
of opiates had a long history and was being studied extensively at the time of discovery, facilitated a rapid development of knowledge about endorphins. Many classically
trained pharmacologists moved into a new field, that of
brain peptides. The discovery of endorphins is therefore
a very important landmark in the development of the neurosciences and consequently, brain peptides in general are
now the focus of contemporary brain research.

Pharmacologic testing of endorphins or synthetic analogues has confirmed that in most respects they are opiate-
like. Thus, they produce analgesia and opiate-like dependence, they inhibit respiration and affect various tissues
in vitro like classical opiates. All these effects are
reversed by naloxone, emphasizing the similarity to

[1]The term neuromessenger is suggested as a general
word for a signal substance released from a nerve, be it
a neurotransmitter, neuromodulator or neurohormone.

opiates. Although there is some subspecialization of
opioid receptors and some receptors show selectivity foɪ
opioid peptides (LORD et al., 1977; TERENIUS, 1977), the
general conclusion which can be drawn from these experiments is that opioid peptides act on the same receptors
as opiates, and that in general they can be expected to
produce the same effects.

     Most of the early work on endorphins was based on
the classical pharmacology of opiates as analgesics or
as inhibitors of electrically stimulated contractions of
sensitive smooth-muscle preparations (GOLDSTEIN, 1976).
However, opiates are also known to be psycho-active agents
with profound general effects on the mood and behavior
in animals and man (JAFFÉ, 1975). It was also observed
that nerve fibers with opioid peptides were spread over
wide areas of the CNS (ELDE et al., 1976), suggesting that
they might not only serve as endogenous regulators of
pain. It was therefore proposed that endorphins could
be involved in mental disorders (TERENIUS et al., 1976;
BLOOM et al., 1976; JACQUET & MARKS, 1976), in drive-
reward regulation (BELLUZI & STEIN, 1977) and in the regulation of sexual behavior (MEYERSON & TERENIUS, 1977).
Testing of the hypothesis that endorphins may influence
the mental state in the human being were soon to come
(TERENIUS et al., 1976; GUNNE et al., 1977). This communication critically summarizes the present status of the
knowledge of the role and importance of endorphins in clinical pain syndromes, with special reference to methodological and principal problems. Since this area has been
reviewed recently (TERENIUS, 1978), only the more recent
significant developments will be covered.

                SOME CHARACTERISTICS OF ENDORPHINS AND
                    ENDORPHIN PRODUCING SYSTEMS

     The endorphins have been found in many different
organs. In terms of function one may distinguish between
endorphins produced in neurons and those produced in non-
neuronal tissue. Chemically, it seems as if we should
deal with two separate systems, one with the enkephalins
of 5 amino acids, the other with beta-endorphin, a longer-
chain peptide with 31 amino acids. Shorter fragments of
beta-endorphin with opioid activity have been isolated,
but whether they are present in significant amounts under
physiologic conditions is not known. In all probability,
the enkephalins and beta-endorphin are the most important
opioid peptides. Their distribution is outlined in Table 1.

Thus, endorphins may be anything of neurotransmitter candidates, neuromodulators, neurohormones or hormones, dependent on the site of production. It is clear that functional studies will be complicated by this diversity.

As suggested from Table 1, the brain has the two
different endorphin systems, with enkephalins and beta-
endorphin as the predominant peptides, respectively.
The best evidence is perhaps the lack of overlap in the
CNS of fibers with enkephalins and those with beta-endorphin (BLOOM et al., 1978; WATSON et al., 1977). Furthermore, there is evidence that the peptide precursor to the
enkephalins is different from that for beta-endorphin
(LEWIS et al., 1978). If this can be proved, the two
opioid systems would be truly separate, biochemically as
well as topologically. One may question whether there
may be still further complexity. In fact, there are two
enkephalins, each with 5 amino acids but with different
C-terminal amino acids, leucine and methionine, respectively. These two peptides seem to occur in different
proportions in different parts of the rat CNS (HUGHES
et al., 1978). However, so far it has not been possible
to establish with immunohistochemical means whether some
neurons contain only one enkephalin. It seems as if they

TABLE 1

Principal endorphin systems.

Neuronal Systems	Brain	Enkephalin
		Beta-Endorphin
	Ganglia	Enkephalin
Nonneuronal Systems	Pituitary	Beta-Endorphin[1]
	Gastro-intestinal tract	Enkephalin
	Adrenal	Enkephalin

[1]It has been questioned whether beta-endorphin is
formed in human pituitary gland (LIOTTA et al., 1978).

co-exist in one neuron, but in varying relative amounts.
In addition to these peptides of known structure, various
investigators have found evidence for the existence of
other opioid substances. For instance, cerebrospinal
fluid (CSF) contains two fractions, denoted I and II with
opioid activity (TERENIUS & WAHLSTRÖM, 1975). A compound
which has high affinity to morphine-directed antibodies
and which reacts with opioid receptors has also been characterized and found to have a unique distribution with
high levels in the cerebellum (GINTZLER et al., 1976,
1978). These substances which have not yet been characterized chemically may or may not be related to the known
endorphins.

The difference in chemical composition of the enkephalins and beta-endorphin is certainly not trivial. The
enkephalins are very unstable if introduced into the CNS
while beta-endorphin is considerably more stable (FELD-
BERG & SMYTH, 1976). The enkephalins are therefore likely
to be degraded rapidly after release while beta-endorphin
would give rise to sustained effects. In mechanistic
terms, beta-endorphin might act as a neurohormone or a
neuromodulator capable of reaching distant synapses while
enkephalin would give a local effect. Another great difference is that any degradation of the enkephalin molecule
will cause an almost complete drop in activity (FREDERICK-
SON, 1977) while fragmentation of beta-endorphin can occur
to the enkephalin pentapeptide sequence with retention
of activity. However, the larger size of the beta-endorphin molecule also means that it may carry several messages, in different sequences of the molecule. The differences between the two opioid peptide systems seem to
be fundamental and may have far-reaching consequences.

The differences may also extend to differences in
receptor activation. It is well-known that there must
be more than one opioid receptor. Binding analysis points
to at least 3 different binding sites (Table 2). Firstly,
there is a marked difference between the opioid peptides
and morphine-like alkaloids, the former only having considerable affinity for the delta-sites. Secondly, beta-
endorphin, but not the enkephalins seems to bind to some
sites with high affinity for naloxone. The functional
importance of these sites is not known and direct studies
with labelled beta-endorphin would be required to study
the properties of these sites. In terms of affinity,
beta-endorphin is always more avidly binding than the
enkephalins. The following conclusions seem warranted.
Although the beta-endorphin fibers in the CNS are

relatively few and appear to project diffusely from hypothalamic to thalamic areas (BLOOM et al., 1978; WATSON et
al., 1977) the high receptor affinity and metabolic sta-
bility of beta-endorphin could give it an important physiologic role. The relative closeness of the beta-endorphin
nerve terminals to the third ventricle and the aqueduct,
also suggests a distribution via CSF. Thus, one might propose a neuroendocrine role for the beta-endorphin system.

It was mentioned earlier that endorphins are present
outside the central nervous system (Table 1). The function of the endocrine systems with beta-endorphin (pituitary) and enkephalins (adrenal medulla) is unknown. Release of beta-endorphin from the rat pituitary seems to
follow almost molecule for molecule, the release of ACTH
and is consequently influenced by stress, glucocorticoid
steroids, etc. (GUILLEMIN et al., 1977). The existence
of beta-endorphin in human plasma is a subject of controversy (SUDA et al., 1978). It has even been suggested
that human pituitary beta-endorphin might be an extraction artifact (LIOTTA et al., 1978). If it exists, it
is not clear whether it reaches the brain. Contradictory
results from animal experiments give no lead. In man,
plasma levels of immunoassayable beta-endorphin do not
correlate with those in the CSF (JEFFCOATE et al., 1978).
Furthermore, there is no decrease in CSF levels of beta-
endorphin in hypopituitarism and no clinical studies suggest hyperalgesia in these disorders. However, there

TABLE 2

Relative affinities of opioids for different binding
sites.

Opioid	Affinity		
	Site μ	Site δ	Site "Antagonist"
Alkaloid agonists	High	Low	Low
Alkaloid antagonists	High	Low	High
Enkephalin	High	High	Low
Beta-Endorphin	High	High	Intermediate

Modified from LORD et al. (1977); TERENIUS (1977).

are reports that systemic beta-endorphin produces analgesia in man and that it has behavioral effects (CATLIN
et al., 1978). This is of course no proof for a physiologic function of pituitary beta-endorphin in the regulation of human CNS functions. It has been observed clinically that destruction of pituitary function actually
may give pain relief and in fact this procedure is done
successfully to treat severe pain (MORICCA, 1974). This
suggests that if anything, the human pituitary enhances
rather than attenuates pain sensitivity.

     The situation for the adrenal system is not very
much clearer. It has been assumed that the enkephalins
are too unstable in blood to produce any systemic effects.
However, recent reports suggest that this may not be true
(KONTUREK et al., 1978). So far, it is not known whether
enkephalin is released at all, or if so, how the release
is stimulated. Since   section of the splanchnic nerve
increases the apparent concentration of enkephalin in
the rat adrenal medulla, an active role for the peptide
is indicated (SCHULTZBERG et al., 1978). It is unlikely
however, that enough enkephalin of adrenal origin would
cross the blood-brain barrier and act significantly on
CNS structures.

THE STUDY OF ENDORPHIN MECHANISMS IN CLINICAL PAIN

     This is not the place for detailed discussions of
the nature of pain. It may be useful, however, to consider
some of the problems involved in the testing of pain and
particularly to test the role of endorphins in pain modulation. Pain may be considered as response to activation
of high-threshold sensory fibers. The emotional component
in a pain reaction is always considerable in clinical
pain. This component is lacking in the experimental situation. A number of studies have shown that morphine, in
clinical doses, is very poorly active against experimental
pain, that is, on the sensory threshold (JAFFÉ, 1975).
Similarly, it is not surprising that naloxone hardly affects
pain thresholds in healthy volunteers in experimental
situations (EL-SOBKY et al., 1976; GREVERT & GOLDSTEIN,
1977). With more sophisticated technology, naloxone has
been shown to reduce pain thresholds in pain insensitive
individuals (BUCHSBAUM et al., 1977). However, experimental test methods cannot reflect adequately what is
probably highly relevant in the clinical situation, the
reaction to pain. Paradoxically, it is probably more
relevant to perform animal experiments than experiments

in human beings to study the significance of endorphins
in the physiological regulation of pain. In the animal,
the pain reaction will be colored by fear and anxiety
and will evoke a flight reaction. Animal experiments have
also more consistently shown an increase in pain sensitivity on naloxone injection (JACOB et al., 1974; FREDER-
ICKSON et al., 1977). After habituation, however, the
naloxone response is no longer significant (GOLDSTEIN
et al., 1976), reemphasizing that when the stimulus is
no longer experienced as really threatening, endorphin
activation will no longer occur.

    Naloxone has been found to worsen postoperative pain
intensity moderately (LEVINE et al., 1978). In this investigation, naloxone was studied in a double-blind fashion
in patients recovering from anaesthesia after a molar
extraction. The situation must be considered as rather
moderate both in terms of pain suffering and in the arousal
of anxiety and fear. It is possible that in a more severe
and threatening situation, the naloxone test would have
given more distinct results. Ethical considerations will
make it difficult to make progress in this direction.
The important point here is to emphasize that the relative
lack of effects of naloxone in experimental or near-experi-
mental situations should not be taken as evidence against
the importance of the endorphins in the combat of pain.

    The difficulties in studying endorphins indirectly
by using naloxone led us very early to search for more
direct methods. It was considered likely that the CNS
endorphin activity would be reflected in the concentrations of endorphins in the cerebrospinal fluid (CSF).
Therefore, a method was developed which allowed the measurement of endorphins by a receptorassay. This procedure
would have the advantages of being direct, easy to quan-
tify, and objective. It might also have disadvantages,
such as being inadequate in reflecting CNS endorphin activity. The best way to test the validity of endorphin
analysis for functional studies was by pilot experiments
(TERENIUS & WAHLSTRÖM, 1975). Although the case number
was small, the observations were clear-cut, patients with
chronic neurogenic pain showed low CSF endorphin levels.
This finding was later substantiated in a large series
of patients. The assay has therefore been maintained
as an empirical tool for functional studies in pain syndromes as well as in psychiatric disorders. The assay
has been described in detail elsewhere (TERENIUS et al.,
1978), and here only some of its essential characterictics
will be discussed.

Experimental Procedure for CSF Endorphin Analysis

1. 12.5 ml CSF sample obtained from the patient in the
    morning after a night's rest.

2. Sample centrifuged for 5 minutes, clear fluid frozen
    at -20°C or lower until analysis.

3. Careful thawing and filtration through a PM10 ultra-
    filter. Filter washed with 1 ml saline per 4 ml CSF.

4. Five ml combined filtrate run through a Sephadex
    G 10 column (2x50 cm).

5. The Fraction I (15 ml; 1.3 V_O, V_O = void volume)
    and the Fraction II (20 ml; 2 V_O) are collected
    and lyophilized.

6. The fractions are tested for competitive affinity
    against dihydromorphine $-^3H$ in a receptorassay
    (TERENIUS, 1974). A mixture of 0.2 mg synaptic
    plasma membrane protein from rat brain, test frac-
    tion and label is incubated in 0.4 ml buffer, pH
    7.4, in polypropylen centrifuge tubes (Beckman
    340196). Following incubation, the tubes are cooled
    and centrifuged for 5 minutes in a Microfuge (Beckman).
    The tubes are inverted and any remaining fluid is
    removed by centripetal forces induced by a rotary
    wheel. The tips of the tubes are cut and counted
    for radioactivity.

FIGURE 1

THE RADIORECEPTORASSAY FOR CSF ENDORPHINS

An outline of the individual steps of the assay is
given in Figure 1. All handling of the CSF samples prior
to the analysis is done with a minimum of heat. If possible, samples are stored at -90°C. Thawing is done very
carefully and the samples are never allowed to reach a
higher temperature than +4°C. Ultrafiltration is also
done in the cold. After the filtration, further proces-
sing can be done at room temperature. The most critical
point of the assay is the chromatographic separation of
CSF endorphins. The Fractions I and II, which contain
most (> 80%) of the assayable activity, must be very accurately defined. Chromatographic fractions are therefore

obtained by drop-counting and not by chromatography time
and the position of the salt peak is also measured on each
chromatogram. Its position is taken as a landmark for the
positions of the Fractions I and II. The receptorassay
has been improved lately to an extent that variation between replicates is minimal and not much larger than the
pipetting error. The assay is also run on a microscale to
allow maximum sensitivity. All components of the incubation mixture (Figure 1) are incubated in centrifuge tubes
under gentle agitation. Following incubation, the tubes
are briefly cooled on ice and centrifuged in a Microfuge
(Beckman[R]). The tubes are then inverted and put indivi-
dually into the peripheral edge of a wheel which is rota-
ted for one minute at low speed. This procedure removes
any residual supernatant fluid and the membrane pellet
and its bound radioactivity is obtained by simply cutting
the tip of the tube. Thus, following the initial pipetting, no further transfer of material is done with very
good reproducibility as a consequence.

The main disadvantage of the assay is that it is
time-consuming. The average number of samples tested
per week and operator is only 10. Since several of the
steps are so slow, it is difficult to achieve any increase
in output. Otherwise, a receptorassay should have the
advantage of giving biologically active material. Since
the material is derived from CSF, which is probably very
close in its composition to the extracellular fluid of
the brain, the measure levels might also rather directly
relate to the actual levels at the receptor and give some
indication of the receptor occupation. If such a direct
relationship exists it can be calculated that the receptor
occupation can hardly surpass 10 percent in a normal individual.

A definite risk with a receptorassay is that some
of the measured activity is in fact not due to endorphins
at all but rather to some component which destroys recep-
tor-binding unspecifically. This possibility does not
seem so likely. The situation would have been very different if serum or urine were to be tested.

It is presently not known what chemical composition
characterizes the Fractions I and II (TERENIUS et al.,
1978). It would be of great interest to know if they
derive from the enkephalin or beta-endorphin systems or
if they do not relate at all. Despite considerable effort
we have so far been unable to unravel the chemical structures. The basic problem is the low concentrations of
active material and the limited supply of CSF which has

been available to us. Because of these difficulties,
we have recently investigated whether there is any co-
variation in levels of Fractions I and II and enkephalins,
respectively, as measured in a radioimmunoassay. Despite
reports in the literature (SARNE et al., 1978; AKIL et
al., 1978) of considerable concentrations of enkephalins
in human CSF, our findings are negative for most studied
cases. (We found that one exceptional patient had measurable methionine- and leucine-enkephalin as indicated by
the proper retention times in HPLC analysis and positive
identification with radioreceptorassay and radioimmunoassay.) Enkephalins, if present in CSF, would be obtained
with about 50% recovery in our Fraction II. Tests with
a number of different enkephalin antisera have yielded
generally negative results. One antiserum, raised against
leucine-enkephalin, gives positive results in several
cases, however, activity could not be due to leucine-enkephalin since other antisera which would react with this
compound give negative results. No correlation between
Fractions I or II on one side, and the radioimmunoactive
material on the other, is apparent. It is interesting
to note that one group, which has reported on enkephalin-
like material in the human CSF, recorded activity against
a leucine-enkephalin directed antiserum (SARNE et al.,
1978). Beta-endorphin has been reported by two groups
to occur in human CSF (JEFFCOATE et al., 1978; HOSOBUCHI
et al., 1978). With our procedures, beta-endorphin if
present, will not interfere with the determination of
Fraction I and II, since it elutes ahead of both of them.

THE CSF ENDORPHIN ASSAY AND PAIN SYNDROMES

The early observation of low CSF endorphin Fraction
I in chronic neurogenic pain syndromes was born out in
a larger series of patients (SJÖLUND et al., 1977; ALMAY
et al., 1978 and unpublished). No corresponding correlation was observed for Fraction II. The possibility that
measurements of Fraction I endorphin levels might be used
to discriminate between different types of pain syndromes
was investigated. A series of consecutive cases with
chronic pain admitted to the University Hospital in Umeå,
Sweden was studied (ALMAY et al., 1978). The pain syndrome had lasted for at least 6 months. No patient had
previously been taking narcotic analgesics or had any
history of alcohol or drug abuse. The patients were subjected to neurological and psychiatric evaluation. The
neurological investigation attempted to define whether the
pain origin was neurogenic or not (by X-ray, biopsy).

The psychiatric evaluation was quite extensive, using
items of the CPRS scale (ÅSBERG et al., 1978) covering
items for depression, anxiety and retardation. All evalua-
tions were made blind and then compared. The pain syndrome
was then classified as "organic" (i.e., somatogenic, a
syndrome with clear neurogenic involvement or due to an
organic lesion and with a clear and distinct pain description) or "psychogenic" (a syndrome where there is no evidence for an organic component and the pain description
being vague and emotionally strongly colored and the
patient showing neurotic or hysterical signs). The distinction between "organic" and "psychogenic" will always
have to be more subjective than between neurogenic and
non-neurogenic pain syndromes. The results of this investigation are summarized in Table 3. It seems as if the
neurogenic pain syndrome is associated with low CSF Fraction I levels, and that measurements of this fraction
might be used for discriminative purposes. It is of
interest that Sicuteri and co-workers (1978) recently
made similar observations with regard to patients with
essential headache. These patients had low endorphin
Fraction I levels, particularly in association with pain
attacks.

TABLE 3

Pain syndromes and Fraction I endorphin levels in CSF
(from ALMAY et al., 1978). Figures denotes number of
cases.

Pain Syndrome	Fraction I (pmol / ml)	
	≤ 0.6	> 0.6
A. Organic	13	1
B. Psychogenic	7	16
C. Neurogenic	12	2
D. Non-neurogenic	2	21
E. None (Healthy Volunteers)	3	16

Difference A - B, $\chi^2 = 11.3$ $p < 0.001$

C - D, $\chi^2 = 18.1$ $p < 0.001$

It is not difficult to understand that reduced activation of endorphins, either due to loss in sensory input
as in the neurogenic pain syndrome or to intrinsic factors
such as in essential headache, might lead to a pain syndrome. In fact, it has independently been suggested
mainly based on electrophysiologic evidence that neurogenic pain syndromes are more likely associated with a
defect in pain inhibitory processes than in sustained
activation of pain generating processes (MELZACK & LOESER,
1978). Our series did not incorporate patients with cancer
pain or arthritic pain. A few cases with arthritic pain
have been studied separately and found to have endorphin
levels in the normal range, indicating that pain derives
from the pain-generating mechanisms. The most intriguing
cases are the psychogenic pain patients. Here, endorphin
levels tend to be normal or slightly above the normal
range (ALMAY et al., 1978). High levels have also been
recorded in patients with the primary diagnosis of endogenous depression (TERENIUS et al., 1977). Incidentally,
such patients are commonly insensitive to experimental
pain (VON KNORRING, 1975) and it has been suggested that
there is a link between depressive syndromes and chronic
pain (ENGEL, 1959; STERNBACH, 1974). In line with this
clinical observation is the positive correlation between
endorphin levels and the depth of depression (ALMAY et
al., 1978). It is too early to explain why these patients
develop pain syndromes at all. One possibility is to consider the pain reaction as a compensatory feeling due to
an overprotection of pain-signalling systems. Pain then
becomes a projection rather than a primary senation.

Two series of experiments were conducted to study
the functional importance of the CSF endorphins. Fraction
I was found to correlate both with experimental pain
thresholds and pain tolerance limits in the pain patients
(VON KNORRING et al., 1978a). Pain was here induced by
electrical stimulation of two fingers. A positive correlation was also found between Fraction I and the average
evoked responses to visual stimuli (V.AER). Patients with
low Fraction I tended to respond to increasing visual
stimulus intensities by increasing EEG responses (augment-
ing) while cases with high Fraction I tended to show no
or decreased EEG responses (reducing) in a comparative
test situation (VON KNORRING et al., 1978b). In summary
then (Table 4), Fraction I endorphins show correlations
with a number of clinical variables and experimental test
variables. These correlations indicate that endorphins
may serve as a biochemical marker for pain and depressive
syndromes and provide a link between them. However, a

unitary explanation for these syndromes is still not possible to offer. Work in progress aims at identifying how
various therapeutic regimens will affect endorphin levels
and the clinical course of the pain syndromes.

It may now be stated that stimulation-produced analgesia is at least partly explainable in terms of endorphin activation. As summarized in Table 5, intracerebral
stimulation has been found to give naloxone reversible
pain relief and a similar observation has been made for
low-frequency transcutaneous nerve stimulation (lo-TNS
or electroacupuncture). Electroacupuncture will also increase CSF endorphin, Fraction I (SJÖLUND et al., 1977).
However, high-frequency TNS (hi-TNS) seems to work differently since naloxone cannot block the induced pain
relief (SJÖLUND & ERIKSSON, 1979; ALMAY et al. unpublished). The clinical value of these stimulation techniques is already considerable. The results also show
that endorphin systems are potentially very powerful and
suggest that during adequate stimulation conditions they
may serve an important physiological role.

TABLE 4

Fraction I endorphin values and clinical variables in
patients with chronic pain (n=44).

Positive correlation with:	pain threshold
	pain tolerance
	depression scores
Negative correlation with:	duration of pain (organic group)
Correlation with:	EEG responses to visual stimuli (V.AER)
No correlation with:	sex
	age
	self-rated severity of pain

From ALMAY et al. (1978); VON KNORRING et al. (1978a, b).

TABLE 5

Basic differences between conventional high-frequency,
low intensity transcutaneous nerve stimulation (hi-TNS)
and low frequency, high intensity stimulation (lo-TNS,
also called electroacupuncture).

Stimulation mode	Pain relief naloxone-reversible	Effect on CSF endorphin Fraction I
hi-TNS	No	None
lo-TNS	Yes	Increase

Data compiled from SJÖLUND et al. (1977) SJÖLUND & ERIKS-
SON (1979); ALMAY et al. unpublished.

CONCLUSIONS

The role of endorphins in pain mechanisms is gradually becoming clear. The endorphins seem to have little
effect on experimental or acute pain, but probably they
affect the reaction to pain when it becomes protracted.
Several observations link endorphins to psychogenic pain
syndromes and depression. This also serves to illustrate
the importance of the psychic components of the chronic
pain syndromes.

Measurements of endorphins in the CSF may be used
diagnostically to discriminate between neurogenic and non-
neurogenic pain syndromes. Lower than normal endorphin
levels in neurogenic pain syndromes suggest that inadequate activation of endorphin systems is involved in the
aetiology of the disease.

The potent nature of the endorphin systems is illustrated by the strong pain relief which can be achieved
by proper electrical or mechanical stimulation. One important goal for the future is to find endogenous mechanisms for activating and inhibiting the endorphin system.

REFERENCES

AKIL, H., RICHARDSON, D. E., HUGHES, J. & BARCHAS, J. D.
    (1978) Science 201, 463-465.
ALMAY, B. G. L., JOHANSSON, F., VON KNORRING, L., TERENIUS,
    L. & WAHLSTRÖM, A. (1978) Pain 5, 153-162.
ÅSBERG, M., MONTGOMERY, S. A., PERRIS, C., SCHALLING, D.
    & SEDWALL, G. (1978) Acta Psychiatr. Scand. Suppl.,
    271.
BELLUZI, J. D. & STEIN, L. (1977) Nature 266, 566-568.
BLOOM, F., SEGAL, D., LING, N. & GUILLEMIN, R. (1976)
    Science 194, 630-632.
BLOOM F., ROSSIER, J., BATTENBERG, E. L. F., BAYON, A.,
    FRENCH, E., HENRIKSEN, S. J., SIGGINS, G. R., SEGAL,
    D., BROWNE, R., LING, N. & GUILLEMIN, R. (1978) Adv.
    Biochem. Psychopharmacol. 18, 89-109.
BUCHSBAUM, M. S., DAVIS, G. C. & BUNNEY, W. E., JR. (1977)
    Nature 270, 620-622.
CATLIN, D. H., HUI, K. K., LOH, H. H. & LI, C. H. (1977)
    Comm. Psychopharmac. 1, 493-500.
EL-SOBKY, A., DOSTROVSKY, J. O & WALL, P. D. (1976) Nature
    263, 783-784.
ENGEL, G. L. (1969) Amer. J. Med. 26, 899-918.
FELDBERG, W. & SMYTH, D. G. (1976) J. Physiol. 260,30P-31P.
FREDERICKSON, R. C. A. (1977) Life Sci. 21, 23-42.
GINTZLER, A. R., LEVY, A. & SPECTOR, S. (1976) Proc. Natl.
    Acad. Sci. U.S.A. 73, 2132-2136.
GINTZLER, A. R., GERSHON, M. D. & SPECTOR, S. (1978)
    Science 199, 447-448.
GOLDSTEIN, A. (1976) Science 193, 1081-1086.
GOLDSTEIN, A., PRYOR, G. T., OTIS, L. S. & LARSEN, F.
    (1976) Life Sci. 18, 599-604.
GREVERT, P. & GOLDSTEIN, A. (1977) Proc. Natl. Acad. Sci.
    74, 1291-1294.
GUILLEMIN, R. VARGO, T., ROSSIER, J., MINICK, S., LING,
    N., RIVIER, C., VALE, W. & BLOOM, F. (1977) Science
    197, 1367-1369.
GUNNE, L.-M., LINDSTRÖM, L. & TERENIUS, L. (1977) J.
    Neural Transm. 40, 13-19.
HOSOBUCHI, Y., ROSSIER, J., BLOOM, F. & GUILLEMIN, R.
    (1978) Soc. Neuroscience Abstr. 4, 410.
HUGHES, J., KOSTERLITZ, H. W. & SMITH, T. W. (1977) Br.
    J. Pharmacol. 61, 639-647.
JACOB, J. J., TREMBLAY, E. C. & COLOMBEL, M. C. (1974)
    Psychopharmacologia 37, 217-223.
JACQUET, Y. F. & MARKS, N. (1976) Science 194, 632-634.
JAFFÉ, J. H. (1975) in The Pharmacological Basis of
    Therapeutics (Goodman, A. & Gilman, A., eds.)

pp. 284-324. Macmillan, London.
JEFFCOATE, W. J., MCLOUGHLIN, L., HOPE, J., REES, L. H.,
    RATTER, S. J., LOWRY, P. J. & BESSER G. M. (1978)
    Lancet ii, 119-121.
KONTUREK, S. J., PAWLIK, W., WALUŚ, K. M., COY, D. H. &
    SCHALLY, A. V. (1978) Proc. Soc. Expt. Biol. Med.
    158, 156-160.
LEVINE, J. D., GORDON, N. C., JONES, R. T. & FIELDS, H. L.
    (1978) Nature 272, 826-827.
LEWIS, R. V., STEIN, S., GERBER, L. D., RUBENSTEIN, M. &
    UDENFRIEND, S. (1978) Proc. Natl. Acad. Sci. U.S.A.
    75, 4021-4023.
LIOTTA, A. S., SUDA, T. & KRIEGER, D. T. (1978) Proc. Natl.
    Acad. Sci. U.S.A. 75, 2950-2954.
LORD, J. A. M., WATERFIELD, A. A., HIGHES, J. & KOSTER-
    LITZ, H. W. (1977) Nature 267, 495-499.
MELZACK, R. & LOESER, J. D. (1978) Pain 4, 195-210.
MEYERSON, B. J. & TERENIUS, L. (1977) Europ. J. Pharmacol.
    42, 191-192.
MORICCA, G. (1974) Adv. Neurol. 4, 707-715.
SARNE, Y., AZOV, R. & WEISSMAN, B. A. (1978) Brain Res.
    151, 399-403.
SCHULTZBERG, M., LUNDBERG, J. M., HÖKFELT, T. TERENIUS,
    L. BRANDT, J., ELDE, R. P. & GOLDSTEIN, M. (1978)
    Neuroscience. In press.
SICUTERI, F., ANSELMI, B., CURRADI, C., MICHELACCI, S. &
    SASSI, A. (1978) Adv. Biochem. Psychopharmacol. 18,
    363-366.
SJÖLUND, B., TERENIUS, L. & ERIKSSON, M. (1977) Acta
    Physiol. Scand. 100, 382-384.
SJÖLUND, B. & ERIKSSON, M. B. E. (1979) Adv. Pain Res.
    Ther. (Liebeskind, J. C., ed.) 3. In press.
STERNBACH, R. A. (1974) in Pain Patients: Traits and
    Treatment. Academic Press, New York.
SUDA, T., LIOTTA, A. S. & KRIEGER, D. T. (1978) Science
    202, 221-223.
TERENIUS, L. (1974) Acta Pharm. Tox. 34, 88-91.
TERENIUS, L. & WAHLSTRÖM, A. (1975) Life Sci. 16, 1759-
    1764.
TERENIUS, L., WAHLSTRÖM, A., LINDSTRÖM, L. & WIDERLÖF, E.
    (1976) Neurosci. Lett. 3, 157-162.
TERENIUS, L. (1977) Psychoneuroendocrinology 2, 53-58.
TERENIUS, L., WAHLSTRÖM, A. & ÅGREN, H. (1977) Psycho-
    pharmacologia 54, 31-33.
TERENIUS, L. (1978) Ann. Rev. Pharmacol. Toxicol. 18,
    189-204.
TERENIUS, L., WAHLSTRÖM, A. & JOHANSSON, L. (1978) in
    Endorphins in Mental Health Research (Usdin, E.,
    Bunney, W. E., Jr. & Kline, N. W., eds.) Macmillan,

     New York. In press.
VON KNORRING, L. (1975) in The Experience of Pain in Pa-
     tients with Depressive Disorders. A Clinical and
     Experimental Study. Umeå University Medical Dis-
     sertations, New Series, no. 2, Umeå.
VON KNORRING, L., ALMAY, B. G. L., JOHANSSON, F. &
     TERENIUS, L. (1978a) Pain. In press.
VON KNORRING, L., ALMAY, B. G. L. & TERENIUS, L. (1978b)
     Neuropsychobiology. In press.
WATSON, S. J., BARCHAS, J. D. & LI, C. H. (1977) Proc.
     Natl. Acad. Sci. U.S.A. 74, 5155-5158.

THE OBESITY OF MIDDLE AGE: A COMMON VARIETY OF CUSHING'S
SYNDROME DUE TO A CHRONIC INCREASE IN ADRENOCORTICOTROPHIN
(ACTH) AND BETA-ENDORPHIN ACTIVITY

D. L. MARGULES

Department of Psychology
Temple University
Philadelphia, Pennsylvania 19122

INTRODUCTION

In 1977, the select committee on nutrition and human
needs of the United States Senate held hearings on obesity
(HEARINGS, 1977). They concluded that obesity is a
killer disease affecting at least 30 million Americans
in 1977. One-half of these people are obese to a degree
that shortens life. The frequency of obesity increases
substantially with age. One-third of American men and
40% of American women between the ages of 40 and 49 are
obese. The obese have a substantially reduced life ex-
pectancy, a high incidence of diabetes mellitus, gall
bladder problems, a high incidence of cardiovascular
disease, thin skin that tears easily, a greater likelihood
to die from anesthesia, greater susceptibility to infec-
tion, greater chances of phlebitis, and a higher likelihood of gout. Many obese individuals have a history of
an unsuccessful life struggle to reduce their body weight.
Over 10 billion dollars a year are spent in this country
for obesity treatments that fail. The few individuals who
have some success at weight loss often regain the lost
weight. Those rare individuals that succeed in maintaining the weight loss have a life-long battle to prevent
relapse of the obese condition (HEARINGS, 1977).

STATEMENT OF THE THEORY

If we knew more about the causes of obesity, more
effective therapy would be possible. In this paper I
present a new theory of obesity that explains certain
types of obesity, particularly adult-onset obesity. I
propose that many middle-age obese individuals may suffer
from a non-tumorous type of Cushing's syndrome in which
excessive hormonal activity occurs by means of two pituitary hormones, adrenocorticotropin (ACTH) and beta-endorphin. The stimuli that release these hormones are known
generally as stressors. The excess ACTH activity stimulates the release of glucocorticoid hormones from the
adrenal cortex gland and this predisposes an individual
toward protein-wasting and enhanced gluconeogenesis (the
conversion of body proteins to glucose). ACTH communicates
a message of severe carbohydrate shortage that requires
the sacrifice of proteins such as muscle. The excess
beta-endorphin activity that occurs concomitantly with the
ACTH activity stimulates the release of insulin from the
endocrine pancreas and this primarily promotes lipogenesis
in the adipose tissue. In other words, glucose made
available from protein is converted to fatty acids, which
are stored in adipose tissue as triglycerides. Thus, the
individual literally turns his own proteins into fat by
the action of these two hormones. This provides a means
for an individual to become fat without overeating. In
many obese individuals, however, overeating occurs and
this provides additional glucose and fatty acids that
further enhances the adiposity and also contributes to
the release of more insulin. Thus, snowballing occurs
with stress generating glucose, insulin and eating which
generates more glucose, insulin and eating.

The stimulus for overeating is unknown. Some evidence suggests the overeating is stimulated directly by
an action of beta-endorphin in the brain (MARGULES et
al., 1978). It also is possible that the hormonal state
created by beta-endorphin in the periphery acts back on
the brain via neural and/or hormonal feedback and thus
produces overeating indirectly. Probably both direct and
indirect mechanisms exist because receptors for beta-
endorphin are present in the periphery as well as the
brain.

The theory presented here states that greatly enhanced protein-wasting, gluconeogenesis and lipogenesis
produced respectively by abnormally high ACTH and beta-
endorphin activity are responsible for a certain type

of overeating and obesity in the middle-aged. It should
be pointed out that mildly enhanced gluconeogenesis and
lipogenesis are normal concomitants of the aging process
and occur in older individuals who are not obese (KERWICK,
1965). The aging process involves the conversion of
bodily proteins to fat and this occurs to some extent in
all lean individuals as they age. For example, a 25 year-
old human male with a lean body weight of 70 kg has a body
composition of 14% adipose tissue and 86% lean body mass.
If this male does not gain any additional weight for his
entire life span, his body composition does not stay fixed,
but shifts toward greater proportions of fat. Thus, at
age 40 he will consist of 22% adipose tissue and 78% lean
body mass, and at age 55 he will consist of 25% adipose
tissue and 75% lean body mass (KERWICK, 1965). This same
situation occurs in females who have a larger adipose
organ to start with than men. For example, a 25 year-old
female with a lean body weight of 60 kg will have 26%
adipose tissue. At age 40 this will increase to 32%
adipose tissue, and at age 55 to 38% adipose tissue (KER-
WICK, 1965). In both sexes it is necessary to lose weight
with aging in order to maintain the lower proportion of
body fat that is characteristic of the younger years. Very
few individuals accomplish this. Most do not even maintain
their lean adult body weight. The tendency of age to increase body fat can be viewed as part of the normal aging
process. Thus, the obese may have an exaggerated version
of a normal process.

    This process is very similar to what occurs in Cush-
ing's syndrome. In this paper I propose that age-related
protein-wasting, gluconeogenesis, and lipogenesis may act
in concert to produce a mild form of Cushing's syndrome.
This non-tumorous form of Cushing's syndrome may be more
widespread than formerly believed. Moreover, it may have
implications for a better understanding of the aging
process.

                        CUSHING'S SYNDROME

    In 1932 (CUSHING, 1932), Cushing first described a
syndrome that consists primarily of centrally-induced
obesity. The chief set of symptoms consisted of a highly
specific pattern of fat distribution, rather than a diffuse
distribution over the entire body. Major accumulations
of adipose tissue occurred on the face ("moon face"), the
upper back ("buffalo hump") and the trunk and girdle areas
("pot belly"). In addition, Cushing described several

symptoms of protein-wasting such as thin skin (as indicated
by purple stretchmarks), osteoporosis or wasting of the
bone matrix, and diminished size and strength of striated
muscles. Other symptoms included impaired glucose tolerance leading to diabetes mellitus in some cases, hypertension and cardiovascular disease, plethora or excess
blood volume, hirsutism or excess hair, and susceptibility
to infections (CUSHING, 1932). It seems to me that this
syndrome has a remarkable similarity to the syndrome seen
in the ordinary obesity of the middle aged.

Cushing attributed the syndrome to a tumor, a basophil adenoma of the pituitary that chronically secretes
an excess of ACTH. Later work (ALBRIGHT, 1942-43) showed
that ACTH itself was not responsible for the syndrome,
but rather that it was due to a stimulating effect of ACTH
on the cortex of the adrenal gland. The chronically elevated ACTH acted to stimulate a chronic release of high
levels of glucocorticoids from the adrenal cortex. It was
these prolonged and abnormally high levels of glucocorticoids that were directly responsible for the Cushing syndrome of symptoms, including the protein-wasting of the
skin, muscle and bone, produced by the excessive gluconeogenesis, which converts the protein matrix of these tissues
into glucose. The glucose is stored with the help of insulin, for the most part in adipose tissue, by conversion
of glucose to fatty acids and triglycerides. This leads
to excess weight gain (adiposity), impaired glucose tolerance, diabetes and hypertension.

                EXPERIMENTAL MODEL OF CUSHING'S SYNDROME

Experimental animals treated chronically with high
levels of glucocorticoid hormones show increases in adipose tissue, but only when insulin is injected simultaneously. If insulin is omitted, the animals lose weight,
nitrogen and fat (HAUSBERGER, 1958). Thus, excess insulin
is necessary for those Cushing's symptoms that favor adiposity. Insulin here serves a permissive role rather than
a causative role. This leads me to wonder what the stimulus
must be for the production of excess insulin in obese
individuals, particularly individuals who have become fat
without overeating. A very good candidate for an insulin
stimulant is beta-endorphin, which is released concomitantly with ACTH (GUILLEMIN et al., 1977). Recently,
beta-endorphin has been shown to stimulate the release of
insulin and glucagon from the endocrine pancreas in vitro
as long as there was some glucose in the medium (IPP,

VOBBS & UNGER, 1978). In the glucose-free preparation, beta-endorphin stimulated the release of glucagon only (IPP, VOBBS & UNGER, 1978). This condition would not exist in vivo, particularly with the individual in a gluconeogenic state. Thus, the animal research suggests that Cushing's syndrome involves the participation of the endocrine pancreas as well as the adrenal cortex, in order for adiposity to occur.

OTHER TUMORS THAT PRODUCE CUSHING'S SYNDROME

Two other tumors have been associated with Cushing's syndrome. One secretes ACTH but it is located at various sites other than the pituitary. These carcinomas of the lungs, pancreas, thymus and other organs are known as ectopic sources of ACTH. There is reason to believe that these tumors also produce excess beta-endorphin. The other type of tumor is located in the adrenal cortex where it produces autonomously an excess of glucocorticoids. It is not known if these tumors also produce beta-endorphin. There is no reason to believe that they would. Thus, patients with glucocorticoid-secreting tumors would be expected by the present theory, to have less insulin and less adiposity than patients with ACTH beta-endorphin secreting tumors. The discoveries of the ectopic ACTH-secreting tumor and the adrenal glucocorticoid-secreting tumor have reinforced the theory that excess glucocorticoids cause Cushing's syndrome. More important, they have so strengthened the connection between tumors and Cushing's syndrome that many physicians assume the existence of a tumor in patients that show the symptoms of Cushing's syndrome. This is a proper assumption, clinically speaking, given the immediate life-threatening risk that tumors present. Thus, the possibility that there might exist a non-tumorous source of excessive ACTH and beta-endorphin receives little if any attention. Yet, this idea is beginning to emerge slowly. Recently, a subset of Cushing's disease patients were identified that showed remission of the disease when treated chronically with cyproheptadine (KRIEGER, 1977). These findings have led to the hypothesis that excess activity in the brain's serotonergic neurons may be the cause of the excess ACTH production in a subset of patients with Cushing's syndrome. This is the first proposal that considers the possibility of a non-tumorous origin for this syndrome.

THE OB/OB MOUSE: A SINGLE GENE DEFECT WITH CUSHING'S
                             SYNDROME

     A very interesting animal model exists that contains,
all at once, a substantially shortened life span, an intense obesity, and a pituitary excess of ACTH from an
early age.   These important characteristics are found
in the C57BL/6J-ob/ob mouse from the Jackson Laboratory.
The excess of ACTH reaches levels as high as 14 times
lean control mice (EDWARDSON & HOUGH, 1975) and the excess is not caused by a tumor.   These mice also have
excess glucocorticoid production in the adrenals and in
the blood serum (NAESER, 1974).   I propose in this paper
that the genetically obese mouse C57BL/6J-ob/ob is suffering from a non-tumorous form of Cushing's disease.
This is the first such proposal to my knowledge.   I also
would like to make the suggestion that the excess glucocorticoids may be responsible for the short life span of
these mice by acceleration of the aging process.   Recently
prolonged administration of glucocorticoids was shown to
accelerate brain aging in rats (LANDFIELD, WAYMIRE &
LYNCH, 1978).

     The origin of the pituitary ACTH excess in the
ob/ob mouse is a single defective gene located on chromo-
some 6 in a position 14 centimorgans from the sig gene.
All the other genes of the ob/ob mouse are normal.   Thus,
all differences between the genetically obese mice and the
lean controls must be due to the action of the one defective gene alone.   This animal model offers one of the
simplest and most powerful experimental models for the
study of obesity and other age-related disorders.

        HYPOPHYSECTOMY AND ADRENALECTOMY AS ATTENUATORS OF
                        CUSHING'S SYNDROME

     If excessive pituitary secretions of ACTH and beta-
endorphin contribute to the development of a Cushing's-
like syndrome, it should be possible to attenuate the
obesity and protein-wasting by hypophysectomy, which
removes a major source of both hormones.   An extensive
and careful series of studies has shown that hypophysectomy completely eliminates further weight gain in the
C57BL/6J-ob/ob genetically obese mouse (PLOCHER & POWLEY,
1977) and in the Zucker genetically obese rat (fa/fa)
(POWLEY & MORTON, 1976).   These animals still remain
obese, however, and do not lose fat accumulated prior to
the surgery.   This failure to return to a lean body weight

after hypophysectomy is not surprising. First of all, the elimination of further weight gain in obese rodents is most impressive. This result suggests that pituitary participation is required for the lipogenesis involved in the continuous accumulation of adipose tissue seen in the obese. Secondly, the failure to lose earlier accumulations of fat suggests that the pituitary also participates in the lipolysis necessary to restore leanness. Hypophysectomy has been shown to retard the rate of lipolysis in rats (SCHLLINGER & GERHARDS, 1974). In all probability, different sets of pituitary hormones are involved in the lipogenesis and lipolysis. In order to further analyze these events, it would be advantageous to be able to manipulate individual hormonal systems within the pituitary.

Adrenalectomy of the ob/ob mouse at two months of age produces a reduction in excess body weight, a lowering of blood glucose, a substantial reduction in blood insulin, an abolition of insulin resistance, and a restoration of the normal feeding response to both a fasting challenge and a glucose load (NAESER, 1973; SOLOMON & MOYER, 1973). These studies suggest that a large part of the problem in the ob/ob mouse is due to excessive adrenal output. They also indicate that adrenal hyperfunction is not the entire story because adrenalectomy does not restore body weight, blood glucose, or blood insulin to the levels of lean control mice. The present theory can account for these failures by the insulin-stimulating action of beta-endorphin on the pancreas (IPP, VOBBS & UNGER, 1978), and by the inferred feeding-stimulating action of beta-endorphin in these rodents (MARGULES et al., 1978). Both of these actions should survive adrenalectomy and may even be enhanced by it because adrenalectomy increases the beta-endorphin content of the pituitary (GUILLEMIN et al., 1977).

                    BETA-ENDORPHIN LEVELS IN THE PITUITARY

The precursor molecule for ACTH is a large polypeptide, known as pro-opiocortin which is produced by messenger RNA in the cells of the intermediate pituitary and scattered cells of the anterior lobe. Pro-opiocortin contains within its structure not only ACTH, but also beta-lipotropin, the immediate precursor to beta-endorphin. Thus, ACTH and beta-endorphin are related, and not only are they produced from the same precursor, but also they are released concomitantly (DUBUC, MOBLEY & MAHLER, 1975). The genetically obese ob/ob mouse has an excess

of beta-endorphin in the pituitary (MARGULES et al.,
1978). This excess also occurs in the blood of these
animals (MARGULES et al., 1978). One consequence of the
excess release of beta-endorphin into the systemic cir-
culation would be the excessive occupation of opioid
receptors, which have been demonstrated to exist on the
stomach, ileum, vas deferens, and some of the other smooth
muscle systems in the periphery. Likewise, opioid receptors are also known to exist in the brain. Naloxone, an
opioid receptor blocker, abolishes the overeating in the
ob/ob mouse, at doses that have no effect on the feeding
of lean littermate control mice (MARGULES et al., 1978).
Thus, it seems that beta-endorphin may function as a hormonal messenger to bring about excessive ingestive behavior
of food in obese mice.

Preliminary data is available that bears on the question of whether the beta-endorphin increase is or is not
related to the excessive accumulation of fat in the ob/ob
mouse. To answer this question the weight gain in ob/ob
mice was restricted by means of food deprivation so that
they remained at body weights equivalent to those of their
lean littermate controls. Despite this reduction of food
intake the mice still accumulated a substantial amount
of adiposity. The beta-endorphin levels in the pituitaries
of these food-deprived obese mice remained at elevated
levels significantly above the levels found in lean mice.
However, they were significantly below the beta-endorphin
levels of full fed obese mice. This suggests that the beta-
endorphin increase is somehow correlated with the amount
of excess accumulation of fat.

BETA-ENDORPHIN LEVELS IN THE BLOOD

Beta-endorphin levels have been found to be elevated
significantly in the blood of patients with Cushing's
disease, Addison's disease and Nelson's disease, compared
to the blood of normals (SUDA, LIOTTA & KRIEGER, 1978).
Highest levels were found in patients with Nelson's
disease, who have both pituitary tumors and the removal
of the two adrenal glands. Adrenalectomy is known to
increase the production of ACTH and beta-endorphin by
elimination of the negative feedback influence of the
adrenal cortex on the pituitary. Intermediate levels of
beta-endorphin were found in patients with Addison's
disease, which is a form of adrenal cortex insufficiency.
Lowest levels of beta-endorphin were found in Cushing's
patients but these levels were still significantly greater

than control levels. Beta-endorphin has also been found
to be elevated in the blood of the obese fatty rat (Zucker
rat) (MARGULES et al., 1978). To date, beta-endorphin
levels have not been measured in the blood of obese humans.
This is a critical test of the hypotheses under consideration. It should be noted that not all obese humans are
expected by the present theory to show increased levels
of beta-endorphin in the blood. Only the subgroup of
obese patients with the symptoms of Cushing's disease
should show this increase. I expect this to be a relatively large subgroup of middle-aged obese individuals.

We have established for the first time a connection
between an excess of pituitary hormone, beta-endorphin
and the overeating problem in the genetically obese
mouse (MARGULES et al., 1978). I have related this animal
model to a human form of obesity as seen in Cushing's
disease. Recently, patients with various varieties of
Cushing's disease have been shown to have excessive amounts
of beta-endorphin in their blood. Normal humans did not
show detectable levels of beta-endorphin in the untreated
state or after vasopressin administration (SUDA, LIOTTA &
KRIEGER, 1978). This raises the important question of
the stimulus events necessary in normal subjects for the
release of beta-endorphin. In general, it has been shown
that stimuli associated with stress, such as electric
footshock, cause increases in beta-endorphin. Such
stimuli are acute rather than chronically or repeatedly
applied. Chronically high levels of ACTH seem to go
along with a concomitant elevation of beta-endorphin. In
the genetically obese mouse (ob/ob) such elevations appear
to be present from an early age and continue into the
later life of the animals. This important mutant mouse
is an animal model of a type of Cushing's disease that
is not caused by a tumor. The question remains open
whether chronically occurring stress in middle-aged individuals can duplicate the chronic hormonal condition
of the ob/ob mouse.

THREE ETIOLOGIES FOR INCREASED ACTH AND
BETA-ENDORPHIN ACTIVITY

Three different etiologies may exist for the development of the hormonal activities that favor adult-onset
obesity. The first is a tumor composed of cells that
produce ACTH and beta-endorphin. In this case there is
an elevated number of cells that causes the problem.

The cause of the tumor itself is unknown. Present-day
theories emphasize the role of carcinogenic agents in
the environment. The second mechanism for the production of this type of obesity involves a genetic defect,
as in the ob/ob mouse. No one knows the function of the
single defective gene in this mouse that is responsible
for its obesity and other hormonal problems. Recently I
presented evidence that favors the theory that the defective gene in this mouse is the gene that specifies the
sequence of amino acids in pro-opiocortin (MARGULES, 1978).
Part of this theory predicted that the glycosylated N-
terminal portion of pro-opiocortin, which to date has not
been sequenced, would be shown to be identical with the
alpha subunit of FSH, LH, and TSH. The two etiologies
described above cannot account for the tendency of normal
individuals to increase the proportion of bodily fat as
they age. A third mechanism must be invoked in order to
accomplish this. Therefore, I propose that an age-induced
shift in the cleavage enzymes present in the pro-opiocortin cell may contribute to the production of the mild
Cushing's-like syndrome in non-obese middle-aged adults.
I suspect that this shift may also contribute to the
more intense form of this syndrome that appears in middle-
aged adults who are overweight. At present the best guess
about the cause of this shift in enzyme type relates to
the following considerations:

  1) It occurs as a part of the normal aging process
      in all individuals in any case.
  2) It is exaggerated in certain obese individuals.
  3) Some obese individuals may be genetically-pre-
      disposed to this exaggeration.
  4) Excess stress, chronically applied, may contribute
      to the problem in both the genetically-predis-
      posed and the non-predisposed individual.

                            SUMMARY

    The common obesity of middle age presents a set of
features that strongly resembles the cardinal symptoms
of Cushing's syndrome: obesity of the face (moon face),
upper back (buffalo hump) and trunk (pot belly) accompanied
by signs of protein-wasting. In non-obese individuals
who remain at a constant weight throughout life, the proportion of adipose tissue increases with age at the expense
of lean tissue loss. Thus, a mild version of Cushing's
syndrome may be part of the normal aging process. A more
intense version of this process may occur in overweight

adults. Excess and chronic activity of two pituitary hormones may contribute to this adiposity. Both hormones are
produced in the same pituitary cell by cleavage from a
common large precursor known as pro-opiocortin. One hormone is adrenocorticotrophin (ACTH), which stimulates the
release of the glucocorticoid hormones. These hormones
promote the conversion of bodily proteins to glucose (gluconeogenesis). The other pituitary hormone is beta-endorphin, a stimulant of appetite that causes the release of
insulin. This pancreatic hormone promotes the conversion
of glucose and fatty acids to triglycerides (lipogenesis).
Three different etiologies are suggested for the excessive
and chronic action of these two pituitary hormones: tumors
that increase the number of cells that synthesize pro-
opiocortin; mutant strains that produce excessive amounts
of ACTH and beta-endorphin such as the genetically obese
mouse (ob/ob) and rat (fa/fa); and an age-determined shift
in the type of cleavage enzymes present in the pro-opiocortin cell that favors ACTH and beta-endorphin production.

ACKNOWLEDGMENT

Supported by grant BNS77.22630 from the National
Science Foundation.

REFERENCES

ALBRIGHT, F. (1942-43) Cushing's syndrome. Harvey Lect.
    38, 123.
CUSHING, H. (1932) The basophil adenomas of the pituitary
    body and their clinical manifestations (pituitary
    basophilism). Bull. Johns Hopkins Hosp. 50, 137.
DUBUC, P.V., MOBLEY, P.W. & MAHLER, R.J. (1975) Elevated
    glucocorticoids in obese hyperglycemic mice. Hormones
    and Metabolic Research 7, 102.
EDWARDSON, J.A., HOUGH, C.A.M. (1975) The pituitary-
    adrenal system of the genetically obese (ob/ob)
    mouse. Journal of Endocrinology 65, 99-107
GUILLEMIN, R., VARGO, R., ROSSIER, J., MINICK, S., LANG,
    N., RIVIER, C., VALI, W. & BLOOM F.E. (1977) Beta-
    endorphin and adrenocorticotropin are secreted con-
    comitantly by the pituitary gland. Science 197,
    1367-1369.
HAUSBERGER, F.X. & HAUSBERGER, B.C. (1958) Effect of in-
    sulin and cortisone on weight gain, protein and fat
    content of rats. Amer. J. Physiol. 193, 455.

Hearings before the select committee on nutrition and
    human needs of the United States Senate, 95th
    Congress, first session, Feb. 1 and 2, 1977, part 2,
    obesity. Diet related to killer diseases, II. U.S.
    Government Printing Office, Washington, 1977.
IPP, E., VOBBS, R. & UNGER, R.H. (1978) Morphine and beta-
    endorphin influence the secretion of the endocrine
    pancreas. Nature 276, 190-191.
KERWICK, A. (1965) Adiposity. In Handbook of Physiology,
    section 5, Adipose tissue, pp. 617-624. American
    Physiological Society.
KRIEGER, D.T. (1977) Serotonin regulation of ACTH secre-
    tion. Annals of the New York Academy of Science
    297, 527-535.
LANDFIELD, P.W., WAYMIRE, J.C. & LYNCH, G. (1978) Hippo-
    campal aging and adrenocorticoids: Quantitative cor-
    relations. Science 202, 1098-1102.
MARGULES, D.L., MOISSET, B., LEWIS, M.J., SHIBUYA, H. &
    PERT, C.B. (1978) Beta-endorphin is associated with
    overeating in genetically obese mice (ob/ob) and
    rats (fa/fa). Science 202, 988-991.
MARGULES, D.L. (1978) Molecular theory of obesity,
    sterility and other behavioral and endocrine problems
    in genetically obese mice (ob/ob). Neurosciences
    and Biobehavioral Reviews 2(4), in press.
NAESER, P. (1973) Effects of adrenalectomy on the obese-
    hyperglycemic syndrome in mice (ob/ob). Diabetologia
    9, 376-379.
NAESER, P. (1974) Function of the adrenal cortex in obese-
    hyperglycemic mice (ob). Diabetologia 10(5),449-453.
PLOCHER, A. & POWLEY, T.L. (1977) Maintenance of obesity
    following hypophysectomy in the obese-hyperglycemic
    mouse (ob/ob). The Yale Journal of Biology and
    Medicine 50, 291-300.
POWLEY, T.L. & MORTON, S.A. (1976) Hypophysectomy and body
    weight regulation in the genetically obese Zucker
    rat. American Journal of Physiology 230, 982-987.
SCHILLINGER, E. & GERHARDS, E. (1974) Effects of pituitary
    hormones and corticosterone on lipolysis in hypo-
    physectomized rats. Acta Endocrinologia 77, 502-
    508.
SOLOMON, J. & MOYER, J. (1973) The effect of adrenalectomy
    on the development of the obese hyperglycemic syndrome
    on ob/ob mice. Endocrinology 93, 510-513.
SUDA, T., LIOTTA, A.S. & KRIEGER, D.T. (1978) Beta-endor-
    phin is not detectable in plasma from normal sub-
    jects. Science 202, 221-223.

HORMONAL AND OTHER EFFECTS OF NALTREXONE IN NORMAL MEN*

J. VOLAVKA[1], A. MALLYA[1], J. BAUMAN[2], J. PEVNICK[1],
D. CHO[1], D. REKER[1], B. JAMES[1], AND R. DORNBUSH[3]

[1]University of Missouri Columbia-School of
Medicine, at Missouri Institute of Psychiatry,
St. Louis, Missouri
[2]Masters and Johnson Institute, St. Louis,Missouri
[3]New York Medical College, Valhalla, New York

INTRODUCTION

Many experimental approaches have been devised to
study the functions of endorphins. A strategy which has
been used very widely consists in the administration of
an opiate antagonist. It is assumed that the antagonists
displace endorphins from their receptors; once this happens,
any function for which endorphins are needed should be
altered. It has been hypothesized that endorphins play
a role in endocrine functions (see below), pain perception (BUCHSBAUM, DAVIS & BUNNEY, 1977; EL-SOBKY, DOSTROVSKY
& WALL, 1976; GREVERT & GOLDSTEIN, 1978; HOSOBUCHI, ADAMS
& LINCHITZ, 1977), modulation of mood (JONES, 1978), sexual
functioning (GOLDSTEIN & HANSTEEN, 1977), mental health
(WATSON et al., 1978; VEREBEY, VOLAVKA & CLOUET, 1978)
and other areas. These hypotheses were tested by the
administration of the opiate antagonist naloxone. Another
opiate antagonist, naltrexone, has also been used for
similar experiments: it was administered to rats in order
to study the function of endorphins in prolactin release
(GUIDOTTI & GRANDISON, 1978); it was also used in therapeutic experiments in psychotic patients (SIMPSON, BRANCHEY
& LEE, 1977; GUNNE & TERENIUS, 1978; GITLIN & ROSENBLATT,
1978). NUTT et al. (1978) have tested the effects of

* Presented in part at the CINP Congress in Vienna,
July, 1978.

naltrexone on movement disorders. Opiate-naive subjects were used in these studies. The principal advantage of naltrexone against naloxone in human experiments was seen in its longer duration of action after oral administration. Clinical trials of ex-addicts have shown that naltrexone antagonizes the effects of heroin for up to 72 hours and that it is non-toxic in doses up to 200 mg/day (VOLAVKA et al., 1976).

One of the assumptions underlying such experiments is that the antagonist which is being used has essentially no other effects than that of blocking the opiate receptor. Should the antagonist have opiate-like effects of its own, the results of these experiments would be difficult to interpret. Naltrexone has been classified as a relatively pure narcotic antagonist (MARTIN, JASINSKI & MANSKY, 1973), and this classification was appropriate in its original context, i.e. with regard to treatment of heroin addiction. The fact that it caused a non-significant decrease of respiratory rate and pupillary size, and significant decrease of temperature was not considered important in view of the very small size of these effects observed in five ex-addicts (MARTIN, JASINSKI & MANSKY, 1973). The administration of opiates probably alters interactions between the opiate receptor and endorphins (GOLDSTEIN, 1976; EHRLICH et al., 1978). For this reason, these issues are best studied in organisms that have not been exposed to opiates. We have found only two studies of naltrexone in opiate-naive normal humans. Ratings of mood and behavior obtained in one of these studies of naltrexone in opiate-naive normal humans obtained in one of these studies (IRWIN et al., 1974) indicated that naltrexone may have some opiate-like properties. Sleepiness, dysphoria, sexual ideation, penile erection, and an increase of luteinizing hormone were reported after 50 mg of naltrexone (MENDELSON et al., in press).

Effects of opiates and opiate antagonists on plasma levels of hormones may provide clues about the potential role of endorphins in the regulation of endocrine systems. Many endocrine functions are affected by the administration of opiates. These effects are exerted through opiate receptors. Since endorphins are the major endogenous ligands for these receptors, it seems reasonable to hypothesize that they are involved in the regulation of endocrine functions.

The administration of opiates affects the function of the hypothalamic-pituitary-adrenal (H-P-A) axis, the

hypothalamic-pituitary-gonadal (H-P-G) axis, and other
pituitary functions, such as secretion of prolactin
(PRL) (CICERO & DORNBUSH, 1979). We have therefore
decided to study the effects of naltrexone on plasma
levels of adrenocorticotrophic hormone (ACTH), cortisol,
luteinizing hormone (LH), testosterone, and PRL.

METHODS

In our present study, the subjects were 10 healthy
male volunteers (age range 19-31). None of the subjects
had a history of opiate use, as determined by a questionnaire. A urine sample was taken for opiates at the beginning of each session, and no opiates were ever detected.
Each subject had three sessions which were separated from
each other by at least a week. Either placebo, or naltrexone 50 mg or 100 mg was given in tablets by mouth
in each session. These substances were given in a counter-
balanced order. We studied respiration rate (which was
recorded using a strain gauge placed around the subject's
chest), heart rate, blood pressure, oral temperature,
and pupil size (pupil photographs were taken after a 10-
minute adaptation to the dark; a Polaroid camera with
circular flash was used). Electroencephalogram (EEG)
was evaluated by subjecting ten-minute EEG records from
the right and left occipital areas to power spectral
analyses. We also assessed opiate withdrawal using a
questionnaire which consisted of 13 items: stomach
cramps or stomach pain; nausea; bowel movement; eyes
teary; nose runny; yawning; sweating; goosepimples;
chills; muscle, joint or back pain; dizziness; anxiety-
nervousness; loss of appetite. Each item was rated on
a four-point scale (absent, slight, moderate, a lot).
We also studied the subject's sensitivity to pain. This
was determined by having the subjects immerse the nonwrit-
ing hand for two minutes into chipped ice. They indicated
the time of pain onset; this was used as a measure of
pain threshold. Severity of pain and the attitude toward
pain were rated every 30 seconds during the immersion.
Immediately after each pain test, a blood sample was
drawn for the radioimmunoassays of LH, PRL, testosterone,
ACTH and cortisol. Subjects were asked by a male nurse
whether they were sexually aroused and/or had an erection.
All these variables were assessed before the drug administration, then at 1, 2, 4 and 8 hours after the drug.
Short unstructured interviews were done at the same time
periods; an additional interview occurred 24 hours after
the drug administration. All sessions started between

TABLE 1

Multiple regression analysis of pain threshold at 1
hour after naltrexone administration.

Independent Variables	d.f.	R^2 inc	F
Pre-Drug Pain Threshold (PDPT)	1/19	0.05	2.15
Naltrexone 50 mg	1/15	0.01	0.36
Naltrexone 100 mg	1/15	0.03	1.37
Naltrexone 50 mg x PDPT	1/15	0.00	1.29
Naltrexone 100 mg x PDPT	1/15	0.20	15.90

8 and 9 a.m.   Double-blind procedures were followed both
in data acquisition and analyses.

Multiple hierarchical regression analyses were used
to test the significance of the drug effects while account-
ing for the variance of the baseline (pre-drug) values.
These procedures were analogous to analyses of covariance
for repeated measures (KERLINGER & PEDHAZUR, 1973). An
example of such analysis is given in Table 1.

The independent variables were entered into the
analyses in the order in which they are listed. Note
that the drug variables (having the same d.f.) were entered as a set, and that each F ratio of the set may be
viewed as a test of the difference between its associated
drug and the placebo condition. The R^2 increments express the proportions of variance contributed by the independent variables to the dependent variable. The last
two independent variables are the interactions between
the pre-drug values of the pain threshold and the drug
effect. The interaction for 100 mg of naltrexone is significant ($p < 0.001$). The variance associated with inter-
subject differences was accounted for prior to entering
the independent variables listed in the Table. All hormonal analyses were performed by radioimmunoassay. Testosterone was assayed following organic solvent extraction and chromatography on celite columns. An internal
standard was used to correct for recovery. Rabbit-anti-
testosterone supplied by Dr. Burton Caldwell and ^{3}H-tes-
tosterone (Schwarz-Mann) were used in an adaptation of
the method of FURUYAMA et al. (1970).

Luteinizing hormone was assayed (SAXENA et al., 1968)
directly in serum with standards and antiserum supplied

by NIAMDD, National Institutes of Health.

Prolactin was assayed (JACOBS, MARIZ & DAUGHADAY, 1972) in serum by a radioimmunoassay kit supplied by CIS Radiopharmaceuticals. In a preliminary experiment, correlation of PRL values of 36 samples assayed by this procedure and by reagents supplied by NIAMDD was 0.984.

Reagents for the radioimmunoassay of ACTH (ORTH, 1974) in EDTA plasma were also supplied by CIS Radiopharmaceuticals. Standards are supplied as synthetic human ACTH 1-39 in hormone-free plasma. Rabbit anti-ACTH was produced against porcine ACTH, and porcine ACTH labelled with ^{125}I is used as the radioactive tracer. The sensitivity of the method is 5 pg/ml.

Serum cortisol was measured without prior extraction by a solid phase method utilizing highly specific antiserum covalently bonded to polyethylene tubes (ROLLERI et al., 1976). Reagents were supplied by CIS Radiopharmaceuticals.

All samples were assayed in duplicate. Quality control pools at low, normal and high levels were run in each assay. Repeat assays were performed on samples if duplicate determinations had a coefficient of variation greater than 5%, if quality control values in the assay fell more than 2 standard deviations from the mean, if the correlation coefficient of the standard curve was less than 0.950, and, in the case of testosterone, if recovery was less than 50%.

RESULTS

The respiration rate was lower after naltrexone than after placebo (Table 2). The decrease became significant at the first hour after the administration of 100 mg and this effect lasted for at least 4 hours. The decrease after 50 mg seemed less pronounced, but no significant differences between 50 and 100 mg were detected for respiratory rate. Temperature was significantly lower after each dose of naltrexone than after placebo (Table 2). The global scores on opiate withdrawal questionnaire were higher after each dose of naltrexone than after placebo at 4 hours after administration, and remained elevated at 8 hours following the 100 mg dose. No significant differences between the effects of 50 and 100 mg of naltrexone on temperature or on withdrawal questionnaire

TABLE 2

The effects of naltrexone on respiration rate and oral temperature.

		Placebo		Naltrexone 50 mg		Naltrexone 100 mg	
		$\overline{X}$	sd	$\overline{X}$	sd	$\overline{X}$	sd
Respi-	Pre-Drug	15.9	3.4	16.0	3.9	16.0	3.7
ration	Post 1 hr	16.2	3.8	15.1	3.7	14.3a	4.0
(breaths/	Post 2 hr	16.4	2.9	15.7	3.2	14.7a	3.8
min)	Post 4 hr	17.5	4.0	16.1a	3.1	15.8b	3.9
	Post 8 hr	16.7	4.0	15.7	4.2	15.8	3.3
Oral	Pre-Drug	98.0	0.6	97.9	0.5	97.8	0.5
Temper-	Post 1 hr	98.0	0.6	97.7	0.6	97.7	0.4
ature	Post 2 hr	98.1	0.6	97.6b	0.6	97.4b	0.3
(F°)	Post 4 hr	98.4	0.5	98.0a	0.5	97.8b	0.6
	Post 8 hr	98.5	0.5	98.0b	0.6	98.0b	0.4

Significant differences from placebo: a ($p<0.05$); b ($p<0.01$).

were detected. Nausea, abdominal cramps and/or decreased appetite occurred in six of the 10 subjects after 100 mg, in one subject after 50 mg, and in two subjects after placebo. These complaints usually started between 8 and 24 hours after 100 mg of naltrexone. No difference between placebo and naltrexone was detected in reports of sexual arousal or erection. Pupillary diameter showed a very slight non-significant decrease after naltrexone. The average frequency of the EEG alpha activity was significantly slower after naltrexone than after placebo; details are published elsewhere (VOLAVKA et al., 1979).

The pain threshold was affected by naltrexone. There was significant interaction between baseline (pre-drug) values of pain threshold and the drug effect. An example is in Table 1 which uses values obtained before and 1 hour after the drug administration. Analogous regression analyses were performed for observations obtained at 2, 4 and 8 hours after the drug administration; these interactions were significant at 2 and 4, but not at 8 hours.

    To evaluate these interactions, the four post-drug
pain threshold values (obtained at 1, 2, 4 and 8 hours)
were averaged for each subject, and plotted as a function
of the pre-drug value. This was done separately for
placebo, 50 and 100 mg of naltrexone. A regression line
was then computed for each of those three plots. The
three regression lines were then superimposed in one single
plot (Figure 1). This interaction means that the subjects
showing higher threshold (i.e., less sensitivity to pain)
before the drug administration are more likely to experience the analgesic effect of naltrexone than those with
low initial threshold (i.e. those who are more pain-sensitive to begin with). Pain severity ratings did not reveal
any difference between naltrexone and placebo.

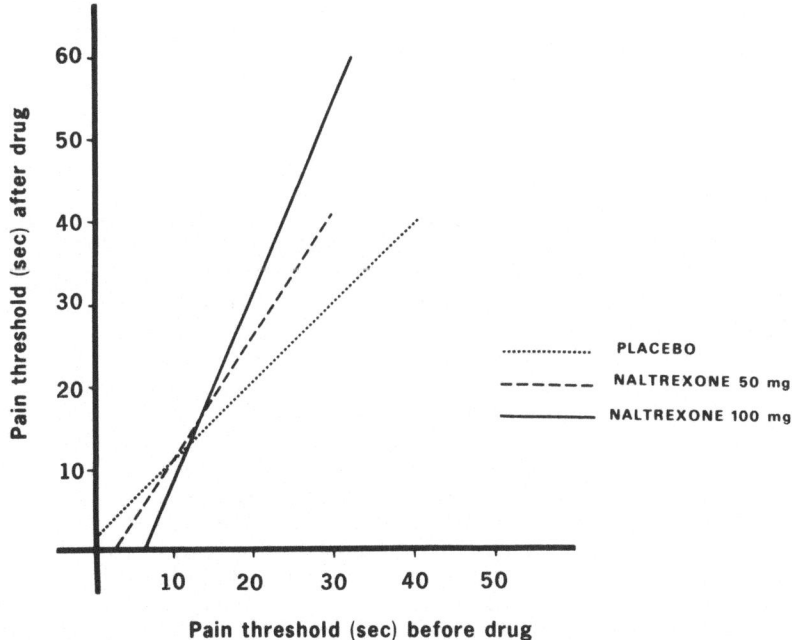

Fig. 1: Post-drug pain threshold as a function of pre-
         drug pain threshold. Each regression line was com-
         puted on the basis of 10 observations. There is a
         significant interaction between treatments and pre-
         drug pain threshold (see text for details).

TABLE 3

The effects of naltrexone on ACTH and cortisol plasma levels.

		Placebo		Naltrexone 50 mg		Naltrexone 100 mg	
		$\overline{X}$	sd	$\overline{X}$	sd	$\overline{X}$	sd
ACTH	Pre-Drug	29.7	25.9	31.9	25.4	22.4	16.2
pg/ml	Post 1 hr	31.9	38.7	49.3	42.1	44.1	37.3
	Post 2 hr	24.7	14.2	35.0	27.6	27.5	21.2
	Post 4 hr	24.5	14.3	27.4	23.7	26.2	22.7
	Post 8 hr	24.8	16.0	32.9	32.1	24.6	16.2
CORTI-	Pre-Drug	9.6	2.6	10.7	3.7	10.4	4.1
SOL	Post 1 hr	9.3	2.5	12.4	4.4	14.0	6.4
µg/dl	Post 2 hr	9.9	4.2	13.1a	5.3	13.0a	5.5
	Post 4 hr	10.3	3.3	10.1	3.6	10.6	3.1
	Post 8 hr	6.5	2.6	10.3a	6.1	11.0b	4.6

Significant differences from placebo: a ($p < 0.05$); b ($p < 0.01$)

The effects of naltrexone on ACTH and cortisol plasma levels are displayed in Table 3. The administration of 50 mg of naltrexone was followed by a 55% increase of ACTH in the first hour; the values then gradually decreased. Naltrexone 100 mg elicited a 97% increase of ACTH levels in the first hour, again followed by a gradual decrease. However, the effects of naltrexone on ACTH were not significantly different from placebo (note the very large standard deviations). Both doses of naltrexone elicited a significant increase of cortisol levels; the increase started in the first hour, and became statistically significant at the second hour after the dose (Table 3). The difference between placebo and naltrexone was again significant at 8 hours after the administration. No significant differences between the cortisol effect of 50 and 100 mg of naltrexone were detected. The effects of naltrexone on ACTH and cortisol may have been mediated by the stress of the experimental situation. To explore this possibility we computed correlation coefficients between the cortisol plasma levels and the pain severity ratings, pain threshold ratings, and global score on the withdrawal questionnaire. This was done separately for

each of the four post-drug observations. Of the 40 correlation coefficients examined, only 4 were significant
at 5% level (all 4 were negative). This means that the
reported pain or discomfort was not positively related
to the plasma levels of cortisol.

The effects of naltrexone on plasma levels of LH,
testosterone and PRL are summarized in Table 4. The
increase of LH is accompanied by an increase of testosterone. No reliable differences between the effect of
50 and 100 mg of naltrexone were detected. The levels
of PRL showed a significant increase after 50 mg of naltrexone. A non-significant increase of PRL was observed
after 100 mg. The effects of the two naltrexone doses
were not significantly different from each other.

                        DISCUSSION

These data indicate that naltrexone has opiate-like
as well as antagonist properties in opiate-naive humans.
The reduction of breathing rate, temperature and EEG alpha
frequency, as well as the effects on pain threshold suggest
opiate-like activity. The presence of clinical symptoms
usually associated with opiate withdrawal suggests that
naltrexone also had antagonist activity. However, these
symptoms are not specific for opiate withdrawal.

Our observations of pain threshold effects differ
from those published by BUCHSBAUM, DAVIS & BUNNEY (1977).
They found that after naloxone, pain-insensitive subjects
showed hyperalgesia whereas the pain-sensitive ones showed
hypoalgesia. Since naltrexone apparently has much more
agonist activity than naloxone, we have seen little or
no hyperalgesia; however, the hypoalgesic effect of naltrexone was more pronounced in our pain-insensitive subjects. This discrepancy in observations may be due to
the difference between naloxone and naltrexone, or to
the differences in methodology. We agree in that the
initial values strongly interact with the drug effects
on pain perception. A slight analgesic effect of naltrexone was observed in the rat but not in the mouse
(BLUMBERG & DAYTON, 1972). We are not aware of any report
of analgesic activity of naltrexone in humans.

Our results indicate that naltrexone has an effect
on the pituitary-adrenal axis. Could the increased level
of cortisol (and the non-significant increase of ACTH)
be interpreted as an agonist or as an antagonist effect?

TABLE 4

The effects of naltrexone on LH, testosterone and PRL plasma levels.

		Placebo		Naltrexone 50 mg		Naltrexone 100 mg	
		$\bar{X}$	sd	$\bar{X}$	sd	$\bar{X}$	sd
LH mIU/ml	Pre-Drug	6.8	3.7	5.0	2.7	7.1	5.0
	Post 1 hr	8.5	3.7	10.4	5.1	11.0	2.9
	Post 2 hr	7.0	4.5	10.2b	4.3	11.1b	4.5
	Post 4 hr	5.9	3.1	9.4b	4.3	11.2b	4.3
	Post 8 hr	9.3	2.7	10.7	4.8	12.2a	3.8
TESTOS-TERONE ng/dl	Pre-Drug	557.2	185.8	626.1	173.5	489.3	133.7
	Post 1 hr	603.7	179.0	641.7	145.7	608.6	190.1
	Post 2 hr	626.1	230.4	683.0	159.8	614.9	137.3
	Post 4 hr	525.0	163.7	590.2a	131.8	592.7a	78.7
	Post 8 hr	544.3	207.6	714.0a	165.9	610.3a	143.5
PROLAC-TIN ng/ml	Pre-Drug	5.2	1.8	5.3	1.4	4.2	1.8
	Post 1 hr	5.1	1.8	7.0a	3.1	6.3	2.9
	Post 2 hr	4.7	1.4	6.4	2.1	5.8	2.1
	Post 4 hr	5.5	1.4	6.3	1.9	6.6	3.2
	Post 8 hr	5.5	1.5	6.9a	1.6	6.6	2.5

Significant differences from placebo: a($p < 0.05$); b($p < 0.01$).

It is well known that morphine increases corticosteroid
secretion in the rat, mouse and dog (GEORGE & KOKKA, 1976).
However, studies in humans have failed to demonstrate
this effect of narcotics (HELLMAN et al., 1975; MENDELSON
et al., 1975). EISENMAN et al. (1958) have shown that
urinary 17-ketosteroid excretion in man was decreased
by morphine; the levels increased during withdrawal.
Abrupt withdrawal from narcotics results in an elevation
of plasma levels of adrenal corticosteroids in rat and
man (SLOAN, 1971). Naloxone-precipitated withdrawal has
the same effect in the rat (KOKKA & GEORGE, 1974). We
have observed that naloxone, a narcotic antagonist free
of opiate-like effects, elicited a highly significant
increase of ACTH and cortisol plasma levels in normal
males (VOLAVKA et al., 1979).   These results
suggest that the naltrexone-elicited increase of ACTH
and cortisol levels was an antagonist effect.

That being so, these results indicate that endorphins
may be involved in the regulation of ACTH (and cortisol)
secretion. Beta-endorphin and ACTH are probably secreted
concomitantly by the adenohypophysis (GUILLEMIN et al.,
1977). Glucocorticoids were shown to regulate endorphin
production in cultured pituitary tumor cells (SABOL, 1978).
Beta-lipotropin may be the common precursor of ACTH and
beta-endorphin (MAINS, EIPPER & LING, 1977). These findings indicate that the mechanism involved in the secretion
of both ACTH and beta-endorphin are very closely related
to each other, or that they are identical. Moreover,
ACTH was shown to have a partial antagonist-like selectivity for opiate receptors (TERENIUS, 1976). Our findings
suggest the possibility of an inhibitory endorphin feedback control of ACTH secretion. As naltrexone displaces
both the endorphins and ACTH from the opiate receptor,
the ACTH (and cortisol) secretion is increased.

A similar inhibitory feedback control of LH secretion
by endorphin was proposed by BRUNI et al. (1977) who observed that methionine-enkephalin depressed serum level
of LH of the rat whereas naloxone increased it. There
is a general agreement that narcotics depress serum LH
levels in all species examined; this effect seems to be
mediated through the hypothalamus, perhaps by inhibiting
the synthesis or secretion of luteinizing hormone-releasing
hormone (CICERO et al., 1977). Met-enkephalin was found
to inhibit (in vitro) dopamine-induced release of LH-
RH from the hypothalamus (ROTSZTEJN et al., 1978). Naltrexone increased the LH levels in normal men (MENDELSON et
al., in press); the same effect was observed after naloxone

VOLAVKA et al., 1979). Our present LH data thus confirm and amplify the results reported by MENDELSON et al. (in press).

However, we cannot confirm MENDELSON'S findings of sexual arousal after naltrexone. Our finding that naltrexone increases PRL was surprising since GUIDOTTI & GRANDISON (1978) have reported decreased PRL levels after naltrexone in the rat. We feel that our results might be explained as an opiate-like effect. PRL decrease was elicited by naloxone in the rat (BRUNI et al., 1977) and the monkey (GOLD et al., 1978), whereas beta-endorphin and morphine elevate PRL levels in the rat (RIVIER et al., 1977). Moreover, serum PRL levels in the rat are depressed following withdrawal from chronic morphine treatment (LAL et al., 1977). JANOWSKY et al. (1978) and VOLAVKA et al. (1979) have not observed any effect of naloxone on PRL in man. We feel that the discrepancies between the observations of PRL in lower animals and man are too great to be explained as a function of different experimental techniques or doses. It seems that the effects are species-specific.

Collectively, our results suggest that naltrexone may not be the most suitable drug for the exploration of the endorphin system. The mixture of agonist and antagonist effects of naltrexone in opiate-naive organisms makes the results difficult to interpret: it is not always easy to decide whether the observed effect results from the displacement of endorphins, or from the agonist-like activity of naltrexone.

REFERENCES

BLUMBERG, H & DAYTON, H.B. (1972) Narcotic antagonist studies with EN-1639A (N-Cyclopropylmethylnoroxy-morphone hydrochloride). In Fifth International Congress on Pharmacology, Volunteer Abstracts. San Francisco, July, p. 23.

BRUNI, J.F., VAN VUGT, D., MARSHALL, S. & MEITES, J. (1977) Effects of naloxone, morphine and methionine enkephalin on serum prolactin, luteinizing hormone, follicle stimulating hormone, thyroid stimulating hormone and growth hormone. Life Sci. 21, 461-466.

BUCHSBAUM, M.S., DAVIS, G.C. & BUNNEY, W.E. JR. (1977) Naloxone alters pain perception and somatosensory evoked potentials in normal subjects. Nature 270, 620-622.

CICERO, T.J., BADGER, T.M., WILCOX, C.E., BELL, R.D. &

MEYER, E.R. (1977) Morphine decreases luteinizing hormone by an action on the hypothalamic-pituitary axis. J. Pharmacol. Exp. Ther. 203, 548-555.

CICERO, T.J. & DORNBUSH, R. (1979) in Research Advances in Alcohol and Drug Abuse (Kalant, O., Kalant, H. and Israel, Y., eds.) Vol. 5 (in press). Plenum Press, New York.

EHRLICH, Y.H., BONNET, K.A., DAVIS, L.G. & BRUNNGRABER, E.G. (1978) Decreased phosphorylation of specific proteins in neostriatal membranes from rats after long-term narcotic exposure. Life Sci. 23, 137-146.

EISENMAN, A.J., FRASER, H.F., SLOAN, J. & ISBELL, H. (1958) Urinary 17-Ketosteroid excretion during a cycle of addiction to morphine. J. Pharmacol. Exp. Ther. 124, 305-311.

EL-SOBKY, A., DOSTROVSKY, J.O. & WALL, P.D. (1976) Lack of effect of naloxone on pain perception in humans. Nature 263, 783-784.

FURUYAMA, S., MAYES, D. & NUGENT, C.A. (1970) A radio-immunoassay for plasma testosterone. Steroids 16, 415.

GEORGE, R. & KOKKA, N. (1976) in Tissue Responses to Addictive Drugs (Ford, D.H. and Clouet, D.H., eds.) pp. 527-540. Spectrum, New York.

GITLIN, M. & ROSENBLATT, M. (1978) Possible withdrawal from endogenous opiates in schizophrenics. Am. J. Psychiatry, 135, 377-378.

GOLD, M.S., REDMOND, D.E. JR., DONABEDIAN, R.K. (1978) Prolactin secretion, a measurable central effect of opiate-receptor antagonists. Lancet Feb. 11, 323-324.

GOLDSTEIN, A. (1976) Opioid peptides (endorphins) in pituitary and brain. Science 193, 1081-1086.

GOLDSTEIN, A. & HANSTEEN, R.W. (1977) Evidence against involvement of endorphins in sexual arousal and orgasm in man. Arch. Gen. Psychiat. 34, 1179-1180.

GREVERT, P. & GOLDSTEIN, A. (1978) Endorphins: Naloxone fails to alter experimental pain or mood in humans. Science 199, 1093-1095.

GUIDOTTI, A. & GRANDISON, L. (1978) in The Endorphins. Advances in Biochemical Pharmacology (Costa, E. and Trabucchi, M., eds.) Vol. 18. Raven Press, New York.

GUILLEMIN, R., VARGO, T., ROSSIER, J., MILNICK, S., LING, N., RIVIER, D., VALE, W., BLOOM, F. (1977) Beta-endorphin and adrenocorticotropin are secreted concomitantly by the pituitary gland. Science 197, 1367-1369.

GUNNE, L.M. & TERENIUS, L. (1978) in Endorphins in Mental
     Health Research (Usdin, E., Bunney, W.E. Jr. and
     Kline, N.S., eds.). Macmillan Press, New York.
HELLMAN, L., FUKUSHIMA, D.K., ROFFWARG, H. & FISHMAN, J.
     (1975) Changes in estradiol and cortisol production
     rates in men under the influence of narcotics. J.
     Clin. Endocrinol. Metab. 41, 1014-1019.
HOSOBUCHI, Y., ADAMS, J.E. & LINCHITZ, R. (1977) Pain
     relief by electrical stimulation of the central gray
     matter in humans and its reversal by naloxone.
     Science 197, 183-186.
IRWIN, S., KINOHI, R.G., COLLET, P.M. & BOTTOMLY, D.R.
     (1974) Acute time-dose-response effects of cyclazo-
     cine, naltrexone, and naloxone in man. Proc. Com-
     mittee on Problems of Drug Dependence. Cambridge,
     Ma.
JACOBS, L.S., MARIZ, I.K. & DAUGHADAY, W.D. (1972) A
     mixed heterologous radioimmunoassay for human
     prolactin. J. Clin. Endocrinol. Metab. 34, 484.
JANOWSKY, D., JUDD, L., HUEY, L., ROITMAN, N., PARKER,
     D. & SEGAL, D. (1978) Negative naloxone effects on
     serum-prolactin. Lancet Sept. 16, 637.
JONES, R. (1978) in Endorphins in Mental Health Research
     (Usdin, E., Bunney, W.E. Jr. and Kline, N.S., eds.).
     Macmillan Press, New York.
KERLINGER, F.N. & PEDHAZUR, E.J. (1973) Multiple Regres-
     sion in Behavioral Research. Holt, Rinehart, Win-
     ston, New York.
KOKKA, N. & GEORGE, R. (1974) in Narcotics and the Hypo-
     thalamus (Zimmerman, E. and George, R., eds.) pp.
     137-157. Raven Press, New York.
LAL, H., BROWN, W., DRAWBAUGH, R., HYNES, M., BROWN, G.
     (1977) Enhanced prolactin inhibition following
     chronic treatment with haloperidol and morphine.
     Life Sci. 20, 101-106.
MAINS, R., EIPPER, E. & LING, N. (1977) Common precursor
     to corticotropins and endorphins. Proc. Natl. Acad.
     Sci. U.S.A. 74, 3014.
MARTIN, W.R., JASINSKI, D.R. & MANSKY, P.A. (1973) Nal-
     trexone, an antagonist for the treatment of heroin
     dependence. Arch. Gen. Psychiat. 28, 784-791.
MENDELSON, J.H., MEYER, R.E., ELLINGBOE, J., MIRIN, S.M.
     & MCDOUGLE, M. (1975) Effects of heroin and metha-
     done on plasma cortisol and testosterone. J. Pharma-
     col. Exp. Ther. 195, 296-302.
MENDELSON, J.H., ELLINGBOE, J., KEUHNLE, J.C. & MELLO,
     N.K. Effects of naltrexone on mood and neuroendo-
     crine function in normal adult males. Psychoneuro-
     endocrinology. In press.

NUTT, J.G., ROSIN, A.J., EISLER, T., CALNE, D.B. & CHASE,
    T.N. (1978) Effect of an opiate antagonist on move-
    ment disorders. Arch. Neurol. 35, 810-811.
ORTH, D.N. (1974) in Methods of Hormone Radioimmunoassay
    (McJaffe, B. and Behrman, M.R., eds.). Academic
    Press, New York.
RIVIER, C., VALE, W., LING, N., BROWN, M. & GUILLEMIN, R.
    (1977) Stimulation in vivo of the secretion of pro-
    lactin and growth hormone by beta-endorphin. Endo-
    crinology 100, 238-241.
ROLLERI, E., ZANNINO, M., ORLANDINI, S., & MALVANO, R.
    (1976) Direct radioimmunoassay of plasma cortisol.
    Clin. Chim. Acta 66, 319.
ROTSZTEJN, W.H., DROUVA, S.V., PATTOU, E. & KORDON, C.
    (1978) Met-enkephalin inhibits in vitro dopamine-
    induced LHRH release from mediobasal hypothalamus
    of male rats. Nature 274, 281-282.
SABOL, S.L. (1978) Regulation of endorphin production
    by glucocorticoids in cultured pituitary tumor cells.
    Biochem. and Biophys. Res. Comm. 82, 560-567.
SAXENA, B.B., DERMURA, H., GANDY, H.M. & PETERSON, R.E.
    (1968) Radioimmunoassay of human follicle-stimula-
    ting and luteinizing hormone in plasma. J. Clin.
    Endocrinol. Metab. 28, 519.
SIMPSON, G.M., BRANCHEY, M.H. & LEE, J.H. (1977) A
    trial of naltrexone in chronic schizophrenia. Curr.
    Ther. Res. 22, 909-913.
SLOAN, J.W. (1971) in Narcotic Drugs, Biochemical Pharma-
    cology (Clouet, D.H., ed.) pp. 262-282. Plenum
    Press, New York.
TERENIUS, L. (1976) Somatostatin and ACTH are peptides
    with partial antagonist-like selectivity for opiate
    receptors. Eur. J. Pharmacology 38, 211-213.
VEREBEY, K., VOLAVKA, J. & CLOUET, D. (1978) Endorphins
    in psychiatry. Arch. Gen. Psychiatry 35, 877-888.
VOLAVKA, J., RESNICK, R.B., KESTENBAUM, R.S. & FREEDMAN,
    A.M. (1976) Short-term effects of naltrexone in
    155 heroin ex-addicts. Biological Psychiatry 11,
    679-685.
VOLAVKA, J., BAUMAN, J., MALLYA, A. & CHO, D. Effects of
    naloxone in normal men. In preparation.
WATSON, S.J., BERGER, P.A., AKIL, H. MILLS, M.J. & BARCHAS,
    J.D. (1978) Effects of naloxone on schizophrenia:
    reduction in hallucinations in a subpopulation of
    subjects. Science 201, 73-76.

MEASUREMENT OF β-ENDORPHIN-LIKE IMMUNOREACTIVITY IN CSF

AND PLASMA OF NEUROPSYCHIATRIC PATIENTS

H.M. EMRICH*, V. HÖLLT, W. KISSLING, M. FISCHLER,
H. HEINEMANN**, D. v. ZERSSEN AND A. HERZ

Max-Planck-Institut für Psychiatrie,
Munich, F.R.G. and
**Bezirkskrankenhaus Kaufbeuren, F.R.G.

INTRODUCTION

The hypothesis of a possible role of endorphins in the pathogenesis of psychoses has its origin in two types of observations. Firstly from the euphorogenic action of opiates and, possibly, endorphins (KLINE & LEHMANN, 1978) in normal subjects, leading to the idea that endogenous opioids may be responsible for mood changes in affective psychoses (TERENIUS et al., 1977) and, secondly, from the fact that certain partial opiate agonists, such as cyclozocine and nalorphine, induce hallucinations and derealization experience in healthy volunteers which can be immediately reversed by the specific opiate antagonist naloxone (JASINSKI et al., 1967). From this latter finding it was concluded that productive symptoms (e.g. hallucinations and delusions) in schizophrenic patients might be induced by abnormally functioning opioids with a spectrum of action similar to that of partial opiate agonists (summary: EMRICH, 1978; SNYDER, 1978; TERENIUS, 1978; VEREBEY et al., 1978).

Consequently, GUNNE et al. (1977) searched for antipsychotic effects of the specific opiate antagonist naloxone and observed in an open pilot study an immediate reversal of auditory hallucinations in 4 patients with

*Supported by the Fritz-Thyssen-Foundation, Cologne

chronic schizophrenia following an intravenous injection
of 0.4 mg naloxone. However, further investigations (for
a summary see EMRICH et al., 1979) using a double-blind
design and somewhat higher dosages of up to 1.6 mg could
not reproduce these findings, whereas in studies using
still larger doses from 4.0-24.8 mg (EMRICH et al., 1977;
WATSON et al., 1978; EMRICH et al., 1978) a small but significant antipsychotic action of naloxone was observed.
On the other hand, from the weakness of this effect at
high naloxone dosages a central role of the endorphinergic
system in the pathogenesis of schizophrenia appears ques-
tionable.

    On the other side, there are two further hints from
other experimental approaches indicating a pathological
significance of endorphins in the pathogenesis of schizophrenic psychoses. One is the finding of PALMOUR (1978),
that in the dialyzate of schizophrenic patients there
exists a high amount of leucine-β-endorphin. The other
hint comes from measurements of endorphin levels in the
CSF. TERENIUS et al. (1976), using the radioreceptorassay,
studied two CSF fractions, separated by gel-chromatography,
and observed, especially in Fraction I, an increase of endorphins in patients with schizophrenia, which was partially reversible by neuroleptic therapy.

    However, by use of the radioreceptorassay, it is not
possible to determine the chemical nature of the opioid
substance.

    An approach to get some more information regarding
these findings has been performed by HÖLLT et al. (1978a,
b), using a radioimmunoassay of β-endorphin.

    The present paper gives a synopsis of some of the
preliminary data which have been obtained by use of this
method.

                            METHODS

    In the CSF study the following types of diagnoses
were included: I: schizophrenia (acute and chronic
cases), II: lumbago, III: herniation of the nucleus pulposus, IV: encephalitis/meningitis, V: neurological disorders (multiple sclerosis, lateral sclerosis, brain
atrophy, etc.), VI: a control group consisting of patients who were punctured because meningitis/encephalitis
was suspected but turned out to have a normal CSF (the

patients of group II and III were punctured for myelography because herniation of the nucleus pulposus was suspected in both groups). All patients with schizophrenia (group I), with the exception of two acute cases (open circles in Fig. 1), were treated with neuroleptic drugs. With the exception of patients of group II and III, who were pretreated with 10 mg diazepam, the patients had no special pretreatment for the lumbar tap. The other patients were free of psychotropic drugs. The CSF samples (2-5 ml) were frozen immediately after the lumbar tap.

In the plasma study the following diagnoses were included: I: schizophrenia without neuroleptic treatment (acute and chronic cases), II: schizophrenia with neuroleptic treatment (acute and chronic cases), III: different types of neuroses (anxiety neurosis, phobia, neurotic depression, etc.), IV: mania and endogenous depression. The neurotic patients (group III) and part of the schizophrenic patients (group I) were drug-free. The other patients were treated in the conventional manner with neuroleptic or anti-depressive drugs. In a second series of experiments, patients with schizophrenia (acute and chronic cases), different types of neuroses and affective psychoses were selected for a re-evaluation of the findings in the first series. In a further study on plasma levels of β-endorphin-like immunoreactivity 3 patients with endogenous depression were selected who were treated with electroconvulsive treatment. Blood samples were taken immediately before and 10 min after electroconvulsion. In all these plasma studies, 9.5 ml venous blood was extracted on puncturing and transferred to a vessel containing 20 mg EDTA, cooled in an ice/water mixture. Within 10 minutes plasma was separated in a cooled centrifuge at 1000 g, frozen, and stored at -70°C.

The radioimmunoassay for β-endorphin has been described in detail elsewhere (HÖLLT et al., 1978). Antibodies against human β-endorphin which exhibit a high avidity for the C-terminal of the peptide were raised in rabbits following the injection of thyroglobulin-coupled human β-endorphin (βh-E) as the immunogen. Methionine-enkephaline, leucine-enkephaline, α-, γ-endorphin, and ACTH peptides do not cause interference in the radioimmunoassay. β-Lipotropin, however, on a molar basis shows a 50% cross-reactivity. The antiserum exhibits the same avidity for human β-leu^5-endorphin and β-met^5-endorphin. β-Endorphin was extracted from human plasma by the following procedure: 1 ml human plasma was mixed with 70 mg silicic acid. After washing procedures with water and

and ether β-endorphin was desorbed from the silicic acid
with a 1.5 ml acetone/0.1 N. HCl (20/80 vol/vol) mixture.
70+5% of β-endorphin was recovered by this extraction procedure. For the RIA the acidic acetone extracts were lyo-
philized in 2.2 ml Eppendorf plastic tubes; thereafter, 0.1
ml β-endorphin antiserum 0.05 ml ^{125}I-endorphin and 350 μl
of RIA buffer was added and the mixture incubated for 36 h
at 2°C. Separation of free and antibody-bound β-endorphin
was achieved by charcoal. Standard amounts of β-endorphin
were added to β-endorphin free plasma samples (obtained
after absorption of β-endorphin by silicic acid). The detection limit of the RIA for β-endorphin after extraction
was between 20 and 40 pg/ml (6-12 fmole/ml), depending on
the precision of the assay technique. The recovery of
β-endorphin from CSF samples was of a similar magnitude
as that from plasma. For convenience, however, a direct
measurement was preferred: 1 ml CSF was incubated together with 0.100 ml of β-h-endorphin antiserum and 0.05 ml
[125]I-β-endorphin at 2°C for 36 h. Control experiments in

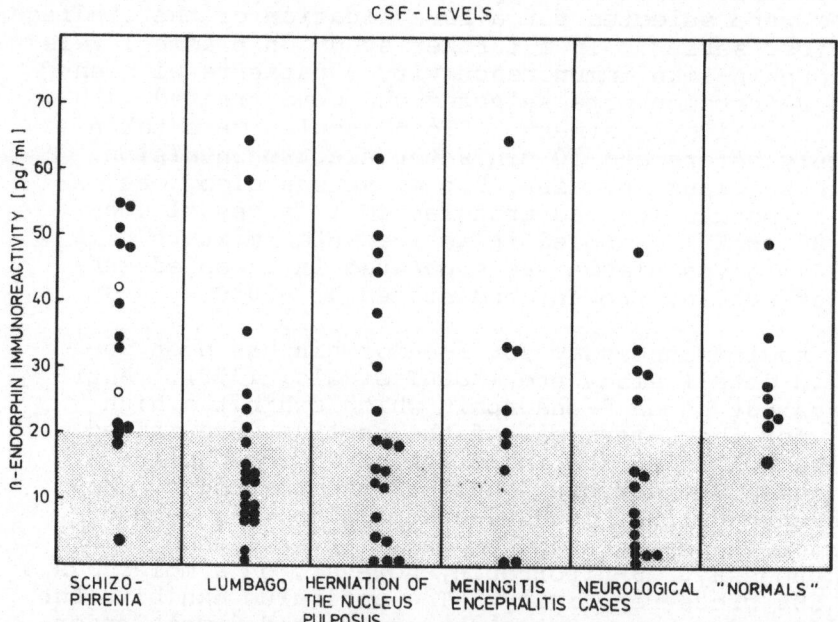

Fig. 1:   β-Endorphin-like immunoreactivity in CSF of
     patients with 6 different types of diagnoses (for
     details see text).

which ^{125}I-endorphin was added to CSF and chromatographi-
cally separated on a sephadex G 50 superfine column
0.9 x 95 cm (equilibrated and eluted in the RIA buffer)
showed no evidence for critical incubation damage after
48 h incubation. Separation was also performed with charcoal, the sensitivity limit of the RIA was about 20 pg/ml
CSF (or 6 fmole/ml) arbitrarily estimated as the concentration of β-endorphin leading to 15% displacement of
^{125}I-β-endorphin from the antiserum.

RESULTS AND DISCUSSION

Fig. 1 shows CSF-levels of β-endorphin-like immunoreactivity in 6 different diagnostic groups:

1. Schizophrenic patients (two of them without neuroleptic therapy, highly acute cases: open circles).
2. Patients with lumbago, punctured for myelography,
   pretreated with 10 mg diazepam.
3. Patients with a herniation of the nucleus pulposus of
   the vertebral disc, pretreated with 10 mg diazepam
   and later operated on.
4. Patients with meningitis/encephalitis.
5. Patients with mixed neurological diagnoses: multiple
   sclerosis, lateral sclerosis, cerebral atrophy, etc.
6. Medical patients who were punctured because of the
   suspicion of a meningitis but whose CSF was normal
   ("normals").

The grey zone in Fig. 1 represents the detection
limit within which no reliable values can be measured.
The values within this limit are only virtual and do not
have a real significance.

If we compare the data of the schizophrenic patients
with the data of the so-called "normal group" there is
practically no difference. There may be a small tendency
to lower values in the group of patients with lumbago and
herniation of the nucleus pulposus. However, these patients
were pretreated with a minor tranquilizer and we do not
know the influence of these drugs on the endorphin system. If β-endorphin, like ACTH, is regarded as an indicator of stress-phenomena, one would suspect a lowering
of values after pretreatment with a minor tranquilizer.

In the group of cases with meningitis/encephalitis
there is no change at all, whereas in the mixed group of
neurological patients there is also a small tendency to

lower values.

From the normal findings in schizophrenic patients
one may derive that the elevated values of TERENIUS and
co-workers (1976) cannot be due to β-endorphin.

Immunoreactive β-endorphin-like material in plasma
ranges from 20 pg/ml (<40, <50 pg/ml) (limits of determination) to 250 pg/ml (Fig. 2, 3, and 4). The detection
limits in the data of Fig. 2, 3 and 4 are different due
to a dissimilar precision of the assay technique.

In a first series of experiments (Fig. 2) blood-
levels of schizophrenic patients with and without neuroleptic therapy have been compared with the levels in different types of neuroses, affective and organic psychoses.
No obvious differences in blood levels of β-endorphin-like
immunoreactivity occur between neurotic patients and patients with affective and organic psychoses, whereas some
of the data in the two groups of schizophrenic patients,
especially after neuroleptic treatment, appear to be
higher than the values in the other patients. However,

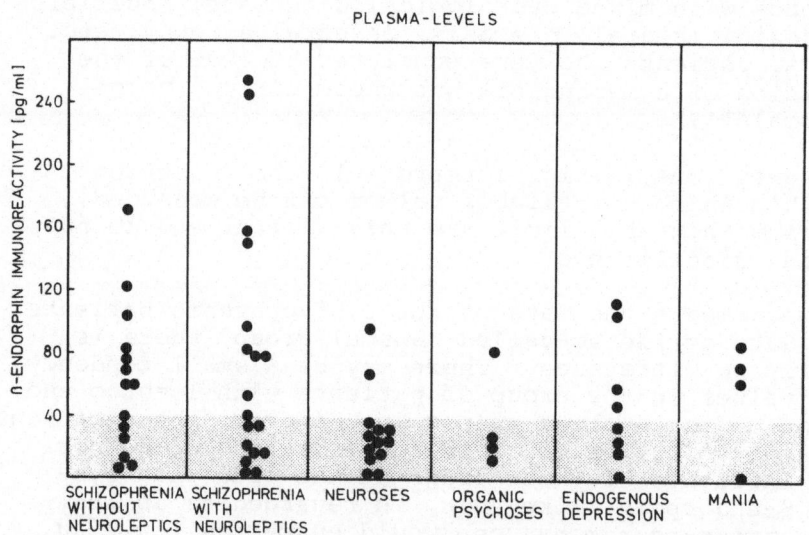

Fig. 2:   β-Endorphin-like immunoreactivity in plasma of
        patients with schizophrenia, different types of
        neuroses, organic psychoses, endogenous depression
        and mania (for details see text).

in the second series (Fig. 3) in which the sensitivity
of the test was somewhat higher, the schizophrenic patients had no higher values as compared with the data of
patients with different types of neuroses and affective
psychoses.

From the present preliminary data no conclusion
should be drawn concerning possible small differences
between these groups.

From these data one can derive that no increase of
β-endorphin immunoreactivity occurs in schizophrenics
which should be present if an abnormally high amount of
leucine-β-endorphin would exist in the plasma of these
patients, since the cross-reactivity of the radioimmunoassay with leucine-β-endorphin is 100%. Thus, our data
do not support the idea of a pathogenesis of schizophrenia
by an abnormally high level of leucine-β-endorphin.

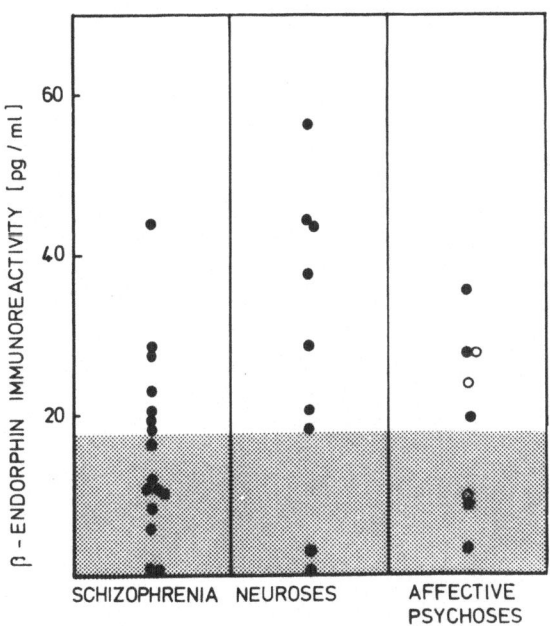

Fig. 3:   β-Endorphin-like immunoreactivity in plasma of
          patients with schizophrenia (acute and chronic cases),
          different types of neuroses and affective psychoses
          (open circles: mania, dots: endogenous depression).

In Fig. 4 the data of plasma levels of β-endorphin-
like immunoreactivity before and 10 min after 9 electroconvulsive treatments in 3 patients with endogenous depression are depicted. In the region >50 pg/ml in which
in this series of experiments reliable measurements were
possible, 7 times an elevation after electroconvulsion
was observed. Concerning this observation, similar findings on ACTH-increase after electroconvulsion has to be
considered (ALLEN et al., 1974). Since β-endorphin immunoreactivity normally goes parallel with ACTH values in
response to electrically induced stress (GUILLEMIN et al.,

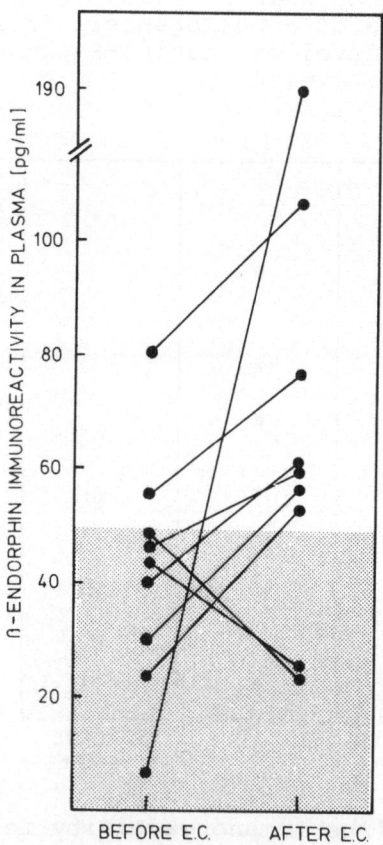

Fig. 4: β-Endorphin-like immunoreactivity in plasma of
    patients with endogenous depression before and 10
    min after electroconvulsive treatment.

1977), the present result has to be expected. On the other side, the possibility that electroconvulsion is effective via an endorphinergic mechanism has to be considered.

A central methodological problem in this work is the inability of the radioimmunoassay employed to discriminate between β-endorphin and β-LPH immunoreactivity, although the cross-reactivity with the enkephalins and α- and γ-endorphin is very low. This is, in fact, a recurrent problem as all radioimmunoassays published to date show comparable cross-reactivity to β-lipotropin (GUILLEMIN et al., 1977; ROSS et al., 1978).

Very recently, SUDA et al., 1978 reported that in contrast to β-lipotropin, β-endorphin is not detectable in plasma from normal human subjects. Preliminary studies, however, performed in our laboratory, provided unequivocal evidence for the presence of both β-lipotropin and β-endorphin in human plasma if 50 ml plasma samples of healthy volunteers were extracted and subjected to gel filtration of a sephadex G 50 superfine column and measured for immunoreactive materials (HÖLLT, in preparation 1979). This column-chromatographic method was not applicable in our clinical study, since the amounts of CSF and plasma required for such a procedure are unobtainable normally. However, for a few selected cases measurements are in progress in which β-lipotropin and β-endorphin are detected separately. The data which are communicated here represent a preliminary attempt to measure actual CSF- and plasma-levels of immunoreactive β-endorphin (comprising β-endorphin and β-lipotropin) in humans under different pathological conditions. However, our data can be interpreted as revealing an upper limit of possibly elevated values of β-endorphin-like immunoreactivity in different types of neuropsychiatric disorders.

ACKNOWLEDGEMENTS

For the supply of CSF-samples thankful appreciation is given to: Prof. Dr. W. KRÄMER, Chief of the Dept. of Neurology, Bezirkskrankenhaus Haar; Dr. H. DWINGER, Dept. of Psychiatry, Bezirkskrankenhaus Haar; Dr. F. SCHMIDT-BERGER, Chief of the Dept. of Neurosurgery, Krankenhaus München-Schwabing; Dr. E. HOLZER, Chief of the 4th Dept. of Internal Medicine, Krankenhaus München-Schwabing; Dr. E. EINHÄUPL, Dept. of Neurology, Klinikum Großhadern, Munich and Dr. D. v. CRAMON, Dept. of Neurology, Max-

Planck-Institut für Psychiatrie, Munich. The authors wish to thank Mrs. F. SAILER and Miss A. WENDL as well as the technical assistants in the Dept. of Clinical Chemistry (Chief: Prof. Dr. D. STAMM) and the nurses of our wards for excellent technical assistance.

REFERENCES

ALLEN, M.J.P., DENNEY, D., KENDALL, J.W. & BLACHLY, P.H. (1974) Amer. J. Psychiat. 131, 1225-1228.
EMRICH, H.M. (1978) Arzneim.-Forsch./Drug Res. 28-II, 1270-1273.
EMRICH, H.M., CORDING, C., PIRÉE, S., KÖLLING, A., v. ZERSSEN, D. & HERZ, A. (1977) Pharmakopsychiat. 10, 265-270.
EMRICH, H.M., CORDING, C., PIRÉE, S., KÖLLING, A., MÖLLER, H.-J., v.ZERSSEN, D. & HERZ, A. (1978) in Endorphins in Mental Health Research (Usdin, E., Bunney, W.E. Jr. & Kline, N.S., eds.). Macmillan Press, New York.
EMRICH, H.M., HÖLLT, V., LASPE, H., FISCHLER, M., HEINE- MANN, H., KISSLING, W., v. ZERSSEN, D. & HERZ, A. (1979) in Recent Advances in Neuropsychopharmacology (Saletu, B., Berner, P. & Hollister, L., eds.). Pergamon Press, Oxford-New York-Frankfurt.
GUILLEMIN, R., VARGO, T., ROSSIER, J., MINICK, S., LING, N., RIVIER, C., VALE, M. & BLOOM, F. (1977) Science 197, 1367-1369.
GUNNE, L.M., LINDSTRÖM, L. & TERENIUS, L. (1977) J. Neural. Transm. 40, 13-19.
HÖLLT, V., PRZEWŁOCKI, R. & HERZ, A. (1978a) Naunyn- Schmiedeberg's Arch. Pharmacol. 303, 171-174.
HÖLLT, V., EMRICH, H.M., MÜLLER, O.A. & FAHLBUSCH, R. (1978b) in Characteristics and Function of Opioids (Van Ree, J.M. & Terenius, L., eds.) pp. 279-280. Elsevier, Amsterdam.
HÖLLT, V. (1979) in prepration.
JASINSKI, D.R., MARTIN, W.R. & HAERTZEN, C.A. (1967) J. Pharmacol. Exp. Ther. 157, 420-426.
KLINE, N.S. & LEHMANN, H.E. (1978) in Endorphins in Mental Health Research (Usdin, E., Bunney, W.E. Jr. & Kline, N.S., eds.). Macmillan Press, New York.
PALMOUR, R.M. (1978) in Endorphins in Mental Health Research (Usdin, E., Bunney, W.E. Jr. & Kline, N.S., eds.). Macmillan Press, New York.
ROSS, M., GHAZAROSSIAN, V., COX, B.M. & GOLDSTEIN, A. (1978) Life Sciences 22, 1123-1130.
SNYDER, S.H. (1978) Amer. J. Psychiat. 135, 645-652.
SUDA, T., LIOTTA, A.S. & KRIEGER, D.T. (1978) Science

202, 221-223.
TERENIUS, L. (1978) in Characteristics and Function of
   Opioids (Van Ree, J.M. & Terenius, L., eds.) pp.
   143-158. Elsevier, Amsterdam.
TERENIUS, L., WAHLSTRÖM, A., LINDSTRÖM, L. & WIDERLÖV,
   E. (1976) Neurosci. Lett. 3, 157-162.
TERENIUS, L., WAHLSTRÖM, A. & ÅGREN, H. (1977) Psycho-
   pharmacology 54, 31-33.
VEREBEY, K., VOLAVKA, J. & CLOUET, D. (1978) Arch. Gen.
   Psychiat. 35, 877-888.
WATSON, S.J., BERGER, P.A., AKIL, H., MILLS, M.J. & BAR-
   CHAS, J.D. (1978) Science 201, 73-76.

SUBJECT INDEX

Acetylcholine, 20, 28, 41, 43, 49, 50, 57, 69, 77, 125, 135, 204

Acetylcholine (muscarinic) receptor, 135, 227, 253

Acetylcholine (nicotinic) receptor, 175-197, 248, 253

  antibodies to, 176, 180, 182, 191, 194

  dephosphorylation of, 187-190, 191-197

  membrane conformation, effect of, 185

  phosphorylation of, 175-187, 191-197

    cations, effect of, 178-186

  purification of, 176-178

Acetylcholine receptors, 87

  sensitivity of, 247, 248, 253

Acetylcholinesterase, 176, 178

N-Acetyltransferase, induction of, 249-251

Action potential, 15, 17-19, 21-23, 46, 48, 49

Adaptation, 76, 88, 91, 92, 94, 95, 97, 105, 107, 200, 247, 249-251, 253-256

Addison's disease, 286

Adenoma of Pituitary, 282, 286, 287

Adenosine, 54, 65-68, 163, 252

5'-Adenosine monophosphate, 54

Adenosine triphosphatase, Na, K,-activated (ATPase), 68, 176, 178, 192, 211

Adenylate cyclase, 44, 46, 47, 52, 53, 55, 65, 79, 89, 93, 96, 133, 138-140, 148, 153, 163, 164, 167, 172, 173, 205-209, 225-229, 235, 236, 243, 249-254

  dopamine stimulated, 79, 94-96, 139, 250, 252-255

  GTP, effect on, 227, 249-251

  norepinephrine stimulated, 139, 140, 153, 250-253

  prostaglandin stimulated, 253

Adipose tissue, 280-282, 285

Adrenal cortex, 200, 201, 207, 215, 283, 286

Adrenalectomy, 284-286

Adrenal medulla, 89, 263, 265, 266

α-Adrenergic receptor, 133, 134, 136, 138-140, 151-156, 227, 252, 253, 256

  adenyl nucleotides, effect of, 136

  cations, effect of, 138, 139

  guanyl nucleotides, effect of, 139, 140

β-Adrenergic receptor, 44, 51, 52, 55, 126, 133-156, 164, 166, 248-253, 256

  adenylate cyclase and, 140

  in brain regions, 142-145

  cations, effect of, 136, 137

  in developing brain, 140, 141

  drugs (acute), effects of, 146

  drugs (chronic), effects of, 146-149